AF782289

Fisheries Management: The Ecosystem Approach to Fisheries

Fisheries Management: The Ecosystem Approach to Fisheries

Editor:

Bruno Augusto Amato Borges

Delve Publishing

www.delvepublishing.com

Fisheries Management: The Ecosystem Approach to Fisheries

Editor: Bruno Augusto Amato Borges

© 2017 Delve Publishing LLC

708 3rd Avenue, 6th Floor

New York, NY 10017

United States of America

www.delvepublishing.com

ISBN: 978-1-68095-855-3

Library of Congress Control Number: 2016951608

This book contains information obtained from highly regarded resources. Reprinted material sources are indicated. Copyright for individual articles remains with the authors as indicated. A Wide variety of references are listed. Reasonable efforts have been made to publish reliable data and views articulated in the chapters are those of the individual contributors, and not necessarily those of the editors or publishers. Editors or publishers are not responsible for the accuracy of the information in the published chapters or consequences of their use. The publisher believes no responsibility for any damage or grievance to the persons or property arising out of the use of any materials, instructions, methods or thoughts in the book. The editors and the publisher have attempted to trace the copyright holders of all material reproduced in this publication and apologize to copyright holders if permission has not been obtained. If any copyright holder has not been acknowledged, please write to us so we may rectify.

Notice: Registered trademark of products or corporate names are used only for explanation and identification without intent of infringement.

Delve Publishing publishes wide variety of books and eBooks. For more information about Magnum Publishing and its products, visit our website at **www.delvepublishing.com**

About the Editor

Bruno Augusto Amato Borges

Bruno Augusto is Aquaculture Engineer from Federal University of Santa Catarina. His focus is on writing projects and development of new technologies in aquaculture. He is currently project service provider, consultant and develops research in the field of Biofloc Technology (BFT) for freshwater fish farming.

Contents

List of Contributors

Jesper Raakjær
Innovative Fisheries Management – an Aalborg University Research Centre, Aalborg, Denmark

Troels Jacob Hegland
Innovative Fisheries Management – an Aalborg University Research Centre, Aalborg, Denmark

Angelo Tursi
Department of Biology, University of Bari Aldo Moro, Via E. Orabona 4, 70125 Bari, Italy

Porzia Maiorano
Department of Biology, University of Bari Aldo Moro, Via E. Orabona 4, 70125 Bari, Italy

Letizia Sion
Department of Biology, University of Bari Aldo Moro, Via E. Orabona 4, 70125 Bari, Italy

Gianfranco D'Onghia
Department of Biology, University of Bari Aldo Moro, Via E. Orabona 4, 70125 Bari, Italy

Ragnar Elmgren
Department of Ecology, Environment and Plant Sciences, Stockholm University, 106 91 Stockholm, Sweden

Thorsten Blenckner
Stockholm Resilience Centre, Stockholm University, 106 91 Stockholm, Sweden

Agneta Andersson
Department of Ecology and Environmental Science, Umea˚ University, 901 87 Umea˚, Sweden

H. E. Markus Meier
Swedish Meteorological and Hydrological Institute, 601 76 Norrkoping, Sweden

Hele´n C. Andersson
Swedish Meteorological and Hydrological Institute, 601 76 Norrkoping, Sweden

Berit Arheimer
Swedish Meteorological and Hydrological Institute, 601 76 Norrkoping, Sweden

Chantal Donnelly
Swedish Meteorological and Hydrological Institute, 601 76 Norrkoping, Sweden

Kari Eilola
Swedish Meteorological and Hydrological Institute, 426 71 Vastra Frolunda, Sweden

Bo G. Gustafsson
Baltic Nest Institute, Stockholm University, 106 91 Stockholm, Sweden

Lech Kotwicki
Department of Marine Ecology, Institute of Oceanology, Polish Academy of Sciences, 55 Powstancow Warszawy Street, 81-712 Sopot, Poland

Tina-Simone Neset
Centre for Climate Science and Policy Research, Department of Water and Environmental Studies, Linkoping University, 601 74 Norrkoping, Sweden

Susa Niiranen
Stockholm Resilience Centre, Stockholm University, 106 91 Stockholm, Sweden

Joanna Piwowarczyk
Department of Marine Ecology, Institute of Oceanology, Polish Academy of Sciences, 55 Powstancow Warszawy Street, 81-712 Sopot, Poland

Oleg P. Savchuk
Baltic Nest Institute, Stockholm University, 106 91 Stockholm, Sweden

Frederik Schenk
Institute of Coastal Research, Helmholtz-Zentrum Geesthacht, Max-Planck-Str. 1, 21481 Geesthacht, Germany

Jan Marcin We˛sławski
Department of Marine Ecology, Institute of Oceanology, Polish Academy of Sciences, 55 Powstancow Warszawy Street, 81-712 Sopot, Poland

Eduardo Zorita
Institute of Coastal Research, Helmholtz-Zentrum Geesthacht, 21502 Geeshacht, Germany

Hui Liu
Yellow Sea Fisheries Research Institute, Chinese Academy of Fishery Sciences, Qingdao, China

Jilan Su
Second Institute of Oceanography, State Oceanic Administration, Hangzhou, China

Julia Olson
NOAA Fisheries, Northeast Fisheries Science Center, Social Sciences Branch, 166 Water Street, Woods Hole, MA 02543, USA

Patricia Pinto da Silva
NOAA Fisheries, Northeast Fisheries Science Center, Social Sciences Branch, 166 Water Street, Woods Hole, MA 02543, USA

Svein Jentoft
Norwegian College of Fishery Science Faculty of Biosciences, Fisheries and Economics University of Tromsø – The Arctic University of Norway, 9037 Tromsø, Norway

Maaike Knol
Norwegian College of Fishery Science Faculty of Biosciences, Fisheries and Economics University of Tromsø – The Arctic University of Norway, 9037 Tromsø, Norway

Vasudha Chhotray
School of International Development, University of East Anglia, Norwich NR4 7TJ, UK

Douglas C K Wilson
Innovative Fisheries Management, Aalborg University Research Centre, Aalborg, Denmark

Rikke Becker Jacobsen
Innovative Fisheries Management, Aalborg University Research Centre, Aalborg, Denmark

Crisol Méndez-Medina
Department of Biodiversity Conservation, El Colegio de la Frontera Sur (ECOSUR), Av del Centenario Km 5.5, Chetumal, C.P. 77014, Quintana Roo, Mexico

Birgit Schmook
Department of Biodiversity Conservation, El Colegio de la Frontera Sur (ECOSUR), Av del Centenario Km 5.5, Chetumal, C.P. 77014, Quintana Roo, Mexico

Susannah R. McCandless
Department of Geography, University of Vermont, 94 University Place, Old Mill Building 200, Burlington VT 05405, USA

Jannike Falk-Andersson
Marine Resource Economics group, Norwegian College of Fishery Science, UiT The Arctic University of Norway, 9037 Tromsø, Norway

Naomi S Foley
Socio-Economic Marine Research Unit (SEMRU), JE Cairnes School of Business and Economics, National University of Ireland Galway, University Road, Galway, Ireland

Claire W Armstrong
Marine Resource Economics group, Norwegian College of Fishery Science, UiT The Arctic University of Norway, 9037 Tromsø, Norway

Sybille van den Hove
Median SCP and Institut de Ciència i Tecnologia Ambientals (ICTA), Universtitat Autònoma de Barcelona, Carrer Vista Alegre 20, 08197 Valldoreix, Spain

Thomas M van Rensburg
J.E. Cairnes School of Business and Economics, National University of Ireland Galway, University Road, Galway, Ireland

Rob Tinch
Median SCP, Calle Vista Alegre, 20 08197 Valldoreix, Spain

Sara Fro¨cklin
Department of Ecology, Environment and Plant Sciences, Stockholm University, 106 91 Stockholm, Sweden

Maricela de la Torre-Castro
Department of Physical Geography and Quaternary Geology, Stockholm University, 106 91 Stockholm, Sweden

Lars Lindstro¨m
Department of Political Science, Stockholm University, 106 91 Stockholm, Sweden

Narriman S. Jiddawi
Institute of Marine Sciences, University of Dar es Salaam, Mizingani Rd., Zanzibar, Tanzania

Troels J Hegland
Innovative Fisheries Management, Aalborg University, Skibbrogade 5, DK-9000 Aalborg, Denmark

Christopher CE Hopkins
AquaMarine Advisers, Granvägen 20, SE-26532 Åstorp, Sweden

Cleber J. R. Alho
Programa de Po´s-Graduac¸a˜o em Meio Ambiente, Universidade Anhanguera-Uniderp, Rua Alexandre Herculano, 1400 – Jardim Veraneio, Campo Grande, MS 79037-280, Brazil

Roberto E. Reis
Laborato´rio de Sistema´tica de Vertebrados, Pontifı´cia Universidade Cato´lica do Rio Grande do Sul, Pontifı´cia Universidade Cato´lica do Rio Grande do Sul, Av. Ipiranga, 6681, Pre´dio 40, sala 151, Porto Alegre, RS 90619-900, Brazil

Pedro P. U. Aquino
Departamento de Zoologia, Universidade de Brası´lia, Campus Universita´rio – Asa Norte, Brası´lia, DF 70910-900, Brazil

Vandick S Batista
Universidade Federal de Alagoas/UFAL, Instituto de Ciências Biológicas e da Saúde/ICBS-LABMAR, Av. Aristeu de Andrade, 452, Farol, CEP: 57051-090, Maceió, Alagoas, Brazil

Liane G Lima
Universidade Federal do Amazonas/UFAM. Av. Gen. Rodrigo Octávio Jordão Ramos, 3000, Mini-Campus. Bairro Coroado I. CEP 69077-000. Manaus, Amazonas, Brazil

Neil J. Holbrook
Institute for Marine and Antarctic Studies, University of Tasmania, Private Bag 129, Hobart, TAS 7001, Australia

Johanna E. Johnson
School of Marine and Tropical Biology, James Cook University, Townsville, QLD 4810, Australia

Preface

With the growth of the fishing industry, a number of changes should be taken to ensure that the activity will grow and prosper for many generations.

The main current discussion with regard to the development and production of fisheries is the concern with the environment coupled with the search for new technologies. So grows the need to discuss the situation and seek improvements aimed at the sustainable development of the activity and prosperity for future generations. This way starts a different approach regarding the ecosystem with respect to fisheries production. The term "Ecosystem Approach to Fisheries" (EAF) was adopted by the FAO Technical Consultation on Ecosystem-based Fisheries Management held in 2002 (FAO, 2003). This term (EAF) is presented by Ward et al. (2002) as "an extension of conventional fisheries management recognizing more explicitly the interdependence between human well-being and ecosystem health and the need to maintain ecosystems productivity for present and future generations, e.g. conserving critical habitats, reducing pollution and degradation, minimizing waste, protecting endangered species". Currently, many terms are used to name and describe the same concept as: ecosystem-based fisheries management (EBFM), ecosystem approach to aquaculture (EAA), environmental management, biodiversity management and ecosystem approach. The ecosystem approach to fisheries seeks to understand the existing relations and their characteristics within the environment, with the objective of improving the management and use of resources by activity. In addition it is extremely important to consider the impact that fisheries production has been producing in each ecosystem and globally, so that decisions involving changes in management can be evaluated. In this context, the main objectives of this book is to present the current situation of the risks and challenges of fisheries production and its impact on the environment, as well as different management techniques already in use and the search for alternatives and improvements.

Editor
Bruno Augusto Amato Borges

INTRODUCTION

Fisheries management draws on fisheries science to formulate rules and regulations that define where, how, and how many fish can be caught each year. The aim is to allow us to harvest as much as possible without destabilizing the delicate marine ecosystem, thus ensuring the availability of fresh-caught fish to future generations. Fisheries managers look at data provided by scientists and use complex models to predict how present fishing efforts will affect the future numbers of various species and distinct genetic populations ("stocks"). If a certain species or stock is identified as being threatened or endangered, fishing may have to be strictly limited or even prohibited in broad areas to allow the fish to recover.

Fisheries management can be thought of as an integrated process to improve the benefits that society receives from harvesting fish from the ecosystem. It includes the activities of (i) information gathering; (ii) analysis; (iii) planning; (iv) consultation; (v) decision-making; (vi) allocation of resources; and (vii) formulation and implementation, with enforcement of regulations or rules which govern fisheries activities. The main aim is to ensure continued productivity of the resources and accomplishment of other fisheries objectives.

In the region, fisheries have been managed mainly from a sectoral perspective. There are few examples of well-managed fisheries with an objective of maximizing the benefits (often considered as economic benefits) while trying to ensure that the catch is commensurate with the natural productivity of the harvest stocks. In many other fisheries the main objective of management has been to reduce conflict and it is often aimed at increasing the overall production.

Existing fisheries in the region may have characteristics such as:

- mainly focused on target species;
- single sector specific (fisheries);
- management actions mainly focus on control of fishing (e.g. gear restrictions and zones);
- stock assessment based; and
- mainly biological management objectives (e.g. increasing production).

Modern fisheries management, as practiced since the early 1940s, is strongly based on the ecosystem theory but focuses primarily on fishing activity and target fish resources. In inland waters, affected earlier and more strongly than marine waters by environmental problems, it developed as an extension of wildlife management and involves a substantial amount of direct intervention on the habitat, species composition, etc. In marine ecosystems, however, because the possibility of direct intervention on the ecosystem is limited, management strategies concentrated on controlling human intervention (fishing) while observing proxies for the state of an otherwise opaque ecosystem and fugitive resources. It is defined as "the integrated process of information gathering, analysis, planning, decision-making, allocation of resources and formulation and enforcement of fishery regulations by which the fisheries management authority controls the present and future behaviors of the interested parties in the fishery, in order to ensure the continued productivity of the living resources". It aims at optimizing the use of fishery resources as a source of human livelihood, food and recreation, dynamically regulating fishing activity, meeting resource-related objectives or constraints, mainly indirectly.

The ecosystem approach for fisheries management is a widely accepted concept and various international instruments require its application. However, there are various interpretations of the ecosystem approach and its application almost always brings about confrontations and resistance among managers, proponents, and stakeholders.

There is no clear and generally accepted definition of fisheries management. We do not wish to get embroiled in a debate about exactly what fisheries management is and isn't, but use here the working definition used in the Technical Guidelines to provide a summary of the task of fisheries management:

"The integrated process of information gathering, analysis, planning, consultation, decision-making, allocation of resources and formulation and implementation, with enforcement as necessary, of regulations or rules which govern fisheries activities in order to ensure the continued productivity of the resources and the accomplishment of other fisheries objectives.

From this description, it can be seen that fisheries management involves a complex and wide-ranging set of tasks, which collectively have the achievement of sustained optimal benefits from the resources as the underlying goal. These tasks are also summarized in Figure 1.

There has been a lot of interest in recent years in moving from fisheries management focused essentially on single-species or single fisheries, to management with an ecosystem orientation. This expanded approach has been termed ecosystem-based fisheries management (EBFM) and was recently discussed at "The Reykjavik Conference on Responsible Fisheries in the Marine Ecosystem" (1-4 October 2001), which was organized jointly by FAO and the Governments of Iceland and Norway. The Conference agreed on the Reykjavik Declaration[1] which included an affirmation "that incorporation of ecosystem considerations implies more effective conservation of the ecosystem and sustainable use" and also a reaffirmation of the principles of the FAO Code of Conduct for Responsible Fisheries.

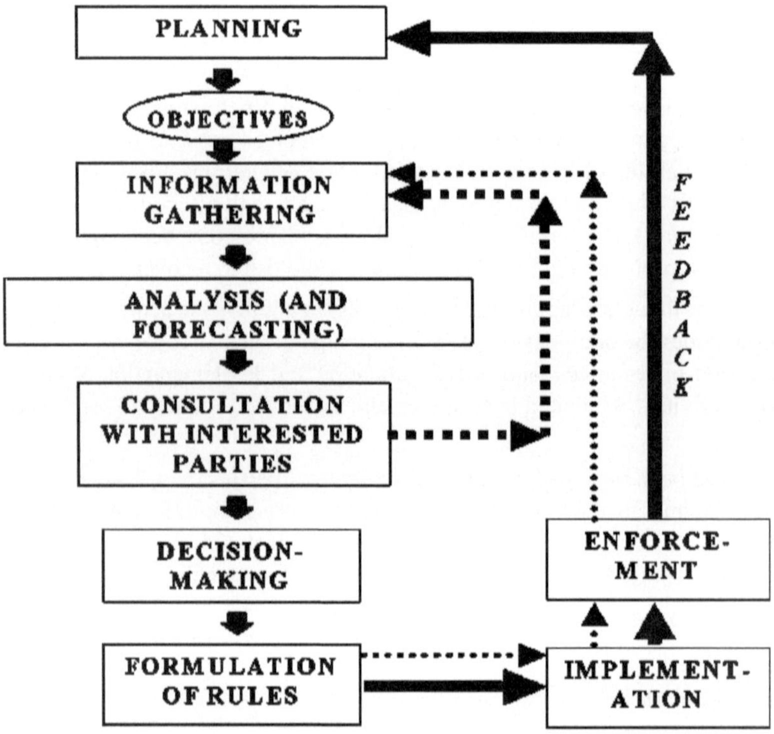

Figure 1 Diagrammatic representation of the functions and responsibilities of a fisheries management authority in relation to fishing, and the inter-relationships between the functions.

What is an Ecosystem?

Ecosystem is a portmanteau word - that is, a word made by jamming two other words together. It originated as a shorthand way of referring to an ecological system, and is now the preferred term.

An ecosystem is simply an easy way to refer to all the life forms (plant and animal) in a cohesive, relatively independent area, and their relationship to one another. In a perfect world, an ecosystem is in balance. That is, the predator species keep the prey species' population in check without completely destroying them, and plant life remains varied without one type predominating and crowding out the rest.

Ecosystems are rarely in balance, and this is often cause for alarm. For instance, pollution runoff from a factory can effect marine life in a lake, causing the fish population to plummet. This destabilizes the entire surrounding ecosystem, and birds and animals that prey on the fish either die off or migrate to areas with more food.

Attempting to counteract damage to an ecosystem requires a complete understanding of all the interrelationships between plants, animals and atmosphere in the system. Since such perfect understanding is rarely possible, unintended consequences are often the result.

For instance, the US government recommended in the 1930s that farmers in the south plant kudzu as a groundcover to help prevent erosion. Now kudzu is a nuisance plant and many thousands of hours a year are logged in trying to keep kudzu from taking over, damaging roadways and buildings and choking out other plant species. A similar takeover of a non-native species is seen in the rabbit problem in Australia. Rabbits have no natural predator native to Australia, so when they were introduced in the nineteenth century, they proliferated like, well, rabbits, to the point where their effects on the ecosystem, particularly farmland, has been devastating.

In nature, no ecosystem can be said to be independent of neighboring ecosystems or indeed, the global ecosystem, since ecosystems are open and plants and animals can move between them. All ecosystems are effected by planet-wide trends that effect the atmosphere, such as pollution and global warming.

One attempt to study an ecosystem in isolation was Biosphere 2, an airtight enclosure that tried to duplicate an ecosystem in balance. Although the experiment was not successful - they were unable to maintain oxygen levels high enough for human health without importing it - the experiment did make clear how fragile the balance of an ecosystem is and how dangerous it can be to severely damage the ecosystem of the entire earth.

What Is Fish Conservation?

Fish conservation is the protection of wild fish populations. Scientists and ecologists who work in fish conservation may work directly with a species of fish, developing policies that protect the future of that species, or they may work to protect and clean up the environment that fish live in. In 2011, fish populations are in decline all over the world, making the conservation of fish species vital.

One component of fish conservation is the responsible management of fisheries. Many species of fish that are caught for food have experienced dramatic population decline in the past few decades. This decline in population can be attributed to a number of different things, including a greater demand for seafood, a change in the temperature of the water, and an increase in pollution. Though the number of fish available has decreased for many fish species, demand for these fish is, in many cases, continuing to rise. Scientists and government agencies must often work together to conserve fish species by imposing strict limits on fishing, both in terms of how many fish can be removed from a given body of water and when those fish can be taken.

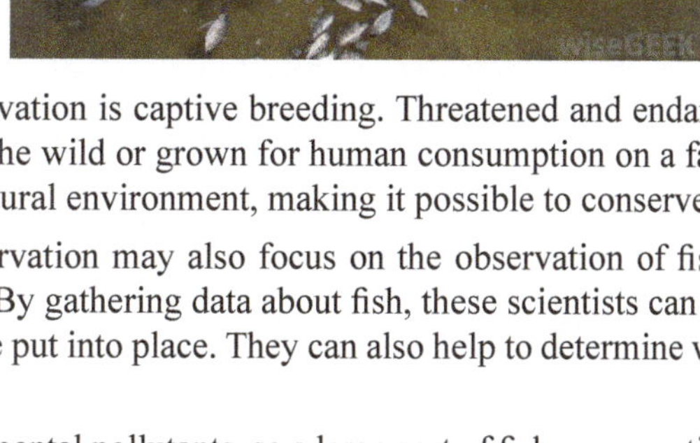

Another component of fish conservation is captive breeding. Threatened and endangered fish species can often be bred in captivity and then either released into the wild or grown for human consumption on a farm. Unlike many animal species, many fish readily adapt to release into the natural environment, making it possible to conserve some species by breeding them.

Scientists who work in fish conservation may also focus on the observation of fish species, aquatic ecosystems, and the environment where the fish are found. By gathering data about fish, these scientists can determine which populations may be in jeopardy so that limits on fishing can be put into place. They can also help to determine what types of environmental factors may be influencing the fish population.

Fish are at great risk from environmental pollutants, so a large part of fish conservation involves protecting the environment. Fish need water not just as a habitat but as a source of oxygen and food as well. When the water in a given area has been polluted, fish may die off or become infertile, decreasing the population of fish in that area. Cleaning up environmental contaminants and putting extra protections in place so that water does not become polluted is a large part of fish conservation.

What Is a Fish Factory?

A fish factory is a fish processing facility where many different species of fish are processed to be sold in the fresh fish and retail market. The fish species that are usually processed in a fish plant include salmon, tuna, herring, mackerel, cod, haddock, hake, and many others. The fish factories may either have special farming facilities to breed their own fish, or they may hire independent fishing trawlers to supply them with fish. They may also have their own trawlers for this purpose.

Since fish are perishable commodities that can deteriorate quickly, it is necessary to preserve them until they can be taken to be processed in a fish factory. Many fishing vessels have freezers to store the fish, and some large fish processing companies even have a fish factory or a cannery row on the fishing vessel itself. Such a factory ship can deliver canned goods directly to the retail industry. Given the rising demand for fish around the world, there has been a corresponding rise in larger and better equipped fish factories.

There are different stages in fish processing. The raw fish is usually first sorted and then it may be preserved as is or it may be skinned and gutted. The fish is then filleted, graded, cut, and trimmed. Depending on how the fish is to be sold, it may be frozen, salted or precooked for preservation. The fish that is to be sold in the retail trade is usually sent to be canned in the cannery row.

Processing fish in this way is not a new concept. There are many instances in history of fishing vessels having specialist fishermen on board to process the fish, or having the fish processed as soon as possible in a coastal fish processing center. Archaeologists have found a Roman fish processing plant in Baelo Claudia in Spain. This fish factory was used to process the Spanish garum fish and it was then exported to Rome.

With increasing fish consumption in modern times, there is a concern about over-fishing and depleting the natural fish populations. To prevent this from happening, many fishing companies either breed their own fish or agree to follow a strict fishing quota. It is also mandatory nowadays for a fish factory to get a food safety certification to show that they are following the correct and hygienic procedures in getting their fish products ready for the consumer market.

The Working Principles of Fisheries Management

The above description presents a complex and possibly confusing picture of all the tasks which need to be considered by the fisheries manager. Some of this complexity can be reduced by attempting to highlight the underlying key issues. There are both benefits and risks in attempting to simplify a subject and over-simplification can lead to neglect of important details. However, simplification can facilitate understanding important principles and highlighting the broad areas which need attention. Arising from the considerations discussed above, a number of key principles can be identified which may serve to focus attention on the starting points for effective fisheries management (Table 1).

Table 1 Suggested fundamental principles of fisheries management

Principle	Management Function	Pertinent Chapters	
1.	Fish stocks and communities are finite and biological production constrains the potential yield from a fishery.	The potential yield needs to be estimated and the biological constraints identified.	1,5
2.	i) Biological production of a stock is a function of the size of the stock andii) it is also a function of the ecological environment. It is influenced by natural or human-induced changes in the environment.	i) Target reference points need to be established through data collection and fisheries assessment; and ii) environmental impacts should be identified and monitored, and the management strategy adjusted in response as necessary.	1, 5
3.	Human consumptive demands on fish resources are fundamentally in conflict with the constraint of maintaining a suitably low risk to the resource. Further, modern technology provides humans with the means, and demand for its benefits provides the motivation, to extract fish biomass at rates much higher than it can be produced.	Realistic goals and objectives must be set. Achieving the objectives will inevitably require controls on fishing effort and capacity.	2 - 5,9 5,6,8

4.	In a multispecies fishery, which description encompasses almost all fisheries, it is impossible to maximise or optimise the yield from all fisheries simultaneously.	Realistic goals and objectives must be established across ecosystems, so as to manage for species and fisheries interactions.	2 - 5
5.	Uncertainty pervades fisheries management and hinders informed decision-making. The greater the uncertainty, the more conservative should be the approach (i.e. as uncertainty increases, realised yield as a proportion of estimated maximum average yield should be decreased).	Risk assessment and management must be done in development and implementation of management plans, measures and strategies.	5,9
6.	The short-term dependency of society on a fishery will determine the relative priority of the social and/or economic goals in relation to sustainable utilisation.	Fisheries cannot be managed in isolation and must be integrated into coastal zone and fisheries policy and planning and national policies.	5-7,9
7.	A sense of ownership and a long-term stake in the resource for those (individuals, communities or groups) with access are most conducive to maintaining responsible fisheries.	A system of effective and appropriate access rights must be established and enforced.	6-8
8.	Genuine participation in the management process by fully-informed users is consistent with the democratic principle, facilitates identification of acceptable management systems and encourages compliance with laws and regulations	Communication, consultation and co-management should underlie all stages of management	7-9

In keeping with the integrated nature of fisheries ecosystems, these principles cannot be considered in isolation in considering how best to manage fisheries: their implications and consequences overlap, complement and confound each other which is what makes fisheries management so demanding and challenging. Nevertheless, the consequences of the principles for fisheries give rise to the fundamental nature and tasks of fisheries management, and hence to the general structure of this Guidebook (Table 1).

Management Plans, Measures and Strategies

There is a lot of terminology floating around in fisheries management that, unless clearly understood, can cause further confusion in an already confusing environment. In addition to the words 'goals' and 'operational objectives', the terms management plans, management measures and management strategies will crop up on many occasions in this Guidebook and we need to clarify what we mean by each of them and how they differ.

The Technical Guidelines on Fisheries Management (FAO, 1997) describe a management plan as "a formal or informal arrangement between a fisheries management authority and interested parties which identifies the partners in the fishery and their respective roles, details the agreed objectives for the fishery and specifies the management rules and regulations which apply to it and provides other details about the fishery which are relevant to the task of the management authority."

As discussed in the previous section, fisheries policy is translated into goals and the goals into objectives that indicate precisely what is expected to be achieved from the fishery. The objectives are achieved through the implementation of a management strategy which will also be a central element of a management plan. The management strategy is the sum of all the management measures selected to achieve the biological, ecological, economic and social objectives of the fishery. It is possible that in a single species fishery a management strategy could consist of a single management measure, such as a specified total allowable catch (TAC), but in practice the great majority of management strategies consist of a number of management measures, encompassing technical, input and output controls and a system of user rights. An effective management strategy, however, should not contain so many management measures that compliance and enforcement become as difficult as to be practically impossible.

Primary Considerations in Fisheries Management

If marine living resources were infinite and indestructible, we could leave people to use and abuse them at will. However, this is not the case and we therefore need to manage fisheries to ensure that the resources are utilized in a sustainable and responsible way, and that the potential benefits are not inefficiently dissipated and possibly totally lost. Fisheries production and yield are constrained by a number of factors which can be classified as biological, ecological and environmental, technological, social and cultural, and economic considerations. There are frequently also considerations imposed by other users of the fishing grounds and neighboring areas.

Biological Considerations

As living populations or communities, aquatic living resources are capable of on-going renewal through the processes of growth in size and mass of individuals and additions to the population or community through reproduction (leading to what in fisheries is often called 'recruitment'). In a population at equilibrium, the additive processes of growth and reproduction on average equal the loss process of total mortality. In an unexploited population, total mortality consists only of natural mortality, made up of processes such as predation, disease, and death through drastic changes in the environment. In a fished population, total mortality consists of natural mortality plus fishing mortality, and a primary task of fisheries management is to ensure that fishing mortality does not exceed the amount which the population can withstand, in addition to natural mortality, without undue harm or damage to the sustainability and productivity of the population. This requires not only that the total population is maintained above a certain abundance or biomass, but also that the age structure of the population is maintained in a state in which it is able to maintain the level of reproduction, and hence recruitment, necessary to replenish the losses through mortality. Further, fishing over a long period on selected portions of a stock, for example large individuals or individuals spawning at a specific time or locality within a wider spawning season or range, can reduce the frequency of the particular genetic characteristics giving rise to that feature or behavior. This has the effect of reducing the overall genetic diversity of the stock. With reduced genetic diversity, the production potential of the population can be adversely affected and it may also become less resilient to environmental variability and change. Fisheries management needs to be aware of this danger and avoid maintaining such selective pressures over a prolonged period.

Achieving an appropriate level and pattern of fishing mortality is hindered substantially by difficulties in estimating population abundance and population dynamics rates and the variability in these rates. The fisheries manager must, however, have sufficient knowledge to make good decisions.

Fishery managers must also respect the stock structure of the resources. Fish populations are frequently made up of a number of different stocks, each of which is genetically largely isolated from the others through behavioral or distributional differences. The different stocks also reflect genetic diversity and if a particular stock is fished to extinction or to very low levels, this genetic diversity may be lost. The stock will not readily be replenished from other stocks, because of the genetic isolation, and therefore the production it was generating will also be lost, leading to a permanent or at least long-term loss of benefits. Fisheries management should therefore attempt to address each stock separately and to ensure sustainable use of each stock and not just of the population as a whole.

Ecological and Environmental Considerations

The abundance and dynamics of a population place an important constraint on fisheries but aquatic populations do not live in isolation. They exist as components of a frequently complex ecosystem, consisting of biological components which may feed on, be fed on by, or compete with a given stock or population. Even those populations which are not directly linked through the food web may indirectly affect each other through their direct interactions with predators, prey or competitors of the other. The physical component of the ecosystem, the water itself, the substrate, inflows of freshwater or nutrients and other non-biological processes may also be very important. Different substrates may be essential for the production of food organisms, for shelter, or as spawning or nursery grounds.

The environment of fish is very rarely static and conditions, particularly of the aquatic environment, can vary substantially over time, from hourly variability, such as the tides, to seasonal variability in, for example, water temperature and currents, to decadal variability as in the occurrence of El Niño events and regime shifts. These changes frequently affect the population dynamics of fish populations, resulting in variability in growth rates, recruitment, natural mortality rates or any combination of these. Such variability can also affect the availability of fish resources to fishing gear, not only affecting the success of the fishing industry, but also the way in which the fishery scientist must interpret catch and catch rate information from the fishery.

Changes in any of the biological, chemical, geological or physical components of the ecosystem can have impacts on the resource population and community. Some of these changes may be beyond human control, such as upwelling processes enriching some coastal ecosystems or large scale temperature anomalies, but they still need to be considered in the management of the resource. Others, such as the destruction of coastal habitats for development, or the direct impact of fishing on the substrate or on other species impacting the resources, are due to human action. In these cases, fisheries management should both take into account their impacts on the resource and, in consultation with other relevant agencies and parties, take steps to minimize their impacts on the fishery ecosystem.

The potential to address the ecosystem considerations will vary depending on whether they are caused by or independent of human action, but in both cases the constraints imposed on the resources and the fishery by biological and non-biological ecosystem factors need to be recognized. At the most fundamental level, these factors in combination with the biology of the species determine the maximum abundance, or carrying capacity, and productivity of the resources. Changes in the ecosystem can affect both and, where they are occurring, need to be considered by the fisheries manager.

Technological Considerations

The fishery manager has very little, if any, ability to influence directly the dynamics of the fish populations or communities which support a fishery. In some case, particularly inland waters, there may be opportunities and a desire to undertake stock and habitat enhancement and in some coastal fisheries, habitat destruction may have had an impact on fish production. This can be done through directly regulating the catch taken, by regulating the amount of effort allowed in the fishery, by specifying closed seasons and closed areas and by regulating the type of gear and fishing methods used. However, there are constraints on how precise the manager can be in setting such regulations. Catch controls are often difficult to monitor and therefore to implement. It is difficult to estimate fishing effort precisely, and normally improving technology and developing skills result in on-going increases in the efficiency of fishing operations, leading to continuing increases in effective effort, unless steps are actively taken to counter these improvements or their consequences. Fishing gear is rarely strongly selective and bycatch of non-target species or unwanted sizes of target species is frequently a problem. The uncertainties in fisheries management are not just at the level of predicting the status and dynamics of the resources, and uncertainties in the real consequences of implementing fishery measures is also a significant problem to the manager.

A fundamental problem in many fisheries is the existence of too much effort. The presence of excess effort will frequently result in on-going pressure on the fisheries manager to exceed the sustainable fishing mortality on a resource. The social and political pressure to provide employment and opportunities for all those with a stake in the fishery is often hard to resist and readily leads to over-exploitation.

Social and Cultural Considerations

Human populations and societies are as dynamic as other biological populations, and social changes take place continuously and on different scales, affected by changes in weather, employment, political circumstances, supply of and demand for fisheries products and other factors. Such changes can affect the appropriateness and effectiveness of management strategies, and therefore need to be considered and accommodated by them. However, again as with biological and technological factors, it can be difficult to identify and quantify the key social and cultural factors influencing fisheries management, generating additional uncertainties for the manager.

A major social constraint in fisheries management is that human societies and behavior are not easily transformed and fishing families and communities may not be willing to move into other occupations, or away from their normal homes when there is surplus capacity in a fishery, even when their quality of life may be suffering as a result of depleted fish resources. The problem is much worse when there are no other opportunities outside of fisheries in which they could earn a basic living. Under such circumstances, the political decision to reduce capacity in the fishery is an extremely unattractive option, as the short-term costs of excluding dependent people from the fishery will be much more visible and hence unpopular than a "hands-off" approach which allows the resource and fishery to dwindle in magnitude and quality under sustained excess fishing mortality. Nevertheless, the ecological, economic and social consequences of the latter choice are far more serious in the longer term. This reluctance or inability to take decisions with serious, immediate social consequences for some has been one of the constraints most responsible for over-fishing around the world.

A key requirement for ensuring that social and cultural considerations are properly considered is to involve the interested parties in fisheries management, keeping them well-informed on the management aspects of the fishery and providing them with the opportunity to express their needs and concerns.

The relative balance between social and economic considerations in a fishery will depend on the priority given by the appropriate authority to social objectives and economic objectives. Social and economic objectives can conflict: for example it is unlikely that maximizing economic efficiency and maximizing employment could be simultaneously pursued within a given fishery, and attempting to do so will result in conflict. A common example of such conflicts is that between a commercial fleet pursuing essentially economic objectives and an artisanal fleet fulfilling primarily social objectives, with both having an impact on the same stock, and possibly also interfering with each other's fishing operations. It is important for the management authority to have identified such potential conflicts and to have resolved them, identifying and specifying compromise objectives that achieve general support.

Economic Considerations

In a fishery for which sustainable economic efficiency had been specified as the sole benefit to be extracted and in which optimum circumstances prevailed, market forces could be anticipated to lead to the desired objective of economic efficiency. However, in reality such optimum conditions are rarely if ever found and uncertainties and externalities distort the natural selection of market forces. Uncertainties include unpredictable variability in resources and other sources of imperfect information, and externalities can include the impacts of other fisheries on the target resources (e.g. taking them as bycatch), subsidies, trade regulations, fiscal regulations and variability in markets and demand. All of these introduce complexity and additional uncertainty into a fishery and, without proper management, will lead to sub-optimum economic performance. It is important for the management authority to consider the broad economic context of a fishery, including relevant macroeconomic factors. As with social considerations, this requires close consultation with the legitimate users who will be the ones most affected by and sensitive to these issues.

At one extreme, although still very common in fisheries especially in many developing countries, are the problems of open access fisheries, in which anyone is allowed entry into a fishery.

Considerations Imposed by Other Parties

Some offshore fisheries operate in effective isolation from any other users and the regional fisheries organizations charged with their management may be able to manage the fisheries without needing to consider conflicts with or interference from non-fishery users. However, the bulk of global fishery landings come from coastal waters and for many if not most of the fisheries producing these landings, other users are an important consideration and frequently a constraint. Other users of the fishing grounds can include, for example, tourism, conservation, oil and gas extraction, offshore mining and shipping, while use of the intertidal and coastal area can include tourism again, aquaculture and mariculture, coastal zone development for housing, business or industry, and agriculture. All of these can impose significant constraints on fishing activities and may be impacted by fishing activities. The manager therefore needs to be aware of such activities and of real or potential impacts in both directions. When developing management strategies and formulating management measures, potential conflicts with other users need to be identified and addressed, and the potential impacts of other users on the efficacy of the management strategy and measures need to be considered. The strategy must be adapted so as to account for and be robust to these impacts.

CHAPTER 1 COMMON FISHERIES POLICY

Jesper Raakjaer and Troels Jacob Hegland

INTRODUCTION

The idea of putting together a special issue of MAST on the issue of regionalisation of the Common Fisheries Policy (CFP), the fisheries policy framework of the European Union (EU), was born in late 2010. Having participated in an EU funded research project looking into how an eco-system based approach to fisheries management could be operationalised in the EU with particular focus on regionalisation, we found that the coming reform of the CFP would be a good opportunity to make a substantial contribution on the topic of regionalisation, which we felt ought to be a central component of the reform discussions.

Consequently, in early 2011 we wrote down some preliminary thoughts on what articles the special issue might contain and took contact to the editorial team of MAST to enquire whether they saw a perspective in our ideas. The response from MAST was very positive. Thus, encouraged by the support from MAST we took contact to some of those of our colleagues, who we felt would be able to contribute, to see if there was any interest in the idea, which indeed turned out to be the case, effectively turning the idea of making a special issue on regionalisation into a project that has consumed considerable time throughout 2011.

A particular challenge of making this special issue has been the fact that while the policy-process leading up to the reform of the CFP, which is expected to find its form later this year, has been on-going, the contributions of this special issue have had to be finalised considerable time in advance of actual publication to allow time for the peer-review process and printing, *et cetera*. In effect, this means that the contributions of this special issue have been finalised in September 2011 and the conclusion in January 2011, which means that it has not been possible to take developments in the policy-process after that into consideration. This is in particular important to point out in relation to David Symes' article, as well as in relation to the overall conclusion of the issue, since both takes stock of the political situation at the time of writing. The last major milestone of the policy-process that the authors of this special issue have had the possibility to take into consideration has been the proposal for a new basic regulation, released by the Commission of the European Communities (Commission) in July 2011 ([Commission 2011]). And, as will be evident

Citation: Jesper Raakjær, Troels Jacob Hegland, Common Fisheries Policy, DOI: 10.1186/2212-9790-11-5.

in Symes' contribution ([Symes 2012]) and the conclusion ([Raakjær et al. 2012]), this proposal raises significant concern as regards to how genuine the interest of regionalisation in the Commission in reality is.

Our hope is that—although the reform of the CFP is likely close to being in place at the time of publication—the special issue can prompt further consideration of regionalisation of the CFP as it is approaching its next crossroad.

In this introduction, we provide initially a little bit of background on why regionalisation might be seen as particularly interesting in the context of the CFP. Subsequently, we introduce the different contributions in brief.

REGIONALISATION AND REFORM OF THE COMMON FISHERIES POLICY

As stakeholders, researchers, administrators, and politicians alike struggle to find long-lasting and innovative 'technical' solutions to put the CFP on a sustainable track and create structures that facilitate the move towards ecosystem-based fisheries and marine management, discussions of changes to the way that the CFP governance system operates, in terms of how it makes decisions and what roles it attributes to various actors, have intensified.

Over the last couple of years, the governance option of regionalising the CFP has become one of the most interesting possibilities—arguably among many interesting topics—in the debate about the future of the CFP. Regionalisation was already considered in connection with the previous reform of the CFP in 2002, which in that respect led to the setting up of seven stakeholder-led, industry-dominated so-called Regional Advisory Councils (RAC).

On a basic level, regionalisation features potentially two elements that may be seen as instrumental in a successful reform of the CFP: the 'moving down' and the 'moving out' of fisheries management and decision-making authorities currently held by the central level EU institutions. 'Moving down' refers to the fact that regionalisation responds to the concern of the limited efficiency and effectiveness of the CFP by relieving the central EU level institutions of tasks by enabling lower level authorities to step in and design more tailor-made management for particular areas. Similarly, 'moving out' refers to the potential of regionalisation leading to increased involvement of stakeholders in the fisheries management process by transferring authorities from pure public institutions to public-private cooperative institutions or the fisheries sector itself, which indicates that regionalisation might have a potential to make the CFP more inclusive than it is at the present.

Although the extent of underperformance of the CFP can be debated, one would be hard-pressed to find voices arguing that the policy has delivered satisfactory results over the years since set up in 1983. Recent reflections on the performance of the CFP ([Sissenwine & Symes 2007]; [Commission 2009]; [Raakjær 2009]) paint a rather dismal picture: Many fish stocks are overfished and some are on the brink of collapse; notably the situation is regarded as worse than in compatible areas elsewhere in the world. Similarly, the EU fisheries sector is generally characterised by poor profitability with sector employment steadily declining. In addition, the EU is facing intensive competition from freshwater and marine aquaculture production making the market extremely competitive.

The Commission's latest CFP Green Paper, which was released 22 April 2009 and kicked of the process of preparing for reform, identified five main structural failings of the policy that the next reform first and foremost had to focus on:

A deep-rooted problem of fleet overcapacity;

Imprecise policy objectives resulting in insufficient guidance for decisions and implementation;

A decision-making system that encourages a short-term focus;

A framework that does not give sufficient responsibility to the industry;

Lack of political will to ensure compliance and poor compliance by the industry. (([Commission 2009]):7)

The Green Paper suggests various ways of addressing the above structural failings as well as discusses other areas of the policy framework that it would possibly also be beneficial to include in the reform, which according to the Commission needs to be 'whole-scale and fundamental' (([Commission 2009]):4) to deliver sustainable fishing. By examining with a critical eye the current style of management, where almost all decisions are taken at the highest political level in Brussels, the Green Paper confirms significant problems facing the CFP in relation to the style of governance employed.

One of the reform options presented in the Green Paper is to resort to more specific regional management solutions and let the EU level decision-making framework, in the shape of particularly the Council of the European Union (Council) and European Parliament (Parliament), focus more on core long-term principles. Such an approach would, according to the Green Paper, likely involve member states organising at the level of marine regions to decide on and implement management solutions subject to standards and control from the EU level. This strategy is perceived to have a number of advantages:

"Putting policy in its right place by letting Council and Parliament focus on principles and delegating implementation decisions [...] would make implementation more sensitive to specific local conditions and give the industry more responsibility in shaping its own destiny. It would enable governments and the industry to adapt the implementation of the policy to their needs and to find the best solutions both technically and economically.(([Commission 2009]): 10)*"*

However, otherwise the Green Paper remained weak in terms of giving specific directions or suggestions on how regionalisation could be achieved in practice. Following its publication in April 2009, stakeholders and other interested parties were invited to take part in a public consultation on the content of reform to which submission of comments had to be made by 31 December 2009. Considering the vague directions of the Green Paper, it was basically left to those being consulted to flesh out the different, possible governance models that could provide for regionalisation.

Although, arguably, this created a very open consultation process, it also catered to a process where the consulted parties had to start from scratch and spend considerable effort figuring out which models were available instead of examining and discussing which of the available models they would prefer. Notably, they had to do this between preparing responses to the many other issues put on the agenda by the Green Paper. The challenge was not lessened by the fact that there was and still is—as recently described by Long ([Long 2010])—uncertainty about the legal limitations imposed by the overall policy framework of the Lisbon Treaty in relation to increasing the regional element of the governance system of the CFP.

The Commission's synthesis of the public consultation was published (as a Commission staff working document) on 16 April 2010 and contained a few paragraphs on regionalisation, which deserves to be cited here:

"Support for a move to some form of increased regionalization is generalised. A mix of terminology is used—e.g. some refer to regionalization at the sea-basin level, others to regions within the Member States. Some insist on subsidiarity where Member States create regional committees for management and a high level of self-regulation. Others propose simple co-operation between Member States on issues of implementation and control, and some see room for delegated decision powers on e.g. access, resource or fleet management."

"A significant number identifies the need for a separate regional body, with varying degrees of powers and responsibilities. Most, including the European Parliament, envisage a mainly advisory body to discuss and prepare proposals for policy and legislation adoption by the EU institutions. Associating the stakeholders and others involved, the regional body would then be used for dialogue and discussion. [...] Others envisage a regional body as the implementing entity for long-term plans with some room for operative regulating powers and implementation decisions. Some advocate devolution of powers (e.g. technical details and effort regulation). Some contributions suggest a combination of functions."

"On the composition most see the regional body as a Member State-led entity, in a number of cases membership of industry and stakeholders is advocated, while in other contributions the stakeholders keep an advisory function through the RAC. The European Commission is envisaged as a member in some contributions while in others as an active observer/collaborator. Some propose a transformation of the RAC into a regional advisory body with both Member States and stakeholders.(([Commission 2010]):6)*"*

Clearly the consultation provides the EU with a clear mandate, if not an obligation, to move toward 'some form of increased regionalisation' of the CFP. However, it is equally clear from the above that the consultation did not provide the Commission with a clear indication of how to move forward; every imaginable model seems to be in play. The Commission also notes the employment of mixed terminology and different conceptions of what regionalisation entails. Nevertheless, there seems to be general agreement on some key aspects, such as keeping overarching principles and policy objectives as a task for the EU level and that micromanagement should be avoided.

As mentioned earlier, already in connection with the 2002 reform steps towards some degree of regionalisation had been taken with the setting up of RACs, predominantly structured along marine regions, and the incorporation in the basic regulation of the stated objective of the EU to move towards 'a progressive implementation of an eco-system-based approach to fisheries management' (([Council 2002]): Article 2(1)) also highlights the centrality of marine ecosystem regions in the CFP of the future.

The link between regionalisation and the thinking behind the ecosystem approach can be exemplified by the Convention on Biological Diversity ([United Nations 1993])—traditionally identified as one of the original sources of the idea of an ecosystem approach to managing the environment, which outlines twelve guiding principles for the implementation of an ecosystem approach. Although the principles are—unlike the convention itself—non-binding on the contracting parties, they provide good guidance in the underlying philosophy of ecosystem-based management. Notably, several of these principles place regionalisation high on the agenda in the effort to turn towards ecosystem-based fisheries or marine management. A handful of the twelve principles are potentially facilitated by regionalising the CFP depending on how it is carried out in practice. However, in the context of the CFP three of the twelve principles are particularly related to the idea of regionalisation and the 'moving down and out' of policy-making:

"Principle 2: Management should be decentralized to the lowest appropriate level."

"[...]"

"Principle 7: The ecosystem approach should be undertaken at the appropriate spatial and temporal scales."

"[...]"

"Principle 11: The ecosystem approach should consider all forms of relevant information, including scientific and indigenous and local knowledge, innovations and practices."

Consequently, as also described by Berghöfer, Wittmer, and Rauschmayer ([Berghöfer et al. 2008]), ecosystem-based fisheries management can be seen as closely linked to the idea of regionalisation, although it cannot be argued that regionalisation is a precondition of ecosystem based management under the CFP, as such.

The implementation of a true ecosystem approach to fisheries management will have to entail taking into account that appropriate management measures in the context of one ecosystem might be sub-optimal or even counter-productive in the context of another ecosystem. Thus a CFP based on the principles of the ecosystem approach will increasingly have to be sensitive to the specific characteristics of the various ecosystems and not least the fisheries taking place within them as an integral part of the ecosystem.

CONTENTS OF THIS SPECIAL ISSUE

Although the possible change of CFP governance towards regionalisation is the overarching theme that binds the different contributions of this special issue together, there are other underlying themes that to varying degrees go through them; some of which we have already touched upon earlier in this introduction. One such sub-theme is the understanding of the EU as increasingly an instance of so-called multi-level governance (most clearly reflected in the contribution by van Hoof, van Leeuwen, and van Tatenhove, see beneath). Another thematic is the implications of the on-going refocusing of management towards a more holistic approach both in fisheries in isolation but also in the context of marine management at large. Similarly, the alleged discrepancy between what is needed and preferable in governance *versus* what is readily possible within the legal provisions both within the CFP but also broader looms underneath the surface in several contributions. As a final example, the issue of differences in terms of readiness to tackle and embrace regionalisation in various regional sea areas or RACs of the EU is also touched upon in more contributions.

In the first article of this special issue, the leading expert on and founding father of the thinking behind CFP regionalisation, David Symes (Symes [2012]), sets the scene for the following contributions by offering a comprehensive overview of the history of regionalisation in relation to EU fisheries management, hereby documenting the apparent need for regionalisation—not as an end in itself but a means of achieving many of the CFP›s objectives and establishing basic legitimacy for the policy, which seems to be all the more needed in the future, but also explaining why the EU remains slow to adopt the idea of regionalisation. Symes rounds off his contribution with a—not all too optimistic—discussion of the implications of the Commission's proposals for reform published in July 2011 *vis-à-vis* regionalisation.

Departing from the same conclusion arrived at by the Commission in its synthesis of the public consultation, namely that the debate on regionalisation is further complicated by the lack of a common terminology, Troels Hegland, Kristen Ounanian and Jesper Raakjær ([Hegland et al. 2012a]) outline in the second article a conceptual framework for discussing regionalisation. The authors provide initially a typology of the many different theoretical benefits that regionalisation might be perceived as having. Subsequently, the discussion of regionalisation is disentangled into three key problem dimensions, being the issues of 'who', 'where' and 'what'. In the final part of the article a selection of five so-called 'archetypes' of regionalisation are outlined. The selected archetypes, which are rough skeletons of possible, regionalised governance models for the future CFP, capture important perceptions of ways to go forward towards regionalisation in practice.

In the third article the same set of authors go on to provide an insight in what regionalisation means to people by presenting material collected primarily by means of interviews and a survey of RAC meeting participants ([Hegland et al. 2012b, in press]). The article documents both the empirical variety of motivations for wanting to regionalise the CFP, as well as the variety of perspectives on how this can be done. Notably, by utilising the developed 'archetypes' of regionalisation and the survey method, Hegland, Ounanian and Raakjær provide an insight into how RAC meeting participants evaluate different regionalisation models when confronted with them.

In the fourth article of the issue, Luc van Hoof, Judith van Leeuwen and Jan P.M. van Tatenhove ([van Hoof et al. 2012, in press]) broaden our perspective from focussing on fisheries management to also considering marine management more broadly by focussing on the CFP and the Marine Strategy Framework Directive (MSFD) together. A major challenge of future EU fisheries management is the integration of fisheries management with broader marine management. The focus on ecosystem based management, which is in itself one of the drivers of regionalisation, is also one of the factors that complicates regionalisation. Departing from a multi-level governance framework for analysis, this article discusses some of these issues by looking specifically into the implications of integrated management for regionalisation.

Based on a survey of RAC meeting participants, Kristen Ounanian and Troels Hegland ([Ounanian & Hegland 2012, in press]) return in the fifth and final article of the special issue to illuminate the current capacities and functions of the RACs. The paper reveals that the RACs possess additional—often not sufficiently recognised—roles and values to the advice they produce, as they facilitate understanding across and within sectors and interest groups and act as key purveyors of information. The previous reform of the CFP introduced the RACs to enhance stakeholder involvement and correct one of the policy's primary deficiencies, its lack of legitimacy. While some have criticize the 2002 reform as not going far enough to alleviate the democratic deficit of the CFP, in certain ways the RACs represent an interim institutional stage, facilitating better information sharing and cultivating stakeholder relationships, thereby to some extent paving the way for further regionalisation.

In a final conclusion ([Raakjær et al. 2012, in press]), the policy implications of the contributions will be discussed, which add to the understanding of the benefits, challenges and drawbacks of regionalisation as a way forward for the Common Fisheries Policy. Importantly, moreover, the conclusion also offers some words of advice *vis-à-vis* the coming reform of the CFP, which—it is the conclusion of the authors—needs to take the possibility of regionalisation seriously—however legally and in other ways complicated that route may seem.

REFERENCES

1. Berghöfer, A, Wittmer, H, and Rauschmayer, F. 2008. Stakeholder Participation in Ecosystem- Based Approaches to Fisheries Management: A Synthesis from European Research Projects. Marine Policy 32: 243–253.
2. Commission, 2009. Green Paper: Reform of the Common Fisheries Policy. COM (2009) 163. Brussels: Commission of the European Communities.
3. Commission, 2010. Synthesis of the Consultation on the Reform of the Common Fisheries Policy. Commission Staff Working Document. SEC (2010) 428.
4. Commission, 2011. Proposal for a Regulation of the European Parliament and of the Council on the Common Fisheries Policy. COM (2011) 425. Brussels: Commission of the European Communities
5. Council. 2002. Council Regulation (EC) No 2371/2002 of 20 December 2002 on the conservation and sustainable exploitation of fisheries resources under the Common Fisheries Policy. Official Journal of the European Communities L 358/59(31.12.2002): 61–80
6. Hegland, TJ, Ounanian K, and Raakjær, J. 2012a. Why and how to regionalise the Common Fisheries Policy: a conceptual framework. Maritime Studies 2012, 11: 7.
7. Hegland, TJ, Ounanian, K, and Raakjær, J. 2012b. What Does 'Regionalisation' Mean? An Exploratory Mapping of Opinions on Reform of the Common Fisheries Policy. Maritime Studies 2012, In press

8. Long, R. 2010. The Role of Regional Advisory Councils in the European Common Fisheries Policy: Legal Constraints and Future Options. The International Journal of Marine and Coastal Law 25: 289–346.
9. Ounanian, K, and Hegland, TJ. 2012. The regional advisory councils: Current capacities and unforseen benefits. Maritime Studies 2012, In press
10. Raakjær, J. 2009. A Fisheries Management System in Crisis - the EU Common Fisheries Policy. Aalborg: Aalborg University Press.
11. Raakjær, J, Degnbol, P, Hegland TJ and Symes, D. 2012. Conclusion. Regionalisation Aborted or Put on Hold. Maritime Studies 2012, In press.
12. Sissenwine, M, and Symes. D. 2007. Reflections on the Common Fisheries Policy. Report to the General Directorate for Fisheries and Maritime Affairs of the European Commission.
13. Symes, D. 2012. Regionalising the Common Fisheries Policy: Context, Content and Controversy. Maritime Studies 2012, 11:6 United Nations, 1993. Convention on Biological Diversity [http://www.cbd.int/ecosystem/principles.shtml]
14. van Hoof, L, van Leeuwen, J and van Tatenhove, JPM. 2012. All at Sea. Regionalisation and the Integration of Marine Policy. Maritime Studies 2012, In press.

CHAPTER 2 FISHERY RESOURCES: BETWEEN ECOLOGY AND ECONOMY

Angelo Tursi, Porzia Maiorano, Letizia Sion, Gianfranco D'Onghia

Keywords: Fishery resources, Ecosystem approach, Management, Ecology, Economy

FISHERY RESOURCES, FISHERIES AND SYSTEM COMPLEXITY

Fishery resources are represented by populations of animal and plant organisms used by man, especially for human food purposes to produce fishmeal for animal husbandry, aquaculture products as well as crafts and decoration of various kinds.

The populations of aquatic organisms are generated and maintained through complex biotic and abiotic interactions within the marine ecosystem. Their texture and evolution are linked to numerous selective factors that act in the ecosystem (hydrographic conditions, productivity, predation, etc.) and human activities over the past two centuries, first of all fishing. Therefore, in the natural system, governed by the laws of biological evolution, relevant actions affecting population structures and communities of organisms are caused by a single species, man. This fact immediately highlights how the ecological dimensions of natural systems intersect with the social and economic ones of human communities (Cataudella and Spagnolo 2011).

Fishing can be defined as the collection or capture of wild organisms from the oceans, seas and inland waters (lakes, rivers, streams, etc.) and is one of humanity's oldest activities. Fishing is an important economic activity which provides food from the sea and thereby creates employment not only for whoever (the fishermen) directly collects that food (fish, crustaceans, molluscs) but also for those who operate in the entire chain, from the construction of vessels and equipment supplies related to the marketing of products. Therefore, fishing is one of those human activities that includes ecological (about populations and aquatic ecosystems), socioeconomic (those working in the sector), technological (boats, motors, tools, etc.) and, consequently, political and administrative concerns. Organisms exploited by fisheries mainly belong to food chains whose first link is microscopic algae with distribution limited to the availability of light, more or less in the first 200 m of depth. Primary productivity due to phytoplankton is related not only to the availability of light but also to that of nutrients (nitrogen, phosphorus, iron, etc.) as well as to particular subsidiary energy sources (upwelling currents, tidal flows,

Citation: Angelo Tursi, Porzia Maiorano, Letizia Sion, Gianfranco D'Onghia, Fishery resources: between ecology and economy, DOI: 10.1007/s12210-014-0372-3.

etc.). Estimates by Pauly and Christensen (1995) indicate that on average 8 % of the global primary productivity of the sea supports global fishing or, in other words, this is the percentage of primary productivity that becomes food for humans. This percentage drops in the open ocean (approximately 2 %) and increases in coastal areas and in areas with rising currents (between 24 and 35 %), confirming the increased productivity of these latter environmental systems in terms of resources exploited by man.

The study of fishery resources implies a substantial leap in the complexity typical of ecological studies of the connection and interaction between the parties (living forms as well as physical, chemical and biological factors, climatic conditions, etc.) which determines a unique and unrepeatable scenario. The study of relations and connections between natural phenomena on different space–time scales requires the integration of different scientific disciplines. Economics and social sciences play an important role in issues related to the management of fisheries' resources.

FISHERY RESOURCE MANAGEMENT: BETWEEN ECOLOGY AND ECONOMY

Over the millennia and until the nineteenth century, man could not even remotely conceive that our actions could result in the depletion or even the extinction of exploited populations and indeed believed that fish stocks were inexhaustible and so all that was necessary to collect a larger amount was to spend more time and improve the means employed in their procurement, i.e., increase the so-called "fishing effort". In fact, while the populations of aquatic species are renewable, they are not inexhaustible. The exploitable fraction of the population is defined as the "stock" and is characterized by its own dynamics related to the contribution of new individuals (recruitment) and their growth as well as the mortality rate of individuals. This latter is both by nature and in relation to the capture of individuals by the fisheries (fishing mortality). The increase in catches occasionally recorded in some areas is generally due to the increase in the fleet and improved fishing technology rather than to greater availability of resources in the sea. According to the latest statistics, the total world production was 158 million tonnes in 2012, of which 91 million tonnes referred to fishery captures and 67 million tonnes to aquaculture production (FAO 2014). The fishing effort maintained at high levels may result in the collapse of one or more resources in the distribution areas. Currently, over 70 % of world fish stocks are fully exploited, about 20 % overexploited and only 10 % of stocks are underexploited by fishing (FAO 2014).

An unregulated fishery causes overfishing of resources both in biological and economic terms. In bio-ecological terms, overexploitation of resources can cause not only the depletion of stocks of the species of interest to fishing activity (target species) but also of many others of lower or no economic value, but nevertheless important ecologically, caught accidentally together with the first. The main objective of fisheries' management has been to safeguard fish stocks or, in other words, find a level of exploitation that could provide the maximum yield by weight in a lasting way or maximum sustainable yield (MSY). In parallel, in economic terms, the main objective is the achievement of maximum sustainable economic performance or MSE (King 1995). In the face of the different issues concerning the fisheries, management objectives at present, in addition to the protection of individual stocks affected by fishing, extend to the protection and preservation of species and the marine ecosystem to maintain economic viability as well as the preservation of jobs.

THE SITUATION IN THE MEDITERRANEAN BASIN

The Mediterranean is a semi-enclosed sea with not only high variability of coastal morphology and a high diversity of habitats and living resources mainly on the continental shelf (on average within 200 m of depth), but also a high diversity of anthropic pressures (civil and industrial discharges, farming and livestock, fisheries, coastal tourism, marine traffic, etc.). Its waters are considered "oligotrophic" that is poor in nutrients, and thus not very productive. Nevertheless, fishing activity has ancient traditions with very high-value resources and diversity of species. In fact, the Mediterranean fishery is known to be multi-specific, because the commercial catches are characterized by a high number of species, contrary to North Atlantic fisheries which maximize the capture of single stocks. In Mediterranean, the monospecific catches occur with small and large pelagic fish. Therefore, this semi-enclosed basin characterized by mainly multispecific catches but also by artisanal or coastal activity results to be a mosaic with many diversified fisheries and gears, along more than 45,000 km of coastline (Lleonart and Maynou2003; Relini 2003).

The fishery resources are constituted by five different groups with different problems and issues: small pelagics, large pelagics, demersals, bivals (mussels, oysters, clams) and others (sponges, coral, algae etc.). Small pelagics provide large catches based mainly on sardine (*Sardina pilchardus*) and anchovy (*Engraulis encrasicolus*). The most abundant species among large pelagics are bluefin tuna (*Thunnus thynnus*) and swordfish (*Xiphias gladius*) principally caught by the industrial fleet working in the Mediterranean. Demersal resources constitute about 55 % of the Mediterranean catch and represent those organisms which live and feed close to the seabed. About a hundred of species are exploited and the most important are: hake (*Merluccius merluccius*), greater forkbeard (*Phycis blennoides)*, blue whiting (*Micromesistius poutassou*), red mullets (*Mullus surmuletus* and *M. barbatus*), red shrimps (*Aristeus antennatus* and *Aristaeomorpha foliacea*), Norway lobster (*Nephrops norvegicus*), rose shrimp (*Parapenaeus longirostris*) and some cephalopods (*Octopus*, *Sepia*, *Eledone,* etc.). The fleets fishing on the small pelagic and demersal stocks is semi-industrial or artisanal (Lleonart and Maynou 2003; Relini 2003).

In the Mediterranean–Black Sea, declining catches were observed, with decreases of 15 and 30 % since 2007. The Global Capture Production of fishery resources (fish, molluscs, crustaceans) was about 1.4 and 1.2 million tonnes in the 2011 and 2012, respectively, observing a variation of less 10.8 % (FAO 2014). Although, this amount seems to be negligible with respect to the world catch, its average economic value is generally greater than that at world level. For example, the Mediterranean fisheries provide about 20 % of the total catch in weight of the European Community but 35 % of the total economic value (Lleonart and Maynou 2003).

From the 1950s, fishing activities have been extended

toward deeper grounds, as deep as 700–800 m in many Mediterranean basins, with the aim of increasing the capture of living resources. This has occurred mainly along the Spanish coasts, in the Tyrrhenian Sea, Sicilian Channel and Ionian Sea.

On the basis of the latest data, most of the stocks of the Mediterranean are either fully exploited, or overexploited. Particularly hake (*Merluccius merluccius*) and red mullet (*Mullus barbatus*) stocks are considered overexploited. The main stocks of small pelagic fish (sardine and anchovy) are assessed as either fully exploited or overexploited. In general, the Mediterranean and Black Sea had 52 % of assessed stocks fished at unsustainable levels, and 48 % fully or underfished in 2011 (GFCM 2011; FAO 2014).

The Mediterranean aquaculture production has increased extensively over the past 10 years (25 % per year). Presently, aquaculture production is more than 200,000 tonnes per year for marine production and more than 500,000 tonnes for brackish water production (without the freshwater production, which is also quite important in countries such as Egypt, Turkey and Italy). It complements fisheries' production and is becoming one of the main sources of aquatic products for many countries of the Mediterranean (GFCM—General Fisheries Commission for the Mediterranean 2011).

With the Italian law relative to the rationalization and development of sea fishing (Law n. 41, 17.02. 1982 *Piano per la razionalizzazione e lo sviluppo della pesca marittima*) no longer in force, systematic studies were promoted by the then Ministry of Merchant Navy (now Ministry of Agriculture, Food and Forestry Policy) to monitor the distribution, abundance and exploitation status of the fishery resources in the Italian seas. Thus, the knowledge on the bio-ecology of many species has been acquired in the Italian basins where fishing techniques and typologies often reflect resource distribution and availability. Some of these studies are still ongoing and they are currently funded by the European Union. In particular, the European Commission has financed the MEDITS project (Bertrand et al. 2000) since 1994 (International bottom trawl survey in the Mediterranean) which regards the demersal resources of the Northern Mediterranean basins and several Member States. Since the year 2000, some EC Directives (e.g., EC Reg. 1543/2000, EC Reg. 1639/2001) have implemented the "Data Collection" program in the Member States, including MEDITS and other projects aimed at the assessment and management of the marine living resources in the European seas.

ECOSYSTEM APPROACH TO FISHERIES (EAF)

Taking into account the fishing effects on organisms, populations and ecosystems, the exploitation of living resources represents one of the environmental issues now tackled on a global scale and according to the principles of sustainability. In fact, most aquatic ecosystems are affected by fishery activities that involve a selective removal of part of the natural production for human subsistence, economic returns and development.

To promote long-term sustainable fisheries, in 1995 the FAO Conference adopted the Code of Conduct for Responsible Fisheries (FAO 1995). The need for an integrated approach for the management of marine resources, in the broadest sense of the word, finds in the Code a summary of how to achieve long-term sustainable use of fisheries' resources in terms of food, employment, recreation and trade as well as ecosystem and socio-economic well-being of populations throughout the world. Even though certain parts of the Code are based on relevant rules of international laws, it is voluntary and provides principles and standards applicable to the conservation, management and development of all fisheries. It also covers the capture, processing and trade of fish and fishery products, fishing techniques and effort, aquaculture and fisheries' research. The Code has been adopted by the over 170 member Governments of the FAO Conference. Its application will be effective when the member Governments translate its principles and objectives into relevant policies and legislation.

The international conventions adopted over the last two decades, including the Code of Conduct for Responsible Fisheries, consider the exploitation of living resources on an ecosystem basis; they stress the need for the adoption of an ecosystem approach to fisheries (EAF). The principles of an EAF are an extension of the conventional principles for sustainable fisheries' development to cover the ecosystem as a whole (Garcia et al. 2003). This implies sustainable management not only of the commercial stocks but also of the whole environmental system which supports their production, including the importance of the economic and social dimension (Spagnolo 2006).

The EAF foresees the integration of different practices and measures to deal effectively with complex situations with respect to a variety of needs and demands, from ecological to socio-economic. The EAF aims to achieve an equilibrium between conservation and sustainable use of biological diversity. In particular, it attempts to satisfy the three components of sustainability, which are the ecological dimension (effectiveness-reproducibility of resources), the economic dimension (efficiency) and the social dimension (equity).

Although the beginning of a Common Fishery Policy (CFP) goes back to the 1980s, the Green Book on the future of the CFP was only presented in 2001 on the basis of the Code of Conduct for Responsible Fisheries. This document identifies the limits of the fishery policy that have been adopted by the European Community and the principles on which to base the reform process of the fishery and aquaculture. In this respect, EC Regulation 2371/2002 strongly promotes the sustainable exploitation of resources from a socio-economic as well as an ecological point of view. Even more importantly, EC Regulation 1967/2006, concerning management measures for the sustainable exploitation of fishery resources in the Mediterranean Sea, puts in place the Action Plan for the Mediterranean Sea as part of the Common Fishery Policy (Cataudella and Spagnolo 2011; Cataudella et al. 2011).

With this Regulation, new rules were established for the protection and conservation of Mediterranean resources in reference not only to commercial species but also to protected species and sensitive habitats. The recommendations of the General Fisheries Commission of the Mediterranean (GFCM) and the International Commission for the Conservation of Atlantic Tuna (ICCAT) as well as the measures of species protection and habitat conservation reported in the Habitat Directive (92/43/EEC) and in the Barcelona Convention are included in the Regulation.

EC Regulation 1967/2006 finally acknowledges that the biological, social and economic characteristics of the Mediterranean fisheries require the Community to establish a specific management framework (EC Reg. 1967/2006).

The regulation of fishing effort is the main measure to favor sustainable fishing in the Mediterranean. In particular, taking into account the differences in the distribution of living resources between the various geographic areas and the consequent different fishing techniques and traditions, the regulations provide for the establishment of Community and National Management Plans, combining the regulation of the fishing effort with specific technical measures (EC Reg. 1967/2006).

Other important measures have been foreseen by the EC Regulation (1967/2006). In particular, the excessive catches of undersized individuals should be avoided. Thus, it is necessary to protect certain areas where juveniles (nursery areas) or adults (spawning areas) congregate. Fishing gears that are too harmful to the marine environment or lead to the depletion of certain stocks should be prohibited or more strictly regulated. To avoid further increases in mortality rates for juveniles and to reduce the amount of discards of dead marine organisms by fishing vessels, it is necessary to increase the selectivity of the currently used gear that is to increase the mesh sizes for trawl nets and bottom-set nets and hook sizes for longlines.

The new regulations will require the establishment of "No-take" marine reserves and management measures to protect the growth and spawning areas as well as the marine ecosystem from destructive fishing effects.

"No-take" marine reserves are considered as an additional form of fisheries' management based mostly on general principles, including ecosystem functions, overall public interest and the need for insurance against the complexity of the environment and socio-economic system. In this regard, starting from 1998, the Ministry of Agriculture, Food and Forestry Policy (MiPAAF) decreed, as part of the protection plan of the fishery resources, the institution of "no-take zones" (Zone di Tutela Biologica, ZTB) where fishing is not allowed and/or regulated during critical phases (spawning, recruitment, feeding, etc.) of the demersal stocks distributed there, with the aim of allowing the renewal of the stocks and their sustainable exploitation in neighbouring areas.

Such no-take zones should correspond to areas which play a fundamental role in the life cycle of one or more demersal species of economic interest. In other words, they would correspond to areas where species concentrate for different biological reasons and whose protection could enhance the management effect through their closure to fisheries. Although some 'no-take zones' were identified along the Italian coast, no data on their effect on the fishery resources are yet available.

The multi-species nature of Mediterranean fisheries, in terms of both species and fishing gears, requires a specific strategy able to combine and integrate the different management measures to preserve flexibility in the fishing activity. In this way, the alternative adoption of a 'closed season' in different areas or catch restrictions and gear limitations together with the main regulation of reducing the fishing effort could avoid the overexploitation of marine resources.

ALTERNATIVE INCOME-GENERATING FISHING ACTIVITIES

Overexploitation of marine resources has lead to the decline of many demersal stocks as well as to habitat degradation, which is particularly severe in marine coastal zones, where human activities have historically been concentrated. In fact, the increasing discharges of waste and materials from the coast, pollution as well as urban and tourism developments have often led to definitive loss of natural habitats that has severe implications in the coastal small-scale fishing activity and consequent economic and social effects. Apart from the regulation of fishing activities previously reported and the protection and conservation of marine habitats, the promotion of alternative actions has been developed in recent years to transfer fishermen from fishery to tourist activities. In particular, the Ministerial Decree 293/99 regulates "fishing tourism" which represents fishing activity for tourist and educational services with which the local fishing community can supplement its income. Fishing tourism is the best way to participate in local development while also integrating socio-economic and cultural activities. Tourist fishing aims not only to discover the resources of the sea, but also to teach people how to experience and respect nature. It is the best way to learn the culture of the sea and of fishing, coastal areas and lagoons, as well as to discover the ancient professions and traditions while also eating excellent fresh fish cooked on board. In fact, aboard a traditional fishing boat everybody can learn from the ancient heritage of fishing techniques and Italian traditions, in the same way as in a museum or a natural park. Moreover, the fishing tourism activity prolonged to different seasons could contribute to the seasonal adjustment of the tourism flow, also reducing resource exploitation in the critical period.

By means of this activity, fishermen could decrease their total fishing effort and be directly involved in the local socio-economic development. The gradual decrease of the main fishing activity produced by the fishing tourism could guarantee sustainable management of resources and/or habitat maintenance either directly through a contribution to conservation and/or indirectly by providing revenue to the local community sufficient for local people to value, and therefore protect, their wildlife heritage as a source of income. An additional important contribution that the fishing world can offer is related to the growth of tourism centered on traditional local products and local cuisine, as well as various activities in which the tourist takes an active role in demonstrational fishing activities, also proposing the start-up of new services, such as fishing tourism, whale–dolphin watching and environmental education.

Human impact leads to decreasing diversity of habitat and species as well as to changing the population and community structure with a severe effect on ecosystem functions.

Although the institution of Marine Protected Areas (MPAs) along the Italian coasts is regulated by the Ministry of the Environment for biodiversity protection and nature conservation and not for fisheries' objectives as in other countries, the MPAs also represent an available and promising management strategy for sustainable use of marine resources. Mediterranean MPAs provide crucial goods and services that support local communities and economies, including recreational and tourist opportunities, producing social and economic development and other benefits. They could also promote environmental education and research. Moreover, MPAs maintain high productivity and high diversity in marine ecosystems, provide a refuge for the conservation of unique species and habitats as well as for exploited species allowing them to recover, increase biomass and restock fishing grounds through the spill-over of egg-larvae, juveniles and adults to replenish commercial stocks in the adjacent fisheries.

Even though each of these areas can have limited relevance when considered in isolation, all of them together (in an MPA network) can contribute effectively to safeguarding the marine environment, also thanks to the spill-over effect for species covered by the analysis.

An MPA must involve local institutions, as well as stakeholders (fishermen, shipbuilders, tourist cooperatives, etc.). The management and the control of the area should benefit from the active involvement of stakeholders (especially fishermen), who should collaborate with the entities in charge of controlling the area (Coastal Guard, etc.), motivated by a direct economic advantage for them. Collaboration between the various actors involved is a necessary condition to promote a sustainable process for fishing, which marries the need for economic development with that of protecting the environment. There is no question that the first step towards establishing the right relationship between the MPA management and the fishing world is that of sharing rules based on shared understanding and shared planning of the goals and potential of the MPA. In any case, managerial uncertainty can also be caused by the efficacy of the checks with regard to the various anthropic impacts, and not only those of fishing, which may take place in the area. In particular, fishermen's involvement can involve: (a) participation in the definition of areas and perimeters, (b) identification and sharing of management practices, (c) implementation of environmental protection activities and surveillance in the MPA territory.

STRATEGIES FOR THE MARINE ENVIRONMENT AND THE ECOSYSTEM APPROACH

Some of the above-reported managerial strategies favour the involvement of users and administrators at the same time. They also have the important impact of embedding ecosystem objectives, which take into account the complexity of the resource–environment system.

As stated in the framework document, it is commonly understood that pressures on natural marine resources and demand for marine ecosystem services are often too high and, therefore, that the need for a reduction in their impact on marine waters, regardless of where their effects are felt, is ever more urgent. On the other hand, the marine environment is a precious asset which needs to be protected, defended and where possible restored, to keep biodiversity and protect the diversity and livelihood of seas and oceans which need to be clean, healthy and productive. To meet these needs, on June 17 2008 the European Parliament and the EU Council issued the framework Directive on the Strategy for the Marine Environment (EC Reg. 2008/56), which was then considered in Italy when issuing law n. 190 on October 13 2010.

The EU Directive is based on an integrated approach and aims to become the environmental pillar of the future marine policy of the EU. The Directive sets the goal of Good Environmental Status (GES) for their marine waters for member States to reach by 2020. To reach GES, each member-State has to develop a marine strategy which must be agreed with neighboring member States and third countries.

Avoiding loss of diversity is the fundamental goal today, not only for the marine framework directive but also for the EU. Keeping biodiversity is a fundamental requirement to reach GES for marine waters. The marine protected areas (MPAs) represent an important tool to guarantee long-term conservation of nature and of ecosystem services. To guarantee biodiversity, the human component, and the tight connection between the coastal marine system and the cultural, social and economic context of the local communities cannot be underestimated. The biological benefits of MPAs could become secondary, if the social and economic situation of their territory is not taken into account.

Furthermore, the framework tries to ensure the sustainable utilization of marine goods and services through an ecosystem approach to managing human activities, i.e., a methodology which sees the human community as an integral part of ecosystems and of the systems which rule them.

Finally, returning to the starting point for this document, the ecosystem approach is the underlying principle of the (EC Reg 2008/56) Marine Framework Directive, which was originally summarized in 12 principles, and can be grouped in the following three key points: (1) The communities who live in an area are responsible for the biodiversity that surrounds them. (2) Sustainability is based on three pillars: environmental, economic and social-cultural. (3) To manage an environment, it is necessary to have an understanding of both scientific aspects and local traditions.

REFERENCES

1. Bertrand JA, Gil de Sola L, Papaconstantinou C, Relini G, Souplet A (2000) An international bottom trawl survey in the Mediterranean: the MEDITS programme. Actes de Colloques IFREMER 26:76–93
2. Cataudella S, Spagnolo M (2011) The multidimensionality of fisheries: a brief introduction. In: Cataudella S, Spagnolo M (eds) The State of Italian marine fisheries and aquaculture, 1. Ministero delle Politiche Agricole Alimentari e Forestali (MiPAAF), Rome (Italy), pp 3–8
3. Cataudella S, Ferraioli O, Lariccia M (2011) L'evoluzione della pesca italiana verso la sostenibilità nel quadro del Codice di condotta per la pesca responsabile (FAO), della Politica Comune della Pesca (PCP) e della Politica marittima integrata. In AA.VV. (2011) Lo stato della pesca e dell'acquacoltura nei mari italiani, a cura di Cataudella S e Spagnolo M. Ministero delle politiche agricole alimentari e forestali. p 11–31
4. EC Reg. 1543/2000 of 29 June 2000 establishing a Community framework for the collection and management of the data needed to conduct the common fisheries policy. http://eur-lex.europa.eu/legal-content/EN/TXT/?uri=CELEX:32000R1543
5. EC Reg. 1639/2001 establishing the minimum and extended Community programmes for the collection of data in the fisheries sector and laying down detailed rules

for the application of Council Regulation (EC). http://eurlex.europa.eu/LexUriServ/LexUriServ.do?uri=OJ:L:2001:222:0053:0115:EN:PDF

6. EC Reg. 2371/2002 on the conservation and sustainable exploitation of fisheries resources under the Common Fisheries Policy, http://eur-lex.europa.eu/legal-content/en/ALL/?uri=CELEX:32002R2371
7. EC Reg. 1967/2006. Misure di gestione per lo sfruttamento sostenibile delle risorse della pesca nel Mar Mediterraneo e recante modifica del regolamento (CEE) 2847/1993 e che abroga il regolamento (CE) 1626/1994. http://eur-lex.europa.eu/LexUriServ/LexUriServ.do?uri=OJ:L:2006:409:0009:0064:IT:PDF
8. EC Reg. 2008/56 Establishing a framework for community action in the field of marine environmental policy (Marine Strategy Framework Directive).http://eur-lex.europa.eu/legal-content/EN/TXT/?uri=CELEX:32008L0056
9. FAO (1995) Code of Conduct for Responsible Fisheries. Rome
10. FAO (2014) The State of World Fisheries and Aquaculture. Rome
11. Garcia SM, Zerbi A, Aliaume C, Do Chi T, Lasserre G (2003) The ecosystem approach to fisheries. Issue, terminology, principles, institutional foundations, implementation and outlook. FAO Fisheries Technical Paper, N. 443. FAO. Rome, p 71
12. GFCM (General Fisheries Commission for the Mediterranean) 2011. Performance Review of the General Fisheries Commission for the Mediterranean and Black Sea [online]. [Cited 28 November 2011]. http://151.1.154.86/GfcmWebSite/GFCM/35/CAF_II_2011_Inf.5_COC_V_Inf.4_GFCM_XXXV_2011_Inf.8.pdf
13. Habitat Directive (92/43/EEC) on the conservation of natural habitats and of wild fauna and flora. http://eur-lex.europa.eu/legal-content/EN/TXT/?uri=CELEX:31992L0043
14. King M (1995) Fisheries biology, assessment and management. In: Fishing News Books. Blackwell Science, p 341
15. Law n. 41, 17.02.1982 Piano per la razionalizzazione e lo sviluppo della pesca marittima.http://www.normattiva.it/uri-res/N2Ls?urn:nir:stato:legge:1982-02-17;41@originale
16. Lleonart J, Maynou F (2003) Fish stock assessments in the Mediterranean: state of the art. Scientia Marina 67(Suppl. 1):37–49
17. Ministerial Decree 293/99 on fishing tourism. http://www.europalex.kataweb.it/article_view.jsp?idArt=24953&idCat=374
18. Relini G (2003) Fishery and Aquaculture Relationship in the Mediterranean: Present and Future. Mediterranean marine science 4(2):125–154CrossRef
19. Pauly D, Christensen V (1995) Primary production required to sustain global fisheries. Nature 374:255–257CrossRef
20. Spagnolo M (2006) Elementi di economia e gestione della pesca. (Ed). FrancoAngeli, Milano, p 279

CHAPTER 3 BALTIC SEA MANAGEMENT: SUCCESSES AND FAILURES

Ragnar Elmgren, Thorsten Blenckner, Agneta Andersson

Keywords: Eutrophication, Persistent organic pollutants, Overfishing, Ecosystem-based management, Climate change

INTRODUCTION

The Baltic Sea including the Kattegat (Fig. 1) is a large, almost non-tidal North European inland sea of 393 000 km^2, with a mean depth of a mere 54 m. It is divided into a series of basins, mostly separated by shallow sounds or sills. Surface salinity gradually declines inwards, from 18–26 in the Kattegat to 2–4 in the innermost Bothnian Bay (using the unit-free Practical Salinity Scale, which is near-identical to older scales with units of g/kg or ‰). There is also permanent salinity stratification with depth, strongest in the Kattegat (bottom salinity 32–34) and weakening inwards to the Gulf of Bothnia. This causes stagnation of the bottom water, and in recent decades has led to widespread deep-water oxygen deficiency, seasonal in Kattegat and the Danish Sounds, near-permanent in the Baltic proper, intermittent in the Gulf of Finland, but not affecting the Gulf of Bothnia. Water renewal is slow, on the order of 50 years for the whole Baltic (description above based on Leppäranta and Myrberg 2009), making it vulnerable to pollution from the surrounding catchment (Fig. 1), with an area four times larger than the Baltic Sea, and a human population of some 85 million (Sweitzer et al. 1996). The waters of the Baltic Sea are generally cold, with the northern areas freezing over every winter, but the surface waters heat up in summer, in warm years to over 20 °C. These gradients cause large differences in ecological conditions along the sea, with biodiversity declining sharply with salinity (Remane and Schlieper 1971; Ojaveer et al. 2010).

The Baltic Sea is surrounded by nine riparian countries with a further five in the catchment, making its management politically complicated. Yet, it is perhaps the sea area with the longest record of scientific research and cooperation (Mills 1989), which from an early stage was expressly intended to support management of living marine resources. The International Council for the Exploration of the Sea (ICES), founded in 1901, was one of the first international intergovernmental scientific organizations and focused on the Baltic and North Seas and their living resources (Rozwadowski 2002). When environmental problems started to be seen as a threat to living marine resources in the late 1960s, an ICES working group documented severe environmental problems in the Baltic Sea (ICES 1970). These led to the formulation and 1974 signing

Citation: Ragnar Elmgren, Thorsten Blenckner, Agneta Andersson, Baltic Sea management: Successes and failures, DOI 10.1007/s13280-015-0653-9.

of the Helsinki Convention for the Protection of the Marine Environment of the Baltic Sea Area, which is governed by the Helsinki Commission (HELCOM, http://www.helcom.fi/; Elmgren 2001). From the beginning, this convention had an emphasis on ecosystem management (Rozwadowski 2002), as was made formally explicit in 1992, when the convention was renegotiated after the fragmentation of the Soviet Union.

Figure 1: Baltic Sea map showing subareas, coastal nations, catchment border, and 60 m isobaths. Gulf of Bothnia = Bothnian Sea + Bothnian Bay.

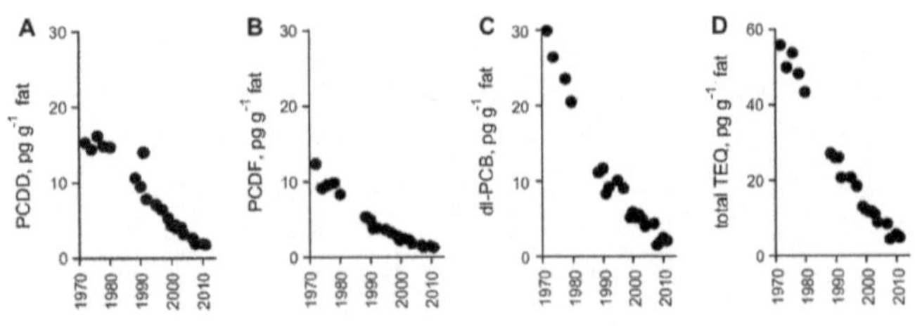

Figure 2: Decrease over time in mother's milk from Sweden, 1972–2011, of **a** polychlorinated dibenzo-p-dioxins, **b** polychlorinated dibenzofurans, **c** dioxin-like polychlorinated biphenyls, **d** sum of toxic equivalents for PCDDs, PCDFs and dioxin-like PCBs ($TEQ_{WHO\text{-}2005}$). Redrawn after Fång et al. (2013, Fig. 1).

Since it was adopted by the Convention for Biological Diversity at the Rio Conference in 1992, ecosystem-based management (EBM) has become the norm for international conventions and organizations, such as ICES and the Food and Agriculture Organization of the United Nations (Leslie and McLeod 2007), as well as in EU's Water and Marine Strategy Framework Directives (Anon. 2000, 2008). Accepting that EBM is the desired goal does not, however, without appropriate institutional basis and funding, guarantee its real life implementation (Österblom et al. 2010). In the early days, "the ecosystem approach" simply meant that the interactions of all components of the ecosystem, including humans, should be considered in the management process (e.g., Jansson 1972; Likens 1992), but starting with the Biodiversity Convention, EBM has come to signify a management process that also has social goals, such as an open and transparent democratic process with strong stakeholder participation, and an equitable distribution of the gains in human well-being resulting from improved management (Levin and Lubchenco 2008). The research on EBM has also increasingly become a search for indices and markers suitable for assessing progress for the political process (e.g., UNEP/GPA 2006; Curtin and Prezello 2010), rather than for a deeper understanding of ecosystem processes (cf. Thrush and Dayton 2010).

The four decades since 1974 seem a suitable period over which to evaluate successes and failures in the Helsinki Convention's application of the ecosystem approach, aimed initially to "protect and enhance the environment of the Baltic Sea" (HELCOM 1993) and, after 1992, to "prevent and eliminate pollution in order to promote the ecological restoration of the Baltic Sea Area and the preservation of its ecological balance" (HELCOM 2008). In the following, we briefly describe some notable successes and failures in the management of the major environmental problems of the Baltic Sea, and assess how well the management process lives up to the ideals of EBM. We also discuss some new challenges for the future.

The two research programs this special issue reports on, Baltic Ecosystem Adaptive Management (BEAM) at Stockholm University and Ecosystem dynamics in the Baltic Sea in a changing climate perspective (ECOCHANGE) at Umeå and Linnaeus Universities, are examples of the type of interdisciplinary research needed to support EBM (Botey et al. 2014).

ENVIRONMENTAL SUCCESSES AND FAILURES OF THE HELCOM PERIOD

Toxic pollutants

The ICES (1970) report considered toxic pollutants as the foremost threat to the Baltic, with particular emphasis on heavy metals and DDT, but acknowledged that there was a large knowledge gap concerning other persistent organic pollutants (POPs). Based on toxicological experience from industrial exposure, heavy metals, also emitted to air and freshwater, were seen as posing a direct threat to humans. Industrial pollution, for example from pulp and paper mills, was commonplace and oil spills were frequent and often large, soiling shores and killing seabirds. Untreated sewage was an obvious problem in many enclosed coastal areas, especially near large cities, and some suspected that the recent increase in the area of oxygen-deficient bottom waters in the Baltic proper was also largely anthropogenic.

After 1974, the first decades of research described the basic functioning of the ecosystem (Jansson 1972; Elmgren 1984), and gave initial estimates of nutrient loads and effects (Larsson et al. 1985). International management was initially largely ineffective, due to political divisions, scientific uncertainty, and weak governance structures. The many problems of the Baltic Sea were tackled separately, often by entirely different management bodies; the fishery even by a separate convention, the Gdansk Convention (signed 1973). Still, use of a few problem pollutants, such as DDT (dichlorodiphenyltrichloroethane) and PCBs (polychlorinated biphenyls), was soon restricted or banned in the Baltic area. Concentrations of DDT, PCBs, dioxins, and dibenzofurans have since decreased rapidly in wildlife, demonstrating the effectiveness of mitigation (Nyberg et al. 2015). After a lag, previously highly threatened Baltic seal and white-tailed eagle (*Haliaeetus albicilla*) populations recovered rapidly, helped also by various conservation measures (Helander et al. 2008; Roos et al. 2012), visibly improving Baltic Sea biodiversity. Nevertheless, concentrations of dioxins and dioxin-like PCBs in fatty fish from parts of the Baltic remain higher than the maximum allowed in foodstuffs by EU. An exception from EU rules permits the sale of such fish for human consumption in Sweden and Finland, where they are the main route of human exposure to dioxin-like compounds. Even so, human exposure, as represented by the toxic equivalent levels of such compounds in mothers' milk in Sweden, has decreased 6.5 % per year over a 40-year period from 1972 (Fig. 2; Fång et al.2013), a truly remarkable improvement. Overall, the management of POPs is the greatest success story in Baltic environmental management but, for some groups of problem compounds, action has still been hesitant and tardy, e.g., for perfluorinated compounds (Roos et al. 2013) and many endocrine-disrupting chemicals (UNEP/WHO 2013).

Industrial pollution has also been curtailed, especially after the collapse of communism in Eastern Europe led to the closure of many heavily polluting and economically unviable plants, thus reducing heavy metal pollution (e.g., Vallius 2014). The HELCOM Joint Comprehensive Program of 1992 was specifically designed to identify and mitigate land-based "hot spots" of pollution (VanDeveer 2011). Still, legacy pollution maintains concentrations of some substances at levels of concern, e.g., PCBs (see above) and arsenic in the Bothnian Bay (Vallius 2014). Through introduction of double-hulled tankers, better surveillance and ship-routes, oil spills have been much reduced in frequency and volume (HELCOM Response 2013). Still, small, often intentional spills remain common and at an unfortunate time and place can kill thousands of seabirds. There have been few major spills from ship accidents in recent years, even though some consider that the risk of such spills may actually have increased due to large increases in ship traffic and the volume of oil transported on the Baltic Sea (Hassler 2010).

Nutrient loads

The gradual accession of all coastal Baltic countries except Russia to the European Union by 2004 made important international environmental agreements legally binding in most of the HELCOM convention area and led to stronger action. Thus nutrient loads to the Baltic have been decreasing since c. 1980, with the largest reductions for phosphorus (Gustafsson et al.2012), as countries have increasingly removed nutrients from municipal sewage, resulting in some notable local improvements in the coastal zone (e.g., Elmgren and Larsson 2001). Yet, in the Baltic proper, open-sea total nutrient concentrations have hardly decreased and anoxic bottom waters have affected record areas and volumes after 2000 (Carstensen et al. 2014). Winter concentrations of inorganic nitrogen have seen a slight decrease since the 1980s, but this may be due more to increased denitrification losses in the increased volume of anoxic bottom water, than to the load reduction (Vahtera et al. 2007). There are reports that primary production during the spring bloom has decreased somewhat, with a shift toward higher summer production (Raateoja et al. 2005, SW coast of Finland), accompanied by larger surface accumulations of cyanobacteria, as recorded by satellite (Kahru and Elmgren 2014, Baltic proper). The agreement on a Baltic Sea Action Plan in 2007 (revised 2013) was widely seen as a major step forward in Baltic Sea management, with nutrient reduction goals that were based on advanced ecosystem modeling (HELCOM 2007, 2013).

Fishing

International Baltic Sea fishery management was initially weak, as exemplified by its inability to prevent overfishing of the Baltic cod (*Gadus morhua*) stock in the late 1980s, which led to large losses in income for the fishery (Sjöstrand 2008). The 2007 cod recovery plan (Anon. 2007), which for once has been adhered to by the politicians, has resulted in a considerable numerical recovery of the eastern cod stock. This success is only partial, however, since in combination with an increased minimum landing length and decreased availability of invertebrate prey through extensive oxygen deficiency, it has resulted in a large population of lean, slow-growing cod of reduced economic value that are harvested soon after reaching legal length (Svedäng and Hornborg 2014).

The Baltic Sea has a long history of mostly unintended introductions of non-indigenous species (Leppäkoski et al. 2002). HELCOM has followed this development, but, as in other marine areas, relatively little has yet been done to still the inflow of alien species. In the future, international agreements on the treatment of ballast water are likely to reduce, but hardly solve the problem, and species already established in the Baltic Sea are unlikely to be eliminated.

Climate change

Invasive species

The Baltic Sea is one of the marine areas with the highest recorded temperature increases during the past century (Belkin 2009; Rutgersson et al. 2014), and this increase is almost certain to continue. A particularly important threshold will be reached once the winter temperature in the open Baltic proper remains above the temperature of maximum density. When this happens, above-halocline temperature convection may be much reduced, lowering both oxygen supply to the deep water and up-mixing of deep water, including pollutants and regenerated nutrients to the surface layer (Hordoir and Meier 2012). Since the causes of climate change are global, there is little HELCOM can do, except follow events and give advice on likely future developments and possibilities for adaptation. Global climate models, and the regional models based on them, generally agree that the Baltic will warm about 2–4 °C by the end of this century, but the projections vary more concerning effects on precipitation (Niiranen et al. 2013), which is likely to be particularly important in the Baltic Sea, where salinity is the ecological master factor (Elmgren and Hill 1997).

EBM must be adapted to the regional conditions, which vary widely in different basins of the Baltic Sea, as described above. Thus, eutrophication, anoxia, and overfishing are dominant problems in the Baltic proper, while arsenic and organic pollutants in fish and wildlife are of greater concern in the Gulf of Bothnia (e.g., Nyberg et al. 2015). On the whole, HELCOM has taken such regional differences into account, but it is worrying that the EU, in implementing the marine strategy directive, is trying to apply a single classification to the whole Baltic Sea area. HELCOM has also inspired and supported research on the Baltic Sea, without which the basis for decisions would have been much weaker, and the Helsinki Convention is now seen as the most successful of all international regional seas conventions.

Holistic assessment

The Baltic Sea is often portrayed as an environmental disaster area, by the media, by non-governmental environmental organisations, and by some scientists. Even HELCOM (2010: Fig. 2.2, panel A), in its first holistic assessment of the ecosystem health of the Baltic Sea presented a map showing the whole Baltic Sea as having only moderate or (mostly) poor or bad ecosystem health, except for a small sliver in good health along the northern half of Sweden's Gulf of Bothnia coast. This map has been produced from good scientific data, according to the principle of the water framework directive of "one out–all out," and is in this sense correct. Yet it fails to convey the enormous value of the Baltic Sea still has as a natural resource for the millions living around it, even in its present somewhat tainted state. It also fails, crucially, to present the real progress that has been made, proving that investing in the environment pays off (Table 1). The remaining problems of the Baltic Sea are real and serious and will take a long time and large resources to mitigate (Table 1), but they would have been much worse without the environmental mitigation work coordinated by HELCOM and ICES. If the public is given the erroneous impression that 40 years of work and investment to protect the environment of the Baltic Sea has been a total failure, its support may falter (Elmgren et al. 2012).

Table 1: Evaluation of Baltic Sea Environmental Management in the HELCOM period

Successes	*Failures/remaining problems*
Helsinki Convention signed and ratified	HELCOM not organized for EBM
DDT, PCBs, TBT and other hazardous organics substances banned	Insufficient action on other endocrine-disrupting and perfluorinated compounds
Concentrations of dioxin-like compounds in wildlife greatly decreased Recovery of seal and eagle populations	Advice on human consumption and exemptions from standards still needed
Industrial heavy metal pollution reduced	Some "hot-spots" and legacy pollution remain
Oil spills reduced in frequency and volume	Seabirds still killed; major spill remains a risk
Nutrient loads considerably reduced, Baltic Sea Action Plan Coastal eutrophication status locally improved	Phosphorus concentration and anoxic volume at all-time high, cyanobacterial blooms larger
Cod recovery plan partial success	Cod concentrated in SW Baltic, grow slowly Fish catches not tailored to ecosystem needs Little effective action to prevent introduction of new invasive species Climate change continues apace
Ecosystem-based management approach enthusiastically adopted	Interconnected problems still managed apart; fisheries handled by a different convention
Water status classification system developed	Classification not adapted to climate change

THE CASE FOR ECOSYSTEM-BASED MANAGEMENT

A food-web perspective

EBM has become the accepted, if not yet the practiced, norm (Österblom et al. 2010; Möllmann et al. 2014). Yet, there are authors who maintain, with some good arguments, that EBM requires too much data and is too expensive to be used as a basis for fisheries management except perhaps for the largest stocks in the wealthiest countries (Hilborn and Ovando 2014), and may not always be needed even there (Cardinale and Svedäng 2011). Are the reasons for advocating EBM for the Baltic Sea really compelling? This requires that we can show that the different major problems of the Baltic Sea are so interconnected, that we cannot expect to deal effectively with them separately. No one disputes that fish in the sea ultimately depend on primary producers, in the open sea the phytoplankton, for their sustenance. Normally there is at least one trophic level between them, zooplankton that are eaten by small fish and fish larvae, but in many cases there are further intermediate trophic links, such as heterotrophic flagellates, ciliates, mysid shrimp, or benthic organisms, which all depend on the same primary production. It is therefore not surprising that catches of Baltic cod reached a maximum in the early 1980s, when Baltic eutrophication had become well established, but before the resulting expansion of oxygen deficiency had stopped recruitment from the spawning areas in Gdansk Bay and the eastern Gotland deep. Total yearly fisheries removals, on the other hand, peaked later, at about 1.5 million tons in 1998, as the depressed state of the overfished cod stock left more of their main fish prey, sprat (*Sprattus sprattus*) and herring (*Clupea harengus*), to the fishery (Zeller et al. 2011). Cod is far more valuable than the other major species in the Baltic fishery, herring and sprat, so the total value of the fishery increased initially as incipient eutrophication increased total fish production, only to fall again when oxygen deficiency became serious enough to stop the eastern spawning areas of cod from producing recruits. Not only the quantity, but also the timing and taxonomy of the phytoplankton matter for fish and fisheries. The classic food chain for effectively supporting fish production is diatoms eaten by crustacean zooplankton eaten by fish. But in the Baltic Sea the main diatom bloom is the nitrogen-limited spring bloom, which is too early for extensive use by most crustacean zooplankton, and ends up largely sinking to the sea-floor (Elmgren 1978). This is followed by a summer-bloom of phosphorus-limited diazotrophic cyanobacteria, which has long been considered as low quality food for zooplankton. Recent research shows, however, that diazotrophic cyanobacteria release much of the nitrogen they fix to the water (Larsson et al.2001; Ploug et al. 2010), effectively fertilizing it for other phytoplankton that bloom alongside them, such as small diatoms and picocyanobacteria. Eaten as part of a diet mixed with other phytoplankton, and the protozoans that feed on them, the diazotrophs can thus support rapid growth of crustacean zooplankton, at exactly the time of maximum food consumption by clupeid larvae and young-of-the-year (Karlson et al. 2015). This means that the summer cyanobacterial bloom is not only a nuisance for the tourist industry, but also supports fish recruitment in the Baltic, and that reducing the bloom may well carry a cost in lowered fish production.

The large freshwater inflow to the Gulf of Bothnia has high concentrations of colored organic matter, making the seawater brownish (Andersson et al. 2015). The organic input stimulates the heterotrophic microbial food web, but lowers pelagic and benthic primary production through shading and competition for production-limiting phosphorus (Ask et al. 2009; Wikner and Andersson 2012). The heterotrophic bacteria-based food web has one to two extra trophic levels (Berglund et al. 2007), making it less efficient than the phytoplankton-based food web. If climate change increases freshwater inflow to the Gulf of Bothnia as projected (Andersson et al.2015), the pelagic food web may become even more bacteria-based, harming fish production and recruitment. On the other hand, a longer productive season and increased nutrient run-off may compensate, making the net effect of climate change difficult to predict (Andersson et al.2015).

In recent years, it has become clear that fisheries management is central to managing the Baltic Sea environment, and that overfishing not only is bad business, and a threat to the fishing industry, but also a very serious environmental problem. Current fisheries management in the Baltic Sea is based on applying the principle of Maximum Sustainable Yield (MSY) to each fish stock separately. This principle has been applied globally for many fish stocks, often with positive results (Worm et al. 2009). Nevertheless, it is questionable on numerous counts. It is economically suboptimal in most cases, since the last fish is generally the most expensive to catch. This means that the maximum economic yield will normally be obtained by catching fewer fish than calculated from MSY. Fish catches need to vary less from year to year if the stock is larger, with more year-classes represented, again implying that catching less than under MSY is better economy. Further, MSY does not take changes in growth, in particular density-dependent growth, into account, with recent dire effects on the Eastern Baltic cod stock (Svedäng and Hornborg 2014). When several stocks with trophic interactions are fished, as in the Baltic Sea, MSY cannot simultaneously be attained for all stocks (e.g., Heath 2012). A multi-species maximum sustainable economic yield model might solve some of these problems, but would still not take the effects of fishing on the entire ecosystem fully into account (cf. Möllmann et al.2014). Currently, the potential risk of trophic cascades from fish to primary producers is not taken into account when setting catch quotas. Such cascades are difficult to prove conclusively, but in the Baltic Sea proper they have been proposed as contributory causes both of increased phytoplankton blooms (Casini et al. 2008, 2012), and the decline of perennial coastal vegetation through over-growth by filamentous annual algae (Eriksson et al. 2011). In addition, overfishing of cod, by allowing stocks of sprat to increase, has probably been a major cause of a decline in growth rate of Baltic herring (Casini et al. 2010), such that herring in most of the Baltic Sea now take about 5–7 years to reach 40 g, whereas this took only 2–3 years around 1980 (ICES 2014). The older the fish, the more time it has had to accumulate organic contaminants (e.g., Kiljunen et al. 2007), indicating that cod overfishing has been a factor increasing human exposure to bioaccumulating POPs from the Baltic Sea, and in the continued ban, mentioned above, on sales of some Baltic fish outside Sweden and Finland. Eutrophication also influences the fate of organic pollutants in the Baltic Sea, with concentrations in fish expected to decrease when eutrophication increases, and vice versa (Larsson et al. 2000).

Classification of water quality

For the implementation of the marine strategy framework directive, EU has developed a water status classification system focused on eutrophication and contaminants. This classification is largely based on estimated historical conditions, but considering ongoing and projected climate change, this may be inappropriate (McQuatters-Gollop 2012). For example, chlorophyll *a* in phytoplankton, which is an established quality factor in the classification, varies greatly with the light climate in the water. Through an increased export of colored dissolved organic matter ("brownification"), climate change is expected to lead to poorer light conditions in the Gulf of Bothnia, which will induce higher chlorophyll content in the phytoplankton community (Andersson and Rudehäll 1993), even without an increase in phytoplankton biomass. It is important that the classification system can adapt to such effects of climate alterations, or the water status may be incorrectly or inappropriately assessed.

Governance

As we have illustrated above, the various environmental problems of the Baltic Sea are indeed so closely intertwined that we can hardly expect to deal successfully with them in a piecemeal fashion. Fish stocks depend on eutrophication, contaminants levels in seafood depend on fisheries management as well as on eutrophication, some negative effects of eutrophication can probably be mitigated by prudent fisheries management, and climate change will affect them all. The synergistic effects of the multiple drivers discussed above may cause unexpected changes in ecosystem functions, often called regime shifts, as reported from the Baltic Sea by Österblom et al. (2007). The potential for such abrupt changes in ecosystem state make it important to address several management sectors simultaneously, as in EBM (Levin and Möllmann 2015). The spatial aspect is of particular importance in geographically large and varied systems like the Baltic Sea, for example when cod management in one basin affects ecosystem dynamics in another basin (Casini et al. 2012).

Overall, we still lack both much of the basic ecological knowledge (Österblom et al. 2010) and the appropriate governance structures (Valman 2013) for effective implementation of EBM of the Baltic Sea. HELCOM was originally organized sector-by sector and the Baltic Sea Action plan was in part intended to re-organize the work to suit an ecosystem approach. However, Valman (2013) failed to find significant institutional change in HELCOM as a consequence of the Baltic Sea Action Plan, or, in fact, in the last 30 years. EBM requires not only ecological data collection, analysis, and modeling capacity, but also coordination between agencies and institutions across sectors and geographical levels. In addition to the top-down governance of the Baltic Sea by nation states through HELCOM and the European Council, there are also emerging bottom-up governance initiatives (Österblom et al. 2010). Experiments with local co-management illustrate the challenges and opportunities of collaborative, more inclusive forms of governance (Sandström et al. 2014). In all, we are still far from making management democratic and equitable, in an area of 14 independent countries, where people have widely diverging attitudes and aspirations and live under quite different economic and social conditions.

OUTLOOK

Prominent among the additional future challenges for Baltic Sea research and EBM is the controversial question of how large populations of top predators, such as seals and great cormorants (*Phalacrocorax carbo sinensis*), that should be maintained in the Baltic Sea, considering the possible cost in lowered fisheries yield, especially if eutrophication mitigation is successful, and results in lower production of commercial fish (cf. Zabel et al. 2003 for seals; KSLA 2013 for seals and cormorants). Climate change and perhaps marine acidification are also likely to change food web function, the interactions among environmental problems and, in the end, fish production (Meier et al. 2012; Lefébure et al. 2013; Niiranen et al. 2013). Environmental management decisions are increasingly being based on scenarios and simulation models, but these are simplifications of nature and no better than the imperfect knowledge on which they are based. Increasingly, managers want to base decisions on results of several independent models (ensemble modeling; Gårdmark et al. 2013; Meier et al. 2013). Models also often fail to incorporate mechanisms leading to sudden, unexpected regime shifts in the managed ecosystems (but see Niiranen et al. 2013). To minimize surprises, we need local and regional experimentation with management methods, coupled with adaptive management, where environmental monitoring is used to warn about sudden changes, or at least allows us to learn from them, so as not to be surprised the next time (Österblom et al. 2010).

For EBM to become successful, it must recognize that we are dealing with complex social-ecological systems that need governance mechanisms adapted to the scale of the problems, whether local, regional or global, with cooperative, multilevel management, partnership approaches, social learning, and knowledge co-production (Berkes 2012). EBM of the Baltic Sea will increasingly have to be extended to include also major human activities in the Baltic drainage basin, such as land use, and their incentive structures. To be realistic, we must also realize that implementing EBM is a 'wicked problem,' without simple, once for all solutions (Jentoft and Chuenpagdee 2009) and will require continuous adjustments, based on continued comprehensive environmental monitoring and gradually improved understanding through adaptive management and modeling (Levin and Möllmann 2015).

CONCLUSIONS

Since the signing of the Helsinki convention 40 years ago, the Baltic Sea's load of toxic pollutants has been greatly reduced; further eutrophication has been halted by greatly reducing nutrient emissions, and the numerical abundance of the main Baltic cod stock has recovered. A loss of the immense value of the Baltic Sea as a natural resource for the citizens of its coastal countries has been prevented. Nevertheless, many serious problems remain: potentially toxic pollutants are still at levels of concern both in wildlife and fish catches, and new ones continue to come into use; undesirable symptoms of eutrophication remain evident in many coastal areas; deep-water oxygen deficiency has record extension, and toxic blooms of cyanobacteria interfere too frequently with tourism and recreation. Little effective action has yet been taken to reduce the influx of non-indigenous, potentially invasive species, and fisheries management, while improved, still falls short. Finally, climate change proceeds at rapid pace, endangering the environmental gains that have been made, and making agreed goals harder to attain. EBM may already be the agreed principle, but it also needs to become the established practice!

ACKNOWLEDGMENTS

The BEAM and ECOCHANGE programmes were funded 2010–2014 by special government grants, supplemented by ordinary University funds. Discussions with the participants in the BEAM and ECOCHANGE programs, and especially Michael McLachlan and Sture Hansson, were very helpful. Carl Folke, Henrik Österblom, and Carl Rolff gave valuable comments on the manuscript. Erik Smedberg prepared Fig. 1, Elena Gorokhova helped with Fig. 2.

REFERENCES

1. Andersson, A., and Å. Rudehäll. 1993. Proportion of plankton biomass in particulate organic carbon in the northern Baltic Sea. *Marine Ecology Progress Series* 95: 133–139.

2. Andersson, A., H.E.M. Meier, M. Ripszam, O. Rowe, J. Wikner, P. Haglund, K. Eilola, C. Legrand, et al. 2015. Projected future climate change and Baltic Sea ecosystem management.*AMBIO*. doi:10.1007/s13280-015-0654-8.

3. Anon. 2000. Directive 2000/60/EC of the European Parliament and of the Council of 23 October 2000 establishing a framework for community action in the field of water policy.*Official Journal of the European Union* L327/1. 22.12.2000.

4. Anon. 2007. Council Regulation (EC) No 1098/2007 of 18 September 2007 establishing a multi-annual plan for the cod stocks in the Baltic Sea and the fisheries exploiting those stocks, amending Regulation (EEC) No 2847/93 and repealing Regulation (EC) No 779/97. *Official Journal of the European Union* L248/1. 22.9.2007.
5. Anon. 2008. Directive 2008/56/EC of the European parliament and of the council of 17 June 2008 establishing a framework for community action in the field of marine environmental policy (Marine Strategy Framework Directive). *Official Journal of the European Union* L164/19. 25.6.2008.
6. Ask, J., J. Karlsson, L. Persson, P. Ask, P. Byström, and M. Jansson. 2009. Terrestrial organic matter and light penetration: Effects on bacterial and primary production in lakes. *Limnology and Oceanography* 54: 2034–2040.
7. Belkin, I.M. 2009. Rapid warming of Large Marine Ecosystems. *Progress in Oceanography* 81: 207–213.
8. Berglund, J., U. Müren, U. Båmstedt, and A. Andersson. 2007. Efficiency of a phytoplankton-based and a bacteria-based food web in a pelagic marine system. *Limnology and Oceanography* 52: 121–131.
9. Berkes, F. 2012. Implementing ecosystem-based management: Evolution or revolution? *Fish and Fisheries* 13: 465–476.
10. Botey, A.P., T. Garvin, and R. Szostak. 2014. Interdisciplinary research for ecosystem management. *Ecosystems* 17: 512–521.
11. Cardinale, M., and H. Svedäng. 2011. The beauty of simplicity in science: Baltic cod stock improves rapidly in a 'cod hostile' ecosystem state. *Marine Ecology Progress Series* 425: 297–301.
12. Carstensen, J., J.H. Andersen, B.G. Gustafsson, and D.J. Conley. 2014. Deoxygenation of the Baltic Sea during the last century. *Proceedings of the National Academy of Sciences of the United States of America* 111: 5628–5633.
13. Casini, M., J. Lövgren, J. Hjelm, M. Cardinale, J.C. Molinero, and G. Kornilovs. 2008. Multi-level trophic cascades in a heavily exploited open marine ecosystem. *Proceedings of the Royal Society B—Biological Sciences* 275: 1793–1801.
14. Casini, M., V. Bartolino, J.C. Molinero, and G. Kornilovs. 2010. Linking fisheries, trophic interactions and climate: Threshold dynamics drive herring *Clupea harengus* growth in the central Baltic Sea. *Marine Ecology Progress Series* 413: 241–252.
15. Casini, M., T. Blenckner, C. Möllmann, A. Gårdmark, M. Lindegren, M. Llope, G. Kornilovs, M. Plikshs, et al. 2012. Predator transitory spillover induces trophic cascades in ecological sinks.*Proceedings of the National Academy of Sciences of the United States of America* 109: 8185–8189.
16. Curtin, R., and R. Prezello. 2010. Understanding marine ecosystem-based management: A literature review. *Marine Policy* 34: 821–830.
17. Elmgren, R. 1978. Structure and dynamics of Baltic benthos communities, with particular reference to the relationship between macro- and meiofauna. *Kieler Meeresforschungen*, Sonderheft 4: 1–22.
18. Elmgren, R. 1984. Trophic dynamics in the enclosed, brackish Baltic Sea. *Rapports et Procès-verbaux des Reunions. Conseil International pour l'Exploration de la Mer* 183: 152–169.
19. Elmgren, R. 2001. Understanding human impact on the Baltic ecosystem: Changing views in recent decades. *AMBIO* 30: 222–231.
20. Elmgren, R., and C. Hill. 1997. Ecosystem function at low biodiversity—The Baltic example. In*Marine biodiversity: Patterns and processes*, ed. R.F.G. Ormond, J. Gage, and M. Angel, 319–336. Cambridge: Cambridge University Press.
21. Elmgren, R., and U. Larsson. 2001. Eutrophication in the Baltic Sea Area: Integrated Coastal Management Issues. In *Science and integrated coastal management*, ed. B. von Bodungen, and R.K. Turner, 15–35. Berlin: Dahlem University Press.
22. Elmgren, R., L. Kumblad, E. Rydin, and F. Wulff. 2012. Political backing to save the Baltic Sea.*Nature* 487: 432.
23. Eriksson, B.K., K. Sieben, J. Eklöf, L. Ljunggren, J. Olsson, M. Casini, and U. Bergström. 2011. Effects of altered offshore food webs on coastal ecosystems emphasize the need for cross-ecosystem management. *AMBIO* 40: 786–797.
24. Fång, J., E. Nyberg, A. Bignert, and Å. Bergman. 2013. Temporal trends of polychlorinated di-benzo-*p*-dioxins and dibenzofurans and dioxin-like polychlorinated biphenyls in mothers' milk from Sweden, 1972-2011. *Environment International* 60: 224–231.
25. Gårdmark, A., M. Lindegren, S. Neuenfeldt, T. Blenckner, O. Heikinheimo, B. Müller-Karulis, S. Niiranen, M.T. Tomczak, et al. 2013. Biological ensemble modeling to evaluate potential futures of living marine resources. *Ecological Applications* 23: 742–754.
26. Gustafsson, B.G., F. Schenk, T. Blenckner, K. Eilola, H.E.M. Meier, B. Muller-Karulis, T. Neumann, T. Ruoho-Airola, et al. 2012. Reconstructing the development of Baltic Sea eutrophication 1850–2006. *AMBIO* 41: 534–548.
27. Hassler, B. 2010. Global regimes, regional adaptation; environmental safety in Baltic Sea oil transportation. *Maritime Policy and Management* 37: 489–503.
28. Heath, M.R. 2012. Ecosystem limits to food web fluxes and fisheries yields in the North Sea simulated with an end-to-end food web model. *Progress in Oceanography* 102: 42–66.
29. Helander, B., A. Bignert, and L. Asplund. 2008. Using raptors as environmental sentinels: Monitoring the white-tailed sea eagle *Haliaeetus albicilla* in Sweden. *AMBIO* 37: 425–431.
30. HELCOM. 1993. *Convention on the protection of the marine environment of the Baltic Sea Area, 1974 (Helsinki Convention)*. HELCOM, December 1993.
31. HELCOM. 2007. *HELCOM Baltic Sea Action Plan.* Helsinki: Helsinki Commission. 103 pp.

32. HELCOM. 2008. *Convention on the protection of the marine environment of the Baltic Sea Area, 1992 (Helsinki Convention)*. HELCOM, November 1993.

33. HELCOM. 2010. Ecosystem Health of the Baltic Sea 2003–2007: HELCOM Initial Holistic Assessment. *Baltic Sea Environment Proceedings* 122.

34. HELCOM. 2013. *Summary report on the development of revised Maximum Allowable Inputs (MAI) and updated Country Allocated Reduction Targets (CART) of the Baltic Sea Action Plan*. HELCOM Ministerial Meeting, Copenhagen, 31 October 2013. 22 pp.

35. HELCOM Response. 2013. Illegal discharges of oil in the Baltic Sea during 2012. HELCOM Baltic Sea Environment Fact Sheets. Online 2014–08–14. http://www.helcom.fi/baltic-sea-trends/environment-fact-sheets/.

36. Hilborn, R., and D. Ovando. 2014. Food for thought. Reflexions on the success of traditional fisheries management. *ICES Journal of Marine Science* 71: 1040–1046.

37. Hordoir, R., and H.E.M. Meier. 2012. Effect of climate change on the thermal stratification of the Baltic Sea: a sensitivity experiment. *Climate Dynamics* 38: 1703–1713.

38. ICES. 1970. Report of the ICES working group on pollution of the Baltic Sea. *Cooperative Research Report, Series A* 15: 1–86.

39. ICES. 2014. *Report of the Baltic Fisheries Assessment Working Group*. Copenhagen: ICES.

40. Jansson, B.-O. 1972. Ecosystem approach to the Baltic problem. *Bulletin of Ecological Research Communications/Swedish Natural Science Research Council* 16: 1–82.

41. Jentoft, S., and R. Chuenpagdee. 2009. Fisheries and coastal governance as a wicked problem.*Marine Policy* 33: 553–560.

42. Kahru, M., and R. Elmgren. 2014. Satellite detection of multi-decadal time series of cyanobacteria accumulations in the Baltic Sea. *Biogeosciences* 11: 3619–3633.

43. Karlson, A.M.L., J. Duberg, N.H. Motwani, H. Hogfors, I. Klawonn, H. Ploug, J.B. Svedén, A. Garbaras, et al. 2015. Nitrogen fixation by cyanobacteria stimulates production in Baltic food webs. *AMBIO*. doi:10.1007/s13280-015-0660-x.

44. Kiljunen, M., M. Vanhatalo, S. Mantyniemi, H. Peltonen, S. Kuikka, H. Kiviranta, R. Parmanne, et al. 2007. Human dietary intake of organochlorines from baltic herring: Implications of individual fish variability and fisheries management. *AMBIO* 36: 257–264.

45. KSLA (Royal Swedish Academy of Forestry and Agriculture). 2013. Seals, cormorants and fisheries. On the effects of seals and cormorants on fish stocks in the Baltic Sea. *Kungliga Skogs och Lantbruksakademiens tidskrift* 152: 1–26 (in Swedish).

46. Larsson, P., A. Andersson, D. Broman, J. Nordbäck, and E. Lundberg. 2000. Persistent organic pollutants (POPs) in pelagic ecosystems. *AMBIO* 29: 202–209.

47. Larsson, U., R. Elmgren, and F. Wulff. 1985. Eutrophication and the Baltic Sea—Causes and consequences. *AMBIO* 14: 9–14.

48. Larsson, U., S. Hajdu, J. Walve, and R. Elmgren. 2001. Baltic nitrogen fixation estimated from the summer increase in upper mixed layer total nitrogen. *Limnology and Oceanography* 46: 811–820.

49. Lefébure, R., R. Degerman, A. Andersson, S. Larsson, L.-O. Eriksson, U. Båmstedt, and P. Byström. 2013. Impacts of elevated terrestrial nutrient load and temperature on pelagic food-web efficiency and fish production. *Global Change Biology* 19: 1358–1372.

50. Leppäkoski, E., S. Gollasch, P. Gruszka, H. Ojaveer, S. Olenin, and V. Panov. 2002. The Baltic—A sea of invaders. *Canadian Journal of Fisheries and Aquatic Sciences* 59: 1175–1188.

51. Leppäranta, M., and K. Myrberg. 2009. *Physical oceanography of the Baltic Sea*. Chichester: Praxis Publishing Ltd.

52. Leslie, H.M., and K.L. McLeod. 2007. Confronting the challenges of implementing marine ecosystem-based management. *Frontiers in Ecology and the Environment* 5: 540–548.

53. Levin, S.A., and J. Lubchenco. 2008. Resilience, robustness, and marine ecosystem-based management. *BioScience* 58: 27–32.

54. Levin, P.S., and C. Möllmann. 2015. Marine ecosystem regime shifts: Challenges and opportunities for ecosystem-based management. *Philosophical Transactions of the Royal Society. B—Biological Sciences* 370: 20130275. doi:10.1098/rstb.2013.0275.

55. Likens, G.E. 1992. *The ecosystem approach: Its use and abuse*. Oldenburg: Ecology Institute.

56. McQuatters-Gollop, A. 2012. Challenges for implementing the Marine Strategy Framework Directive in a climate of macroecological change. *Philosophical Transactions of the Royal Society A—Mathematical Physical and Engineering Sciences* 370: 5636–5655.

57. Meier, H.E.M., H.C. Andersson, B. Arheimer, T. Blenckner, B. Chubarenko, C. Donnelly, K. Eilola, B.G. Gustafsson, et al. 2012. Comparing reconstructed past variations and future projections of the Baltic Sea ecosystem—First results from multi-model ensemble simulations.*Environmental Research Letters* 7: 034005.

58. Meier, H.E.M., H.C. Andersson, B. Arheimer, C. Donelly, K. Eilola, B.G. Gustafsson, L. Kotwicki, T.-S. Neset, et al. 2013. Ensemble modelling of the Baltic Sea ecosystem to provide scenarios for management. *AMBIO* 43: 37–48.

59. Mills, E.L. 1989. *Biological oceanography: An early history 1870–1960*. Ithaca, NY: Columbia University Press.

60. Möllmann, C., M. Lindegren, T. Blenckner, L. Bergström, M. Casisni, R. Diekmann, J. Flinkman, B. Müller-Karulis, et al. 2014. Implementing ecosystem-based fisheries management: from single-species to integrated ecosystem assessment and advice for Baltic Sea fish stocks. *ICES Journal of Marine Science* 71: 1187–1197.

61. Niiranen, S., J. Yletyinen, M.T. Tomczak, T. Blenckner, O. Hjerne, B.R. MacKenzie, B. Müller-Karulis, T. Neumann, et al. 2013. Combined effects of global climate change and regional ecosystem drivers on an exploited marine food web. *Global Change Biology* 19: 3327–3342.

62. Nyberg, E., S. Faxneld, S. Danielsson, U. Eriksson, A. Miller, and A. Bignert. 2015. Temporal and spatial trends of PCBs, DDTs, HCHs, and HCB in Swedish marine biota 1969–2012.*AMBIO*. doi:10.1007/s13280-015-0667-3.

63. Ojaveer, H., A. Jaanus, B.R. MacKenzie, G. Martin, S. Olenin, T. Radziejewska, I. Telesh, M.L. Zettler, et al. 2010. Status of biodiversity in the Baltic Sea. *PlosOne* 5: e12467.

64. Österblom, H., S. Hansson, U. Larsson, O. Hjerne, F. Wulff, R. Elmgren, and C. Folke. 2007. Human-induced trophic cascades and ecological regime shifts in the Baltic Sea. *Ecosystems* 10: 877–889.

65. Österblom, H., A. Gårdmark, L. Bergström, B. Muller-Karulis, C. Folke, M. Lindegren, M. Casini, P. Olsson, et al. 2010. Making the ecosystem approach operational—Can regime shifts in ecological- and governance systems facilitate the transition? *Marine Policy* 34: 1290–1299.

66. Ploug, H., N. Musat, B. Adam, C.L. Moraru, G. Lavik, T. Vagner, B. Bergman, and M.M.M. Kuypers. 2010. Carbon and nitrogen fluxes associated with the cyanobacterium*Aphanizomenon* sp. in the Baltic Sea. *ISME Journal* 4: 1215–1223.

67. Raateoja, M., J. Seppälä, H. Kuosa, and K. Myrberg. 2005. Recent changes in trophic state of the Baltic Sea along the SW coast of Finland. *AMBIO* 34: 188–191.

68. Remane, A., and C. Schlieper. 1971. *Biology of brackish water*, 2nd ed. Die Binnengewässer 25. E. Schweitzerbart'sche Verlagsbuchhandlung (Nägele u. Obermiller). Stuttgart: Wiley.

69. Roos, A.M., B.-M.V.M. Bäcklin, B.O. Helander, F.F. Riget, and U.C. Eriksson. 2012. Improved reproductive success in otters (*Lutra lutra*), grey seals (*Halichoerus grypus*) and sea eagles (*Haliaeetus albicilla*) from Sweden in relation to concentrations of organochlorine contaminants. *Environmental Pollution* 170: 268–275.

70. Roos, A., U. Berger, U. Järnberg, J. van Dijk, and A. Bignert. 2013. Increasing concentrations of perfluoroalkyl acids in Scandinavian otters (*Lutra lutra*) between 1972 and 2011: A new threat to the otter population? *Environmental Science and Technology* 47: 11757–11765.

71. Rozwadowski, H.M. 2002. *The Sea knows no boundaries: A century of marine science under ICES*. International Council for the Exploration of the Sea. ISBN 0-295-98259-4.

72. Rutgersson, A., J. Jaagus, F. Schenk, and M. Stendel. 2014. Observed changes and variability of atmospheric parameters in the Baltic Sea region during the last 200 years. *Climate Research* 61: 177–190.

73. Sandström, A., B. Crona, and Ö. Bodin. 2014. Legitimacy in co-management: The impact of preexisting structures, social networks and governance strategies. *Environmental Policy and Governance* 24: 60–76.

74. Sjöstrand, B. 2008. Fisheries management—A question of good housekeeping. *Havet* 2: 4–5 (in Swedish).

75. Svedäng, H., and S. Hornborg. 2014. Selective fishing induces density-dependent growth.*Nature Communications* 5: Article number 4152. doi:10.1038/ncomms5152.

76. Sweitzer, J., S. Langaas, and C. Folke. 1996. Land cover and population density in the Baltic Sea drainage basin: A GIS database. *AMBIO* 25: 191–198.

77. Thrush, S.F., and P.K. Dayton. 2010. What can ecology contribute to ecosystem-based management? *Annual Review of Marine Science* 2: 419–441.

78. UNEP/GPA. 2006. *Ecosystem-based management: Markers for assessing progress*. The Hague: UNEP/GPA. ISBN 92-807-2707-9.

79. UNEP/WHO. 2013. In *State of the art science of endocrine disrupting chemicals—2012*, ed. Å. Bergman, J.J. Heindel, S. Jobling, K.A. Kidd, and R.T. Zoeller. Geneva: United Nations Environment Programme and The World Health Organization.

80. Valman, M. 2013. Institutional stability and change in the Baltic Sea: 30 years of issues, crises and solutions. *Marine Policy* 38: 54–64.

81. Vahtera, E., D.J. Conley, B.G. Gustafsson, H. Kuosa, H. Pitkänen, O.P. Savchuk, T. Tamminen, M. Viitasalo, et al. 2007. Internal ecosystem feedbacks enhance nitrogen-fixing cyanobacteria blooms and complicate management in the Baltic Sea. *AMBIO* 36: 186–194.

82. Vallius, H. 2014. Heavy metal concentrations in sediment cores from the northern Baltic Sea: Declines over the last two decades. *Marine Pollution Bulletin* 79: 359–364.

83. VanDeveer, S.D. 2011. Networked Baltic Environment Cooperation. *Journal of Baltic Studies*42: 37–55.

84. Wikner, J., and A. Andersson. 2012. Increased freshwater discharge shifts the trophic balance in the coastal zone of the northern Baltic Sea. *Global Change Biology* 18: 2509–2519.

85. Worm, B., R. Hilborn, J.K. Baum, T.A. Branch, J.S. Collie, C. Costello, M.J. Fogarty, E.A. Fulton, et al. 2009. Rebuilding global fisheries. *Science* 325: 578–585.

86. Zabel, R.W., C.J. Harvey, S.L. Katz, T.P. Good, and P.S. Levin. 2003. Ecologically sustainable yield—Marine conservation requires a new ecosystem-based concept for fisheries management that looks beyond sustainable yield for individual fish species. *American Scientist* 91: 150–157.

87. Zeller, D., P. Rossing, S. Harper, L. Persson, S. Booth, and D. Pauly. 2011. The Baltic Sea: Estimates of total fisheries removals 1950–2007. *Fisheries Research* 108: 356–363.

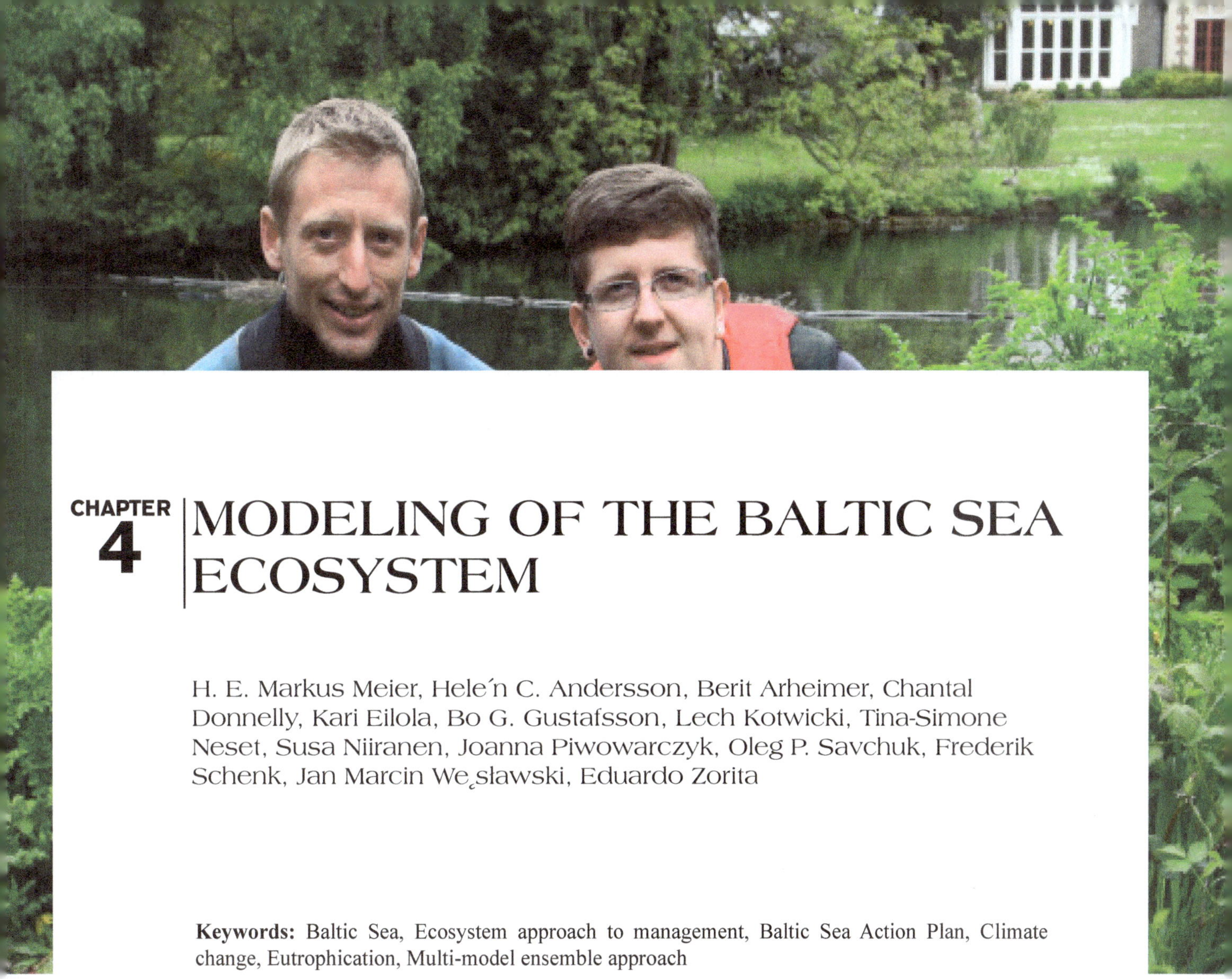

CHAPTER 4 MODELING OF THE BALTIC SEA ECOSYSTEM

H. E. Markus Meier, Hele´n C. Andersson, Berit Arheimer, Chantal Donnelly, Kari Eilola, Bo G. Gustafsson, Lech Kotwicki, Tina-Simone Neset, Susa Niiranen, Joanna Piwowarczyk, Oleg P. Savchuk, Frederik Schenk, Jan Marcin We˛sławski, Eduardo Zorita

Keywords: Baltic Sea, Ecosystem approach to management, Baltic Sea Action Plan, Climate change, Eutrophication, Multi-model ensemble approach

INTRODUCTION

To improve the status of the Baltic Sea environment, the Helsinki Commission (HELCOM) created the Baltic Sea Action Plan (BSAP) by consistent application of the ecosystem approach to management (Backer et al. 2010). In 2007, the BSAP led to international decisions on nutrient load reductions. Policy instruments—like the Marine Strategy Framework Directive, national environmental objectives, and HELCOM's BSAP—do not take the impact of climate change into consideration. For example, Maximum Allowable Inputs (MAIs) of the BSAP are calculated under the assumption that Baltic Sea environmental conditions are in a steady-state, biogeochemically as well as physically, and that it will reach a new biogeochemical steady-state after the internal sinks and sources have adapted to the new loads under the present, prevailing physical steady-state. Within a changing climate this assumption will not hold as the physical environment will change leading to feedbacks upon the biogeochemical cycling, e.g., by enhancing growth rates and mineralization.

Hence, the interdisciplinary ECOSUPPORT project (Advanced modeling tool for scenarios of the Baltic Sea ECOsystem to SUPPORT decision-making) running during 2009–2011 was designed to provide scientifically sound knowledge on how the combination of climate change and nutrient loads from the catchment will impact the marine ecosystem.[1] Since climate change is likely to affect the implementation of policies and environmental objectives, the main aim of ECOSUPPORT was to provide easy access to modeled scenarios of the marine ecosystem, in order to raise awareness of stakeholders and the general public of the impacts of human activities on the ecosystem, as well as for policy decision support.

For this purpose appropriate models need to be developed. Global Earth System Models (ESMs) are fundamental tools to assess future climate change of the twenty-first century (e.g., Solomon et al. 2007). For the upcoming Intergovernmental Panel on Climate Change (IPCC) Fifth Assessment Report, a new generation of ESMs from CMIP5[2] that include complex

Citation: H. E. Markus Meier, Helén C. Andersson, Berit Arheimer, Chantal Donnelly, Kari Eilola, Bo G. Gustafsson, Lech Kotwicki, Tina-Simone Neset, Susa Niiranen, Joanna Piwowarczyk, Oleg P. Savchuk, Frederik Schenk, Jan Marcin Węsławski, Eduardo Zorita, Modeling of the Baltic Sea Ecosystem, DOI: 10.1007/s13280-013-0475-6.

interactions within the Earth system, e.g., by taking feedbacks of the carbon cycle into account, was used. However, despite their complexity and great advances over recent decades, ESMs remain simplified representations of the real Earth system and are associated with a number of errors that reduce their reliability (e.g., Knutti and Sedláček 2012). One of the shortcomings of ESMs is their computational demand that limits the possibility to resolve small spatial scales. However, this information is exactly what is needed by decision-makers (e.g., BACC Author Team 2008). To bridge the gap between global model results and the regional to local information needed for impact studies, regional climate models (RCMs) have been developed as a complementary tool to ESMs allowing increased horizontal resolution and a greater number of explicitly resolved processes (Rummukainen 2010). In the dynamical downscaling approach RCMs are driven with data from ESMs at their lateral boundaries (Räisänen et al. 2004).

To be able to calculate the impact of changing climate on the Baltic Sea ecosystem, so-called Regional Climate System Models (RCSMs) are needed, comprising all relevant components of the Earth system, like atmosphere, sea ice, ocean, land surface physics, atmospheric chemistry, terrestrial and marine biogeochemistry, and marine food-web. Overall, the number of RCSM applications is still very limited. The few world-wide existing regional atmosphere–ice–ocean models are usually not coupled to terrestrial or marine ecosystem model components (Döscher et al. 2002, 2010 and references therein). Further, climate change impact studies for regional seas are very often performed using either results from a regional atmosphere model with sea surface boundary conditions from a global model as driver (Madsen 2009) or atmospheric and hydrological model outputs from global climate models directly (Lasram et al. 2010). Both approaches have significant shortcomings because sea surface boundary conditions taken from global models have significant biases at regional scale (Meier et al. 2011a). Within ECOSUPPORT, the first steps toward a more complex RCSM for the Baltic Sea region were taken. ECOSUPPORT improved the quality of RCSM scenario simulations by (1) developing new component models with increased spatial resolution and refined process descriptions; (2) coupling model components together to study feedback mechanisms within the Earth system comprising the atmosphere–ice–ocean–land surface continuum; (3) applying a holistic, multi-stressor approach that takes the impacts of changing climate, eutrophication, and overfishing into account; (4) assessing long-term changes during 1850–2100 including the transition from oligotrophic to eutrophied states of the Baltic Sea environment; and (5) estimating uncertainties of future projections by applying a multi-model ensemble approach. Finally, ECOSUPPORT disseminated the results from scenario simulations to stakeholders and the public in a more efficient way compared to traditional dissemination techniques by using a novel visualization and communication approach. This paper summarizes major findings of ECOSUPPORT and the impacts on policies.

MATERIALS AND METHODS

Toward Regional Climate System Modeling

Within ECOSUPPORT a hierarchy of regional models was used to downscale results from two General Circulation Models (GCMs) to the spatial scale of the Baltic Sea region including a coupled atmosphere–ice–ocean–land surface model (Döscher et al. 2002; Meier et al. 2011a), two hydrological models of differing complexity (Arheimer et al. 2012; Meier et al. 2012a), one atmospheric chemistry and transport model (Langner et al. 2009), three marine physical–biogeochemical models (Neumann et al. 2002; Eilola et al. 2009; Savchuk et al. 2012), one food-web model for the central Baltic Sea (Niiranen et al. 2012), various statistical fish population models (MacKenzie et al. 2012), and various regional to local scale models of differing parts of the Earth system for the Gulf of Finland, Vistula Lagoon, and Polish Coastal waters, for example, Biological Envelope Modeling (Weslawski et al. 2013; see also Meier et al. 2012c).

An important aspect of the ECOSUPPORT model hierarchy is the proper consideration of the land-sea continuum. ECOSUPPORT introduced for the first time the ability to simulate integrated discharge and nutrients at high-resolution for the entire Baltic Sea catchment using the Balt-HYPE model (Arheimer et al. 2012). The Balt-HYPE model was used to simulate the effects of future climate change and the interaction of climate change with a number of simpler remedial nutrient scenarios. Previously available estimates of the impacts of future climate change to discharge to the Baltic Sea were based on the HBV model forced by today's climate perturbed with a delta-change from the climate scenarios (Graham 2004). New within ECOSUPPORT was that transient estimates of discharge to the Baltic Sea were made using bias-corrected regional, coupled atmosphere–ocean model outputs and that the Balt-HYPE model provides first process based estimates of how nutrient fluxes to the Baltic Sea may change as a result of future climate change (Arheimer et al. 2012).

New Generation of Biogeochemical Models for Baltic Sea Management

In 2007 the agreed reductions of the BSAP were evaluated with the Simple As Necessary BAltic Long-Term large-Scale marine biogeochemical model SANBALTS implemented within the decision support system Baltic Nest (Savchuk and Wulff 2007; Wulff et al. 2007). For a contemporary revision of the BSAP in 2013 and the implementation of the Marine Strategy Framework Directive of EU, the BAltic sea Long-Term large-Scale Eutrophication Model (BALTSEM) has been developed as a next generation marine model in the Baltic Nest system (Savchuk et al. 2012). In addition, two three-dimensional Baltic Sea models have been developed further providing additional information on the sub-basin scale (Neumann et al.2002; Eilola et al. 2009).

Quality Assurance

Within ECOSUPPORT only models of high-quality were used to produce scenario simulations. The quality assurance followed a common protocol. For instance, Eilola et al. (2011) evaluated and compared individually the hindcast results (1970–2005) from three coupled physical–biogeochemical models relative to the seasonal and annual statistics of salinity, temperature, oxygen, nitrate, ammonium, and phosphate estimated from observations in the Baltic Sea. At representative stations, vertically resolved cost functions were calculated to quantify the biases of model means and standard deviations relative to the corresponding values from observations. Also the pools of nutrients in water and sediment, the extension of hypoxic bottom areas as well as cod reproductive volumes were studied and discussed.

Multi-model Ensemble Approach

To estimate uncertainties of future projections caused by biases of the global and regional models, natural variability and unknown scenarios of drivers like greenhouse gas and nutrient load emissions and fisheries, a multi-model ensemble approach was applied. The scenarios cover plausible ranges between the most optimistic and pessimistic cases. For instance, one of the nutrient load scenarios is the BSAP. For the marine biogeochemistry more than 50 transient scenario simulations for the period 1960–2100 were performed. The uncertainty is described by the standard deviation among the projections of the ensemble. For details, the reader is referred to Meier et al. (2012b).

Multi-stressor Approach

The food-web effects of combined multiple-stressors, i.e., climate, nutrient loads and cod fishing, were studied by linking the output from biogeochemical models with an Ecopath with Ecosim (EwE) model for the Central Baltic Sea (BaltProWeb; Tomczak et al. 2012). This model has functional groups from primary producers to seals, including the most important commercial fish of the Baltic, i.e., cod (*Gadus morhua callarias*), herring (*Clupea harengus membras*), and sprat (*Sprattus sprattus*). For future projections (2010–2098), the food-web model was driven by two cod fishing scenarios (intensive and low fishing) in combination with environmental forcing (salinity, temperature, hypoxic area, cod reproductive volume and primary production) as projected by the three coupled physical–biogeochemical models in climate and nutrient load scenarios (for details see Niiranen et al. 2013).

Reconstruction of the Past Baltic Sea Climate Variability

An important aspect of the project was to test the model sensitivity to different drivers on longer time scales by reconstructing the evolution of the Baltic Sea ecosystem from its pristine state around 1850 until today (Gustafsson et al. 2012; Meier et al. 2012c). For this purpose, a key output of the project was to reconstruct the so far longest and most complete set of different forcings from a limited number of available observations. These drivers comprise riverine and atmospheric nutrient inputs (Ruoho-Airola et al. 2012; Savchuk et al. 2012), hydrological data and multivariate daily meteorological fields (Schenk and Zorita 2012). To reconstruct the daily meteorological forcing (High Resolution Atmospheric Forcing Fields, HiResAFF), a nonlinear statistical method was developed. Long historical station records of daily pressure and monthly temperature since 1850 were used to find multivariate analogous target fields within a shorter 50-year regional climate simulation. The advantage of analog-upscaling (Schenk and Zorita2012) lays in the physical consistency of the reconstructed fields including their regional topographic and variable-specific frequency distributions. As the reconstruction makes use of a limited but relatively constant number of predictor stations, HiResAFF provides a homogeneous long-term reconstruction. Hence, it avoids introducing spurious long-term trends by assimilating different station numbers over time as recently shown for the novel twentieth Century Reanalysis since 1871 (Krueger et al. 2013).

Policy Dialog with Stakeholders

The anthropogenic pressure on the ecosystem has a direct link to many of the ecosystem services and thereby impact on economic and societal values that provide human welfare (Fig. 1). For example, changes in the marine habitat may change the stock abundance and distribution of both leisure and commercial fish species and the value of coastal recreational activities due to extensive cyanobacteria blooms. Management strategies may have long-lasting consequences for the environment and habits in the society, like the use of fertilizers, increased meat consumption and the use of phosphate in detergents, affect the marine environment considerably. ECOSUPPORT therefore created a communication platform that enabled system understanding, comparison of the outcome of different management scenarios and assessment of the uncertainties involved in modeled projections.

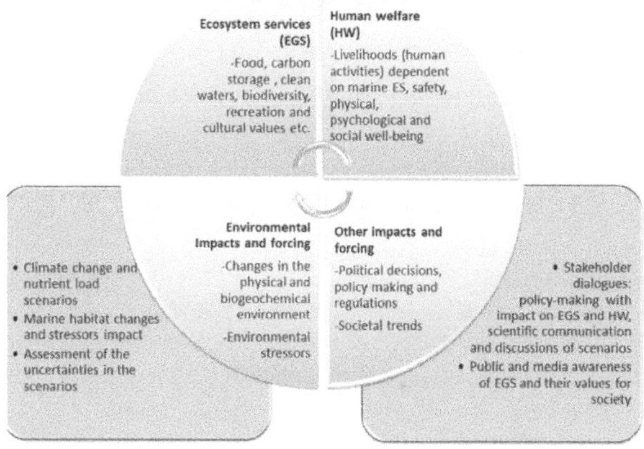

Figure 1: Ecosystem services provide human services. Changes in the physical and bio-geochemical marine environment, as well as policy decisions and societal trends and habits, impact the marine services. ECOSUPPORT aimed at providing awareness and understanding of the human impacts.

Geovisualization

Visualization of scientific results has the ability to provide a rapid understanding of complex and heterogeneous data (Tufte 1997; Ware 2004), and several studies have pointed toward the potential increase in engagement and involvement of the audience in participatory events that feature visualization as a methodology (Nicholson-Cole 2005; Salter et al. 2009; Sheppard et al.2011). Geovisualization takes advantage of human perception capability to find patterns, and allows for large quantities of data and system dynamics to be communicated for a specific spatial context.

For the purpose of involving stakeholders in a dialog between audience and presenter, modules with visual representations of geospatial data for selected parameters for the Baltic Sea Region were, along with short animations, compiled into the interactive visualization software Uniview[3](Fig. 2). The modules were designed to focus on specific issues, including cause, effect and management scenarios for catchment and sea. A narrative was created for each module and the comparative-scenario approach formed a base for further discussions. The interactive presentations were adjusted to suit particular target groups, with the objective to enable data exploration and analysis from different perspectives and to support analytical reasoning and clear understanding of the problems. The interactive application is particularly striking to experience in an immersive GeoDome[4] environment, which has the potential to further enhance perception and understanding (Neset et al. 2010). ECOSUPPORT therefore used an inflatable, portable GeoDome if feasible (Meier and Andersson 2012), and provided the data material additionally for regular flat screens and through web-access.[5] An underlying challenge of this format, especially in communication with the general public, was to emphasize the severity of the presented state-of-the-art scenarios, while stressing that management and societal actions are possible and can make a difference, in an effort to avoid creating feelings of hopelessness and apathy of the audience (e.g., Moser and Dilling 2004).

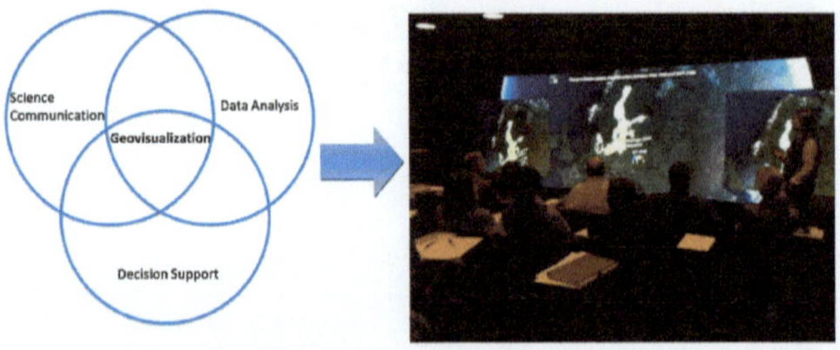

Figure 2: Geovisualization makes use of human perception capabilities when linking large and complex data sets to a geospatial setting. Communication and discussion of ECOSUPPORT scenarios to stakeholders have occurred at a number of occasions, both in the GeoDome and on flat screen. The *photo* is showing discussion and comparison of ECOSUPPORT management scenarios at the Baltadapt workshop, Norrköping 2012.

Challenges and Opportunities in Visualization-Supported Science Communication

To assess the potential of visualization as a tool for science communication and decision support, two separate events were evaluated. Interviews were undertaken with participants of presentations in the GeoDome at the European Strategy for the Baltic Sea Region (EUSBSR) conference in Gdansk 2011. The visualization modules also supported interactive sessions of the BaltAdapt[6] stakeholder workshop in Norrköping 2012, which were recorded and analyzed as well as evaluated through a survey. BaltAdapt is a transnational flagship project, developing a climate change adaptation strategy for the Baltic Sea region, with focus on the marine and coastal environment. Both events brought together stakeholders from the Baltic Sea Region. The EUSBSR conference featured a broad spectrum of issues and sectors, while the BaltAdapt workshop was focused on climate change and agriculture, with participants representing, e.g., farmer's associations, agricultural extension services and agricultural administration, from the national, regional, and local level.

RESULTS

High-Quality Baltic Sea Models

The extensive and detailed model-data comparison for the period 1970–2006 showed that biogeochemical models were capable to reproduce much of the nutrient biogeochemical cycling in the Baltic Sea (Eilola et al. 2011). For example, the Nest component model BALTSEM simulated successfully both the inter-basin spatial gradients and temporal variations at seasonal to long-term scales (Fig. 3; Savchuk et al. 2012). None of the three models was perfect in all aspects. The ensemble means matched the data better than, or as good as, the results of any of the individual models (Eilola et al. 2011). The evaluation of model's performance revealed that, in addition to a number of model-specific needs for improvement, there are a few major model-data discrepancies like (1) the large model-data biases in the Gulf of Bothnia, especially in the Bothnian Bay; (2) the markedly lower rates of primary production compared to those reported from observations; (3) the significant differences in the nutrient turnover time scales between the models, caused primarily by differences in simulated sediment nutrient pools and fluxes.

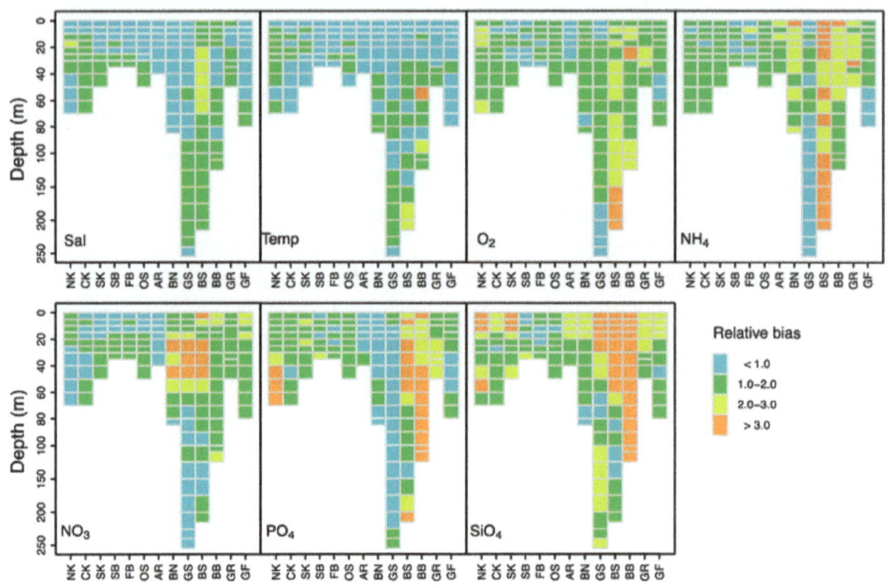

Figure 3: Spatial distribution of the relative bias between simulated and observed dynamics of BALTSEM variables. Comparisons are made

for Sal—salinity, Temp—water temperature, and concentrations of O_2—oxygen, NH_4—ammonium, NO_3—nitrate, PO_4—phosphate, SiO_4—silicate. At the *x*-axis the following Baltic Sea basins are depicted: NK—Northern Kattegat, CK—Central Kattegat, SK—Southern Kattegat, SB—Samsø Belt, FB—Fehmarn Belt, OS—Öresund, AR—Arkona Basin, BN—Bornholm Basin, GS—Gotland Sea, BS—Bothnian Sea, BB—Bothnian Bay, GR—Gulf of Riga, GF—Gulf of Finland.

Projected Changes in the Regional Climate System

According to Meier et al. (2012c) water temperature in the Baltic Sea at the end of the twenty-first century will be higher and salinity and oxygen concentrations will be lower than any values since 1850. Although changes in the Baltic Sea water balance suffer considerably from model uncertainties (Meier et al. 2006), the seasonal dynamics of discharge to the Baltic Sea are expected to increase compared to today's pattern for all ECOSUPPORT projections (Fig. 2 in Meier et al. 2012b). In general, winter discharge to the Baltic Sea increases while summer discharge decreases. The spring flood peak is reduced for all scenarios. It was much more difficult to detect trends in overall nutrient fluxes to the Baltic Sea as a result of a future climate. Interannual variation was much larger than an eventual long-term trend for each climate projection. Nevertheless, there was a consistent change to the seasonal distribution of nutrient concentrations to the Baltic Sea (Arheimer et al. 2012). The seasonal variations of both nitrogen and phosphorous concentrations were dampened with consistent decreases in winter concentrations seen. For phosphorous, there is also an indication that summer concentrations may increase and that more frequent short-term peaks in concentration may occur. The experiments with combinations of remedial measures and climate change in the Balt-HYPE model indicate that there is a possibility to reach all the BSAP targets in the future for most marine basins by the end of this century. For N, the impact of climate change is of the same order as the expected reduction from remedial measures, according to the results of the model experiment. Further, there is a higher probability to reach BSAP targets for P than for N. Thus, climate effects need to be accounted for when estimating the long-term effects of the BSAP.

The future marine ecosystem in the Baltic Sea was projected to change unprecedentedly compared to the past 150 years, and nutrient load reductions and sustainable fishery may be even more important in the future to ensure a healthy marine ecosystem when the stresses from climate change increase (Meier et al. 2012a, b, c; Niiranen et al. 2013). Applying various nutrient load scenarios, it was also shown that under the impact of warming climate hypoxic and anoxic areas will very likely increase or at best only slightly decrease (in case of optimistic nutrient load reduction scenarios) compared to present conditions, regardless of the used global model and climate scenario (Meier et al. 2011b). For the end of the century, prolonged growth and a more than twofold increase in the mean cyanobacteria biomass and nitrogen fixation was found using a coupled biological–physical model with an advanced cyanobacteria life cycle model (Hense et al. 2013). In addition, considerable changes in the spring bloom at least in the northern Baltic Sea are expected as a consequence of the shrinking ice cover in warmer climate (Eilola et al. 2013).

The combination of regional drivers, i.e., cod fishing and nutrient loads, had a large effect on the projected futures of the Central Baltic Sea food-web. In the worst-case scenario, i.e., high cod fishing and nutrient loads a eutrophied and strongly sprat-dominated ecosystem was projected, while the best-case scenario resulted in a cod-dominated ecosystem with eutrophication levels close to present. However, the regional management decisions were not fully able to compensate for some directional climate change effects. For example, cod was negatively affected by worsening reproduction conditions, due to decreasing salinities, and its biomass was projected to decrease during the second half of the twenty-first century across all combinations of fishing and nutrient load scenarios. This indicates that only reduced fishing should be permitted under future climate conditions, if moderate to high cod biomasses are desired (Lindegren et al. 2010). In general, top predatory fish cod was mainly affected by changes in fishing mortality, while phytoplankton and several zooplankton groups responded almost solely to changes in climate and nutrient conditions. The intermediate trophic level groups, i.e., sprat, herring, and*Pseudocalanus acuspes* were most affected by the combination of top-down (cod fishing) and bottom-up forces (nutrient loads and climate drivers). These groups are suggested to have an ecosystem structuring role in the Baltic Sea (Möllmann et al. 2009), indicating the importance of evaluating the interplay of multiple driver effects when projecting the future of the Baltic Sea food-web.

Uncertainties in Projected Biogeochemical Cycles

According to Meier et al. (2011b), uncertainties in projected biogeochemical cycles are dominated by unknown future nutrient loads, biases of the GCMs and biases of the biogeochemical models. We found largely differing sensitivities of the models to changing nutrient loads. Nevertheless, all biogeochemical models suggest that the BSAP will be less effective in future climate than in present climate.

Detection and Attribution in Physical and Biogeochemical Variables

With exception of the somewhat underestimated warming trends in spring, HiResAFF clearly reflect the observed strong seasonal warming trends since 1850. As shown in Gustafsson et al. (2012), the reconstructed trends closely match those derived from coarsely gridded observations like CRUTEM3 over the Baltic Sea since 1871. Besides a high spatial resolution of ~25 km, a novelty of HiResAFF relates to the physically consistent reconstruction of near-surface conditions also over sea-areas where usually little or no observations exist back in time. As an example, Fig. 4 shows smaller winter warming trends over deeper sea-areas like the Baltic Proper.

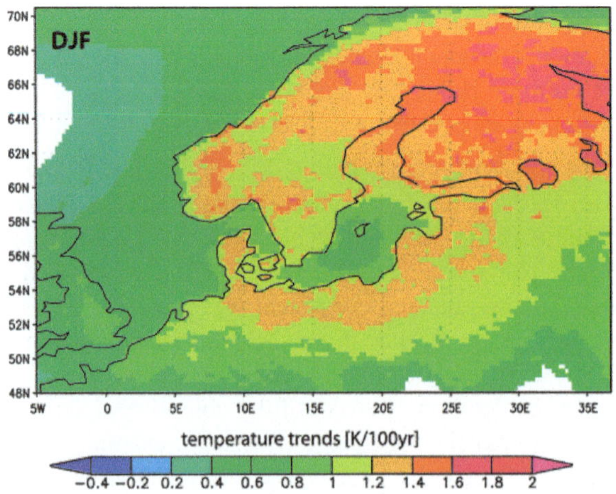

Figure 4: Seasonal near-surface temperature trends in winter 1850–2009 (HiResAFF). Non-significant trends ($p < 0.05$) in *white*.

Impacts of ECOSUPPORT Results on Management

HELCOM's BSAP has been further developed since 2007 and now comprises a number of steps (Fig. 5). The basis of the BSAP has been the definitions of the ecological objectives, agreed upon as the vision of an environmentally healthy Baltic Sea, for example clear water and end to excessive algal blooms. Indicators were developed that would reflect the objectives. Monitoring enables the assessment of the current environmental status, as reflected by the indicators. Quantitative target indicator values for a good environmental status are also established, primarily based on monitoring data and statistical analysis. In the following step, the relationships between pressure (i.e., nutrient loads) and target variables are quantified by means of physical–biogeochemical modeling. The pressure–response relationships differ for the various regions within the Baltic Sea because of differences in, e.g., circulation, ecosystem and nutrient loads. The results of the modeling are basin-wise MAIs of nutrients that will result in a development toward eventually reaching the targets. The MAIs of nutrients, as a first step toward implementation, are allocated as country-wise reduction target where the necessary load reductions are distributed by basin to the contracting countries according to polluter pays principles, and what is considered fair burden and is in agreement with the BSAP. The implementation of nutrient load reductions is planned through national implementation plans.

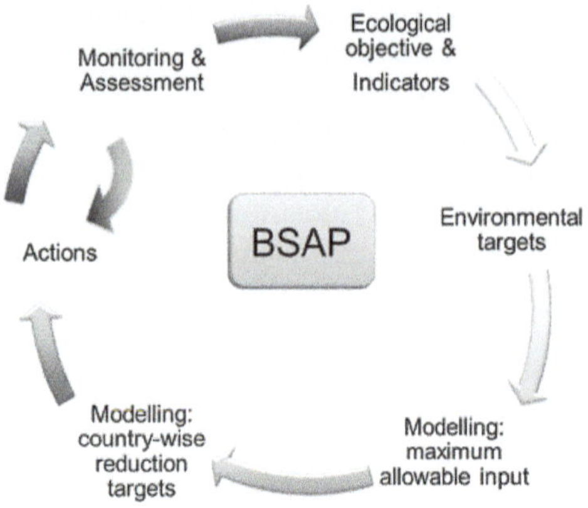

Figure 5: The work process of the Baltic Sea Action Plan.

Leading up to a Ministerial Meeting in 2013, a review of the BSAP agreement from 2007 with its preliminary nutrient reduction targets has significantly gained from ECOSUPPORT developments, e.g., the determination of targets within the TARGREV project (Review of the ecological targets for eutrophication of the HELCOM BSAP, see HELCOM 2013) was improved by using results from the long-term hindcast reconstruction 1850–2007 (Gustafsson et al.2012). From the ensemble mean of the historical simulations guiding targets for nutrients and chlorophyll *a* were derived by estimating indicator levels around 1900 with the levels from the 1970s. Further, the MAI calculation has been strengthened by multi-model validation studies (Eilola et al. 2011) and model developments leading to a new generation of high-quality Baltic Sea models (Savchuk et al. 2012). Although the sensitivity of the models to nutrient load changes largely differed, all models showed the same response qualitatively (Meier et al. 2012b). HELCOM ministerial meetings, both in 2007 and 2010, noted that climate change will have impacts and this should ultimately be reflected in HELCOM policies. Specifically, both climate change aspects as well as ensemble modeling should be reflected in the reviewed BSAP at the 2013 meeting.[7] The HELCOM Executive Secretary emphasized the usefulness of ECOSUPPORT results in the review of the eutrophication segment of the BSAP.[8] Hence, future revisions of the BSAP will hopefully include both changing climate and ensemble modeling to estimate uncertainties of projections.

For a comprehensive summary of ECOSUPPORT results the reader is referred to the AMBIO Special Issue: ECOSUPPORT—Different Ecosystem Drivers Under Future Climate Scenarios in the Baltic Sea (Reckermann 2012 [9]).

DISCUSSION

The ECOSUPPORT results indicate that it is very likely that the work with improving the health of the Baltic Sea will take place in a transient Baltic Sea. As the nutrient reductions will not be as efficient in a future climate, it is important to assess how much additional reductions need to be accomplished in order to reach the goals of BSAP in a changing climate. Further, it is important to understand "when to stop" since improvements in the environment will continue long after the actual reduction took place. The work will undoubtedly be afflicted with uncertainties arising from different sources. The climate change scenarios are uncertain and reflect merely the present state-of-the-art knowledge, and will have to be revised to consider new development, mitigation strategies and technology. The management scenarios are uncertain in a changing climate since both needed reductions and catchment loads are uncertain: warming will, for example, induce higher evaporation and mineralization rates with impact on soil processes and increased precipitation changes in runoff rates, annual cycles and flooding. The processes are considered in the Balt-HYPE model, but process knowledge needs to be further improved (Arheimer et al.2012). Furthermore, the catchment will undergo changes due to changes in agricultural practices, improved technology for land nutrient retention, changing vegetation in a warmer and dryer/wetter climate, demographic changes, etc.—effects not taken into account in the present scenario narrative. Present scenarios also lack understanding of realistic present and future atmospheric deposition of nutrients. When it comes to ecosystem structure and functioning, we will move into a Baltic Sea state where the food-web models cannot be evaluated with present state observation and understanding since changes in, e.g., ocean acidification, lower salinity and impact on invasive species will bring a state unknown to the research community (Niiranen et al. 2012). This can have implications for management actions to protect marine areas and restrict fisheries, but is difficult to foresee with state-of-the-art models and understanding. The ECOSUPPORT efforts not only detected model deficiencies, e.g., for the biogeochemistry in the northern Baltic Sea, which needs to be improved, but also differences in model behavior with implications for management. This relates, e.g., to the changes of the internal loads (nutrient pools) under the nutrient load reduction scenario, which behave differently in the different Baltic Sea models and therefore the models have different response time between abatement and improvement.

There is a need to develop modeling strategies to support adaptive management under combined pressures. ECOSUPPORT was a pioneering effort showing the way and illustrating challenges and opportunities for time-dependent adaptive management. The way to approach the future and handle the uncertainties may be to make the BSAP process operational (Fig. 5). Also in the future the focus will have to be on the monitoring and assessment, in order to evaluate the ecosystem that changes as a response to applied abatement strategies. Models and scenarios should be continuously updated with state-of-the-art understanding and a multi-model, multi-scenario approach (ensemble modeling) would be preferable in order to take uncertainties into account (Meier et al. 2012c). The MAIs and reduction targets will have to be revised to take changes in the physical environment into account. The ECOSUPPORT projections also indicate future regional differences, e.g., the warming over the northern region will be more pronounced than over the southern region, and river runoff is indicated to increase in the northern region and maybe even decrease in southern areas. These changes mean that geographical areas of the Baltic Sea will have different salinity, temperature, and stratification in future compared to present climate. Hence, ecological objectives can either be more easy or difficult to reach than today. This indicates that the country allocations might also have to be changed due to achieved abatements and environmental targets, and should be adjusted to give maximum effect in the most cost-efficient way.

Although ecosystem services and climate change are phenomena that can be defined and quantified using a strictly scientific approach, the decision-making about them is a social process, where scientific information might be of secondary importance. Education and information campaigns could be an entry point to raise public awareness and inform managing strategies and coastal planning.

Although ECOSUPPORT results are highly relevant for marine management, some shortcomings of the approach and future research needs were identified:

- To improve the simulation of biogeochemical cycling, especially in the northern Baltic Sea, existing models should be extended with carbon cycling including dissolved organic carbon, nitrogen, and phosphorus (DOC, DON, and DOP).
- Nutrient retention in the coastal zone is poorly understood. The coastal zone filtering effect should be studied using high-resolution modeling of the coastal ecosystem.
- Models for lower and higher trophic levels should be two-way coupled to study bottom-up and top-down controls of the marine ecosystem.
- To better quantify the carbon and nutrient inputs from land, modeling of the land-sea continuum needs to be further improved, e.g., by including the interactions between climate, land use and socio-economy.
- For a comprehensive risk assessment the multi-stressor approach should be extended and hazardous substances and invasive species should also be taken into account.

ACKNOWLEDGMENTS

This paper is a contribution from the ECOSUPPORT project, which has received funding from the BONUS+ programme funded jointly by European Community's Seventh Framework Programme (FP7/2007-2013, Grant Agreement 217246) and Baltic Sea national funding institutions. The project has received funding from the Swedish Environmental Protection Agency (SEPA, Ref. No. 08/381) with additional support from the Swedish Research Council for Environment, Agricultural Sciences and Spatial Planning (FORMAS) within the projects "Impact of accelerated future global mean sea level rise on the phosphorus cycle in the Baltic Sea" (Grant No. 214-2009-577) and "Impact of changing climate on circulation and biogeochemical cycles of the integrated North Sea and Baltic Sea system" (Grant No. 214-2010-1575) and from the Baltic Ecosystem Adaptive Management program (BEAM, 2010-2014) funded by the Swedish government, as part of its Strategic Research Initiative on Marine Environmental Research. Further, we are in great debt to all institutes and individuals that kindly provided their data to the Baltic Environmental Database (BED) at the Baltic Nest Institute (http://nest.su.se/bed).

REFERENCES

1. Arheimer, B., J. Dahné, and C. Donnelly. 2012. Climate change impact on riverine nutrient load and land-based remedial measures of the Baltic Sea Action Plan. *AMBIO* 41: 600–612. doi:10.1007/s13280-012-0317-0.
2. BACC Author Team. 2008. *BALTEX assessment of climate change for the Baltic Sea basin. Regional Climate Studies*. Berlin: Springer-Verlag. ISBN: 978-3-540-72785-9.
3. Backer, H., J.-M. Leppänen, A.C. Brusendorff, K. Forsius, M. Stankiewicz, J. Mehtonen, M. Pyhälä, M. Laamanen, et al. 2010. HELCOM Baltic Sea Action Plan: A regional programme of measures for the marine environment based on the Ecosystem Approach. *Marine Pollution Bulletin* 60: 642–649.
4. Döscher, R., U. Willén, C. Jones, A. Rutgersson, H.E.M. Meier, U. Hansson, and L.P. Graham. 2002. The development of the regional coupled ocean–atmosphere model RCAO. *Boreal Environmental Research* 7: 183–192.
5. Döscher, R., K. Wyser, H.E.M. Meier, M. Qian, and R. Redler. 2010. Quantifying Arctic contributions to

climate predictability in a regional coupled ocean–ice–atmosphere model.*Climate Dynamics* 34: 1157–1176.

6. Eilola, K., H.E.M. Meier, and E. Almroth. 2009. On the dynamics of oxygen, phosphorus and cyanobacteria in the Baltic Sea: A model study. *Journal of Marine Systems* 75: 163–184.
7. Eilola, K., B.G. Gustafson, I. Kuznetsov, H.E.M. Meier, and O.P. Savchuk. 2011. Evaluation of biogeochemical cycles in an ensemble of three state-of-the-art numerical models of the Baltic Sea. *Journal of Marine Systems* 88: 267–284.
8. Eilola, K., S. Mårtensson, and H.E.M. Meier. 2013. Modeling the impact of reduced sea ice cover in future climate on the Baltic Sea biogeochemistry. *Geophysical Research Letters* 40: 1–6. doi:10.1029/2012GL054375.
9. Graham, L.P. 2004. Climate change effects on river flow to the Baltic Sea. *AMBIO* 33: 235–241.
10. Gustafsson, B.G., F. Schenk, T. Blenckner, K. Eilola, H.E.M. Meier, B. Müller-Karulis, T. Neumann, T. Ruoho-Airola, et al. 2012. Reconstructing the development of Baltic Sea eutrophication 1850–2006. *AMBIO* 41: 534–548. doi:10.1007/s13280-012-0317-x.
11. Hense, I., H.E.M. Meier, and S. Sonntag. 2013. Projected climate change impact on Baltic Sea cyanobacteria. *Climatic Change*. doi:10.1007/s10584-013-0702-y.
12. Knutti, R., and J. Sedláček. 2012. Robustness and uncertainties in the new CMIP5 climate model projections. *Nature Climate Change*. doi:10.1038/nclimate1716Letters.
13. Krueger, O., F. Schenk, F. Feser, and R. Weisse. 2013. Inconsistencies between long-term trends in storminess derived from the 20CR reanalysis and observations. *Journal of Climate* 26: 868–874. doi:10.1175/JCLI-D-12-00309.1.
14. Langner, J., C. Andersson, and M. Engardt. 2009. Atmospheric input of nitrogen to the Baltic Sea basin: Present situation, variability due to meteorology and impact of climate change.*Boreal Environment Research* 14: 226–237.
15. Lasram, F.B.R., F. Guilhaumon, C. Albouy, S. Somot, W. Thuiller, and D. Mouillot. 2010. The Mediterranean Sea as a "cul-de-sac" for endemic fishes facing climate change. *Global Change Biology*. doi:10.1111/j.1365-2486.2010.02224.x.
16. Lindegren, M., C. Möllmann, A. Nielsen, K. Brander, B.R. Mackenzie, and N.C. Stenseth. 2010. Ecological forecasting under climate change: The case of Baltic cod. *Proceedings of the Royal Society B* 277: 2121–2130.
17. MacKenzie, B.R., H.E.M. Meier, M. Lindegren, S. Neuenfeldt, M. Eero, T. Blenckner, M.T. Tomczak, and S. Niiranen. 2012. Impact of climate change on fish population dynamics in the Baltic Sea: A dynamical downscaling investigation. *AMBIO* 41: 626–636. doi:10.1007/s13280-012-0317-y.
18. Madsen, K.S. 2009. Recent and Future Climatic Changes in Temperature, Salinity and Sea Level of the North Sea and Baltic Sea. PhD Thesis, Planet and Geophysics, Niels Bohr Institute, University of Copenhagen.
19. Meier, H.E.M., and H.C. Andersson. 2012. ECOSUPPORT: A pilot study on decision support for Baltic Sea environmental management. *AMBIO* 41: 529–533. doi:10.1007/s13280-012-0317-y.
20. Meier, H.E.M., E. Kjellström, and L.P. Graham. 2006. Estimating uncertainties of projected Baltic Sea salinity in the late 21st century. *Geophysical Research Letters* 33: L15705. doi:10.1029/2006GL026488.
21. Meier, H.E.M., A. Höglund, R. Döscher, H. Andersson, U. Löptien, and E. Kjellström. 2011a. Quality assessment of atmospheric surface fields over the Baltic Sea of an ensemble of regional climate model simulations with respect to ocean dynamics. *Oceanologia* 53: 193–227.
22. Meier, H.E.M., H.C. Andersson, K. Eilola, B.G. Gustafsson, I. Kuznetsov, B. Müller-Karulis, T. Neumann, and O.P. Savchuk. 2011b. Hypoxia in future climates: A model ensemble study for the Baltic Sea. *Geophysical Research Letters* 38: L24608. doi:10.1029/2011GL049929.
23. Meier, H.E.M., R. Hordoir, H.C. Andersson, C. Dieterich, K. Eilola, B.G. Gustafsson, A. Höglund, and S. Schimanke. 2012a. Modeling the combined impact of changing climate and changing nutrient loads on the Baltic Sea environment in an ensemble of transient simulations for 1961-2099. *Climate Dynamics* 39: 2421–2441. doi:10.1007/s00382-012-1339-y.
24. Meier, H.E.M., B. Müller-Karulis, H.C. Andersson, C. Dieterich, K. Eilola, B.G. Gustafsson, A. Höglund, R. Hordoir, et al. 2012b. Impact of climate change on ecological quality indicators and biogeochemical fluxes in the Baltic Sea: A multi-model ensemble study. *AMBIO* 41: 558–573. doi:10.1007/s13280-012-0317-y.
25. Meier, H.E.M., H.C. Andersson, B. Arheimer, T. Blenckner, B. Chubarenko, C. Donnelly, K. Eilola, B.G. Gustafsson, et al. 2012c. Comparing reconstructed past variations and future projections of the Baltic Sea ecosystem—first results from multi-model ensemble simulations.*Environmental Research Letters* 7: 034005. doi:10.1088/1748-9326/7/3/034005.
26. Möllmann, C., R. Diekmann, B. Müller-Karulis, G. Kornilovs, M. Plikshs, and P. Axe. 2009. Reorganization of a large marine ecosystem due to atmospheric and anthropogenic pressure: A discontinuous regime shift in the Central Baltic Sea. *Global Change Biology* 15: 1377–1393.
27. Moser, S., and L. Dilling. 2004. Making climate hot: Communicating the urgency and challenge of global climate change. *Environment* 46: 32–46.
28. Neset, T.-S., V. Wibeck, O. Uhrqvist, and J. Johansson. 2010. Visualizing climate change: The potential of dome presentations as a tool for climate communication. In *Proceedings of Eurographics 2010: Areas Papers*. Norrköping, Sweden. May 2010, pp 31-35.
29. Neumann, T., W. Fennel, and C. Kremp. 2002. Experimental simulations with an ecosystem model of the Baltic Sea: A nutrient load reduction experiment. *Global Biogeochemical Cycles* 16: 1033.

30. Nicholson-Cole, S. 2005. Representing climate change futures: A critique on the use of images for visual communication. *Computers, Environment and Urban Systems* 29: 255–273.

31. Niiranen, S., T. Blenckner, O. Hjerne, and M.T. Tomczak. 2012. Uncertainties in a Baltic Sea food-web model reveal challenges for future projections. *AMBIO* 41: 613–625. doi:10.1007/s13280-012-0324-z.

32. Niiranen, S., J. Yletyinen, M.T. Tomczak, T. Blenckner, O. Hjerne, B.R. MacKenzie, B. Müller-Karulis, T. Neumann, et al. 2013. Combined effects of global climate change and regional ecosystem drivers on an exploited marine food web. *Global Change Biology* 19: 3327–3342. doi:10.1111/gcb.12309.

33. Räisänen, J., U. Hansson, A. Ullerstig, R. Döscher, L.P. Graham, C. Jones, H.E.M. Meier, P. Samuelsson, et al. 2004. European climate in the late twenty-first century: Regional simulations with two driving global models and two forcing scenarios. *Climate Dynamics* 22: 13–31.

34. Reckermann, M. 2012. ECOSUPPORT: Different ecosystem drivers under future climate scenarios in the Baltic Sea. *AMBIO* 41: 533–651.

35. Rummukainen, M. 2010. State-of-the-art with regional climate models. *Wiley Interdisciplinary Reviews: Climate Change* 1: 82–96. doi:10.1002/wcc.8.

36. Ruoho-Airola, T., K. Eilola, O. Savchuk, M. Parviainen, and V. Tarvainen. 2012. Atmospheric nutrient input to the Baltic Sea from 1850 to 2006: A reconstruction from modeling results and historical data. *AMBIO* 41: 549–557. doi:10.1007/s13280-012-0319-9.

37. Salter, J.D., C. Campbell, M. Journeay, and S.J.R. Sheppard. 2009. The digital workshop: Exploring the use of interactive and immersive visualisation tools in participatory planning.*Journal of Environmental Management* 90: 2090–2101.

38. Savchuk, O.P., and F. Wulff. 2007. Modeling the Baltic Sea eutrophication in a decision support system. *AMBIO* 36: 141–148.

39. Savchuk, O.P., B.G. Gustafson, and B. Müller-Karulis. 2012. BALTSEM: A marine model for decision support within the Baltic Sea Region. BNI Technical Report 7. 59 pp. Baltic Nest Institute, Stockholm University. ISBN 978-91-86655-06-8.

40. Schenk, F., and E. Zorita. 2012. Reconstruction of high resolution atmospheric fields for Northern Europe using analog-upscaling. *Climate of the Past* 8: 1681–1703. doi:10.5194/cp-8-1681-2012.

41. Sheppard, S.R.J., A. Shaw, D. Flanders, S. Burch, A. Wiek, J. Carmichael, J. Robinson, and S. Cohen. 2011. Future visioning of local climate change: A framework for community engagement and planning with scenarios and visualisation. *Futures* 43: 400–412.

42. Solomon, S., D. Qin, M. Manning, Z. Chen, M. Marquis, K.B. Averyt, M. Tignor, and H.L. Miller (eds.). 2007. Contribution of Working Group I to the Fourth Assessment Report of the Intergovernmental Panel on Climate Change. 996 pp. Cambridge University Press: Cambridge, UK and New York, NY, USA.

43. Tomczak, M.T., S. Niiranen, O. Hjerne, and T. Blenckner. 2012. Ecosystem flow dynamics in the Baltic Proper: Using a multi-trophic dataset as a basis for food-web modelling. *Ecological Modelling* 230: 123–147.

44. Tufte, E.R. 1997. *Visual explanations*. Cheshire: Graphic Press.

45. Ware, C. 2004. *Information visualization: Perception for design*. Morgan Kaufmann Publishers: Elsevier.

46. Weslawski, J.M., L. Kryla-Straszewska, J. Piwowarczyk, J. Urbanski, J. Warzocha, L. Kotwicki, M. Włodarska-Kowalczuk, and J. Wiktor. 2013. Habitat modelling limitations—Puck Bay, Baltic Sea: A case study. *Oceanologia* 55: 167–183.

47. Wulff, F., O.P. Savchuk, A.V. Sokolov, C. Humborg, and C.-M. Mörth. 2007. Management options and effects on a marine ecosystem: Assessing the future of the Baltic Sea. *AMBIO* 36: 243–249.

CHAPTER 5

VULNERABILITY OF NEARSHORE ECOSYSTEMSUNDER INTENSIVE MARICULTURE DEVELOPMENT

Hui Liu & Jilan Su

Keywords: Large-scale intensive mariculture, Impact on marine ecosystems, Sustainable development, Ecosystem approach to aquaculture

INTRODUCTION

Since the early 1980s, China's economy has developed rapidly with an average annual GDP growth rate over 9 %, according to its National Bureau of Statistics. The growth rate of the 11 provinces and metropolises along the coasts exceeded 10 %. In fact, with only 13 % of China's land area, in 2011, the 11 coastal provinces and metropolises housed about 42 % of its population and produced over 57 % of the national GDP (CCICED Task Force 2013).

Rapid economic growth and urbanization have brought about tremendous pressure on the coastal and marine ecosystems, resulting in the loss or deterioration of many of their services. Land-based pollutant and coastal reclamation are the two most detrimental factors to the health of the ecosystems (CCICED Task Force 2013). According to the official bulletin, in 2009, over half of China's nearshore waters were classified as below the water quality standard and about 20 % as severely below standard. Occurrence of recorded events of harmful algal blooms (HABs) rose sharply to about 80 yearly, covering a total area exceeding 15,000 km^2 (op cit). Between 1990 and 2008, the average reclamation area in China reached 285 km^2 yearly and a further 5780 km^2 was planned for over the next 10 years: The fast pace of reclamation has caused significant coastal landscape fragmentation and loss of biodiversity, as well as destruction of habitats for fish and loss of feeding grounds for shorebirds, leading to sharp decline in fisheries resources and bird species (Wang et al. 2014). In addition, overfishing, rapid expansion of mariculture, and construction of large number of hydraulic works on land all had compounded the impacts on China's coastal and marine ecosystems (CCICED Task Force 2013).

Spurred by both improved culture technology and rise in domestic demand for high-quality seafood, there has been a revolutionary growth of mariculture in China over the last 30 years. China is credited with 63.3 and 54 %, respectively, of the world's farmed marine food fish and aquatic plant production (FAO 2014), which are usually farmed in large-scale aquaculture operations. For such high production, extensive patches of tidal flats and nearshore waters have been converted for mariculture use. However, so far, the ecological impact of

Citation: Hui Liu, Jilan Su, Vulnerability of nearshore ecosystems under intensive mariculture development, DOI: 10.1007/s11356-015-5239-3.

rapid mariculture development and its large-scale operations has not drawn sufficient attention in China. Attention is usually directed at the food provision function of mariculture, while the interference of mariculture with other ecosystem services has not been fully recognized.

In this paper, the rapid development of mariculture in China is reviewed, China's effort in the implementation of ecological mariculture practices is examined, and the vulnerability of marine ecosystems to impacts of mariculture is evaluated, pertaining to large-scale operations. Considering the current scale of mariculture worldwide and its significant upward trend, impacts of the world's intensive mariculture on the ocean ecosystems are far from negligible. Understanding the true state of impact of intensive mariculture on coastal and marine ecosystems will facilitate consideration of precautionary measures to ensure sustainable growth of world mariculture.

MARICULTURE IN CHINA

Along with the world's rapid growth in mariculture over the recent decades, China's mariculture has also undergone unprecedented development (Fig. 1). Since the middle 1970s, both the mariculture production and capture fisheries experienced fast growth. However, after the middle 1990s, the capture fisheries became stagnant while mariculture kept on expanding. The ratio between the mariculture production and fishery catch was almost nil in early 1950s, reached 20 % in 1975, 50 % in 1988, and 100 % in 2006, and stood at 140 % in 2013. This is in sharp contrast to the ratio for Europe (18 %) and for other continents (<15 %) (data source: FAO FishStatJ). In 2012 China's mariculture contributed to 54 and 63.3 %, respectively, of the world's farmed aquatic plants (largely seaweeds) and marine food fish production (FAO 2014). In light of the depletion of wild fish stocks and the continuing rise in domestic seafood demand, mariculture will undoubtedly lead the future growth of Chinese aquaculture in the long run.

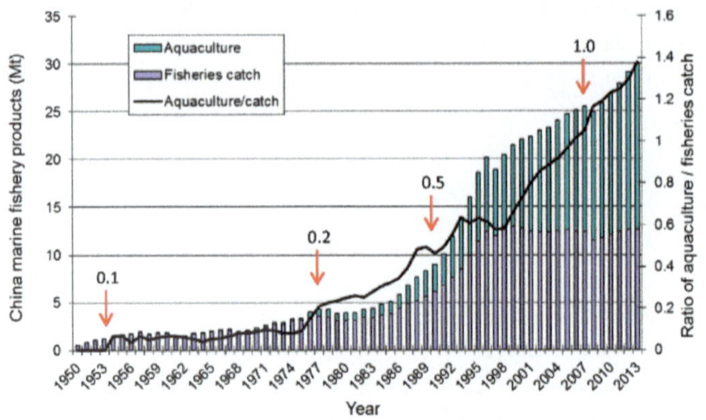

Figure 1: Growth of mariculture production and fishery catch during 1950–2013. The *number above each red arrow* denotes the ratio of "aquaculture/fishery catch" for that year (data source: MOA 1979–2014).

Throughout the 1950~1980, mariculture species in China were largely confined to seaweeds and bivalves, which are extractive species relying on natural supply of nutrients and plankton. Monoculture on longlines over tidal flats was the main type of operation. Commercial culturing of shrimp, finfish, and other high-value species began in the early 1980s, when land-based hatcheries, earthen ponds, and net cages were rapidly set up. The trend of "fishing down and farming up the food web" has been significant in the world fishery industry for decades (Naylor et al. 2000), and China is no exception. In the last 15 years, mariculture production of fed species (finfish and crustaceans) increased sharply from <5 to >15 %, showing a clear upward trend into the future (Fig. 2). This reflects the increased domestic market demand for high-value seafood, corresponding to the economic growth in China. At the same time, this will no doubt bring about increased wastes from mariculture, because, unlike the freshwater finfish which are mostly planktonic filter feeders or omnivorous, marine finfish and crustaceans are generally carnivorous and high in trophic level. Paying attention to this trend is important for rational planning and ecosystem-based management of mariculture.

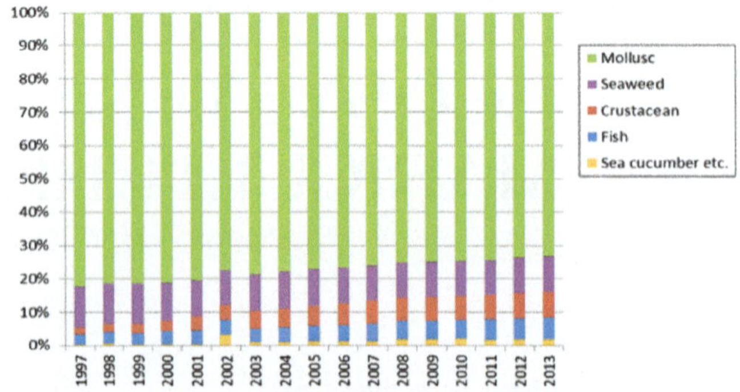

Figure 2: Change of the composition of China's mariculture over time (data source: MOA 1998–2014).

Along with the increase of production, large portions of area from both the intertidal wetlands and nearshore waters of China were used for mariculture (Fig. 3). According to the Chinese fishery statistics (MOA 2014), the total mariculture acreage amounted to 2.32 Mha in 2013, among which 0.8 Mha was converted from intertidal wetlands (nearly one third of China's total) and 0.9 Mha from the nearshore waters (nearly 10 % of China's total) was used for mariculture purpose.

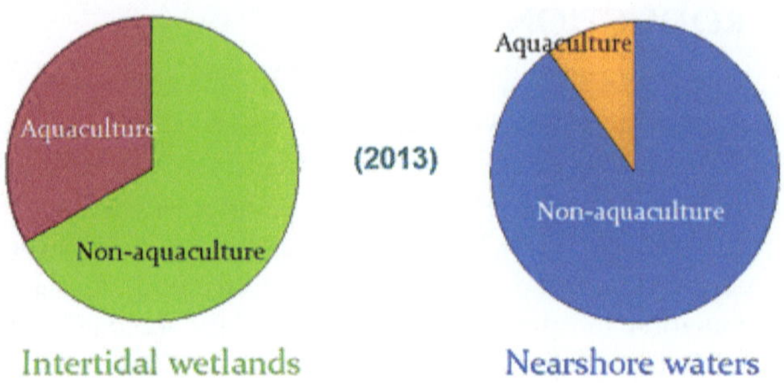

Figure 3: Portion of area used for mariculture in, respectively, intertidal wetlands and nearshore waters of China in 2013. Calculation based on coastal wetland area data from the SFA (2011) and aquaculture area data from MOA (2014).

CHINA'S APPROACH TO ECOLOGICAL MARICULTURE

Aquaculture is basically an economic activity that consumes resources, impacts on biodiversity in different levels and ways, and discharges wastes, hence affecting the environment (e.g., Naylor et al. 2000). With the worldwide rapid development of aquaculture since the 1970s, especially in developing countries, there was first a focus on productivity and benefits, leading to a general scaling up and overstocking. The resulting aquaculture epidemics and mass mortalities raised awareness of ecological carrying capacity of the aquaculture systems (e.g., Grant et al. 2007). This concept soon evolved into ecological aquaculture, with ecosystem approach to aquaculture (EAA) as its implementation strategy (Soto et al. 2008).

Conceptually, EAA embodies the principle of ecosystem-based management by promoting conservation and sustainable use of resources. In implementing EAA, it is necessary to define the ecosystem boundaries in both space and time. For this, three levels of scale have been identified, namely, the farm, the watershed/aquaculture zone, and the global (Soto et al. 2008). In practice, boundaries for the second level, i.e., the watershed/aquaculture zone level, sometimes are not easy to delineate, both managerially and ecologically. For a fjord-like inlet where the physical boundaries are more or less defined, management tools including analysis models for evaluating aquaculture in an ecosystems approach were developed and proved to be useful (Nunes et al. 2011). However, because of the lack of a good understanding of the interaction between a mariculture operation and the marine ecosystems at large, presently, there are no agreed-on guidelines or practical measures for implementing EAA at the level of the watershed/aquaculture zone.

One innovative approach within EAA is the so-called integrated multi-trophic aquaculture (IMTA), in which multi-trophic species with complimentary trophic functions are cultured together. Traditional IMTA has been practiced in the world for thousands of years, including ancient China (Costa-Pierce 2010). Modern polyculture methods in China, as one of the eight techniques embodying China's freshwater aquaculture, were reviewed for the first time in 1959 (CoCFAE 1961).

With the development of intensive aquaculture worldwide since the 1970s, monoculture of omnivorous or carnivorous marine species at high densities, including finfish and crustacean species, was initially carried out in China. The production boom in late 1980s and early 1990s brought about environmental pollution and disease outbreak among the stocks, leading to a crash in harvest. At the same time, technology improvement allowed for restructuring of the existing culture systems. There was then a clear transition in China from intensive monoculture to polyculture. The first successful attempt in marine polyculture was conducted in 1975, when kelp *Laminaria japonica* and mussels *Mytilus edulis* were co-cultured (Xie 1981; Dong 2011). Soon, polyculture in China evolved such that fed (fish and shrimp) and non-fed species (bivalves and seaweeds) were cultured together, where the non-fed species served as a means to recycle wastes and reduce pollution (e.g., Wu et al. 1980; Zhu 1981). Thus, the ecosystem perspective in traditional mulberry-fish pond system from ancient China regained appreciation. Such IMTA mode ecological aquaculture is now commercially practiced in China (Fang 1996; Dong 2011).

A major IMTA demonstration site in China is Sanggou Bay, at the tip of Shandong Peninsula in eastern China, where, according to local statistics, seaweed, shellfish, and finfish are cultured at a ratio of approximately 7:3:1 with an annual seaweed production of about 200 thousand tons (Zhang XJ, Head of Ocean and Fishery Bureau of Rongcheng, private communication). Based on both research findings and technology improvement, the ratio and layout of the seaweed longlines, shellfish lantern nets, and finfish net cages were constantly optimized to minimize the environmental impact from this major IMTA system (Tang et al. 2013, Shi et al. 2013). Although IMTA has become an important component for mariculture in China, there is no official statistics on its scale and total production because it is not yet classified as one standard aquaculture mode.

At the same time, IMTA has been met with worldwide enthusiasm during the last decade, as an effective way to deal with the wastes and improve the output of aquaculture (Barrington et al.2009; Chopin 2012). In the present practice of IMTA, there are various approaches to achieve the basic goals that feed and wastes are recycled within the system. With different species composition and spatial configuration, IMTA has been applied to both pond and open-water mariculture systems. Although there were attempts to study the nutrient balance within an IMTA system from a modeling point of view (e.g., the study of ratios between fed and extractive species in IMTA farms (Reid et al. 2013)), lack of thorough understanding of environmental impact of IMTA remains one major concern (e.g., IMTA White Paper 2010).

From the point of view of EAA, boundary delineation of an IMTA system at the watershed/aquaculture zone level is often not straightforward. This results in rather complicated interaction between the culture system and the natural habitats. For example, large-scale IMTA in shallow bay will significantly alter the tidal currents, resulting in a reduced flushing capability of the bay, as well as insufficient nutrient supply to the bay from the offshore water (Shi et al.2011). In addition, organic wastes do accumulate in the sediments of IMTA area and alter both the benthic and water column chemical parameters (Ren et al. 2014). Given the physical conditions of any particular IMTA site, it is extremely difficult, if not impossible, to recycle all the nutrients and wastes within an IMTA system.

VULNERABILITY OF THE NEARSHORE ECOSYSTEM

As commented in the "Introduction" section, over the last 30 years, the nearshore ecosystems of China have been under multiple pressures of human activities, including land-based pollutants, reclamation, port construction, overfishing, and mariculture. Early concern about the impact from mariculture was limited to its waste discharge pollution and escapees. However, as China's mariculture rapidly expands, there are evidences now that large-scale intensive mariculture may bring about undesirable ecological consequence far beyond the normally perceived boundaries of its operation. The combined impact from mariculture and other human activities has an amplification effect that significantly increases the vulnerability of China's coastal and marine ecosystems. In the following, we will use a few examples to illustrate this point.

Degradation and destruction of natural habitats

As is well known, intertidal wetlands and nearshore waters are critical habitats for many marine organisms, including both large and forage fish. As extensive tidal flats and wetlands are transformed to aquaculture ponds, the natural ebb and flow of tides disappear, along with the local benthic organisms and migrating shore birds (Yang et al. 2004,

Chen and Ouyang 2014). Mariculture not only introduces pollutants into nearshore material cycle, but also interferes with local hydrologic processes which can disrupt the success of recruitment in the early life of fisheries (e.g., Dong et al. 2007). In open-water aquaculture, the environmental impact is heavily influenced by local circulation (Han and Liu 2014). It is noteworthy that mariculture impact on a sheltered bay is usually more significant due to its limited environmental carrying capacity. Large-scale net cages and longline facilities in shallow bays significantly slow down the tidal currents, resulting in weakened flushing capability of the bay, reduced nutrient supply to the inner bay, increased deposition, and reduced productivity of phytoplankton (Ji et al. 1998; Liu et al. 2004; Shi et al. 2011). In addition, accumulation of waste deposits under shellfish longlines or finfish net cages may severely affect the macro-benthos and change their biodiversity, composition, and distribution (Gao et al. 2003; Yang et al. 2007; Wang et al. 2011). In the long run, the establishment of a mariculture system may result in fragmentation of the original natural ecosystems and disappearance or significant change in some of the original biotic communities (Liu and Fang 2014).

Furthermore, mariculture also imposes deleterious effects on the natural fauna and flora through its chemical or pharmaceutical discharge. To eliminate harmful or undesirable organisms in many of the extensive pond mariculture, pellouxite and other disinfectants have been used in high concentrations before or during the culture period (Luo 2012). When the treated water is discharged without further treatment, these chemicals will invariably cause high mortality to the nearby wild fauna and flora.

Although such impacts are well known, in the case of China, the magnitude of these impacts on the nearshore ecosystems needs to be assessed against the scope of the human activities. As discussed in the "Introduction" section, natural wetlands in China suffered great loss and degradation over decades. More than 50 % of the coastal wetlands were lost due to reclamation for various uses, including mariculture (An et al. 2007). The loss of associated ecosystem services due to reduced purification capacity, failure in recruitment of fishery stock, and reduced biodiversity was estimated to amount to 46 billion USD per year (op cit).

For example, the Laizhou Bay of Bohai Sea has always been used as spawning grounds for many fishery species of the Bohai Sea and Yellow Sea (Deng and Jin 2000; Xu and Chen 2010). However, presently, most of the estuaries and wetlands around the bay are severely affected by mariculture activities (Fig. 4). Furthermore, not shown in the figure, there are also large areas of longline mariculture over the shallow waters in the bay. It is possible that most of the nursery grounds for the fisheries are either lost or severely degraded. Thus, in addition to overfishing, reclamation and mariculture activities may also contribute to the collapse of regional fishery (Zuo and Lin 2008; Wang et al. 2014).

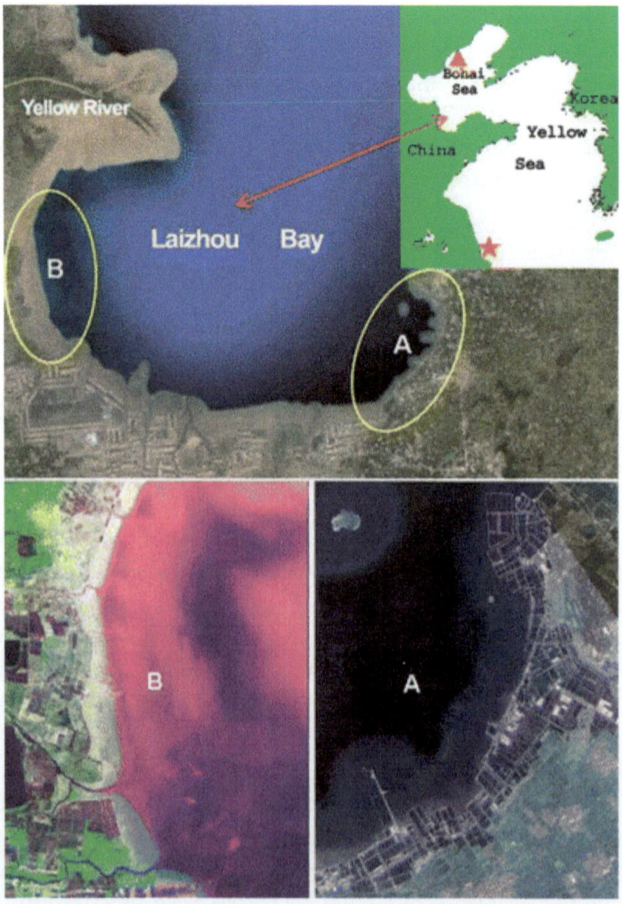

Figure 4: Extensive mariculture facilities along the shores of the Laizhou Bay. In the inserted map on the *top figure*, the *red triangle* and *star* denote, respectively, the location of the large-scale bivalve culture and the extensive Su-Bei tidal shoals.

The loss of coastal wetlands along with its biodiversity is not unique in China. As a matter of fact, there has been extensive loss of tidal flats around the whole Yellow Sea region, associated with economic development in China and the two Koreas. It was found that 28 % of the Yellow Sea tidal flats existing in the 1980s had been reclaimed by the late 2000s (Murray et al. 2014).

Pressure on wild fisheries

There have been very few observations on the direct aquaculture impact on wild fishery, especially regarding a particular species and over a significant length of time. According to a 7-year study in Canada (Loucks et al. 2014), net cage finfish aquaculture produced odors and changed benthic habitat and water quality, which may affect the behavior and movement of wild fishery species (e.g., lobster). It was found that, within 2 years of a finfish aquaculture operation, ovigerous lobsters abandoned the area and overall lobster abundance decreased; the female lobsters returned to this area only when fish farming was discontinued. On the other hand, aquaculture may provide food for wild species. For example, aggregations of wild fish communities were repeatedly observed in the vicinity of aquaculture farms in the Mediterranean (Grigorakis and Rigos 2011).

In addition to habitats, marine capture fisheries and mariculture also compete for other ecosystem services, such as nutrients and feed, broodstock, and larval supply

(Naylor et al.2000). In China, the catch of wild broodstock for hatchery (such as Chinese shrimp*Fenneropenaeus orientalis*) and of young fish for grow-out or fattening (such as sea bass*Lateolabrax maculatus* and freshwater eel *Anguilla japonica*) has been going on for many years (Niu 1982; Gao and Zhang 2002; Zhi et al. 2013). The above-mentioned human activities together imposed devastating effect on the wild stock and also defeated large-scale restocking efforts conducted in China, including artificial reefs and annual release of tens of billions of fishery seedlings (Shan et al. 2012).

Another impact of mariculture on wild fishery is the harvest of forage fish, which is either directly fed to aquaculture or processed into fishmeal. Generally speaking, in China, about one third of forage fish (about 4 Mt) are fed directly to aquaculture species (Mai 2010), although there is large site-specific variability. Annual production of these small forage fish amounted to several million tons, though not all such productions are recorded in the official fishery statistics (op cit). Statistics of specific forage fisheries varied widely over the decades, reflecting overfishing pressure on these species (Fig. 5).

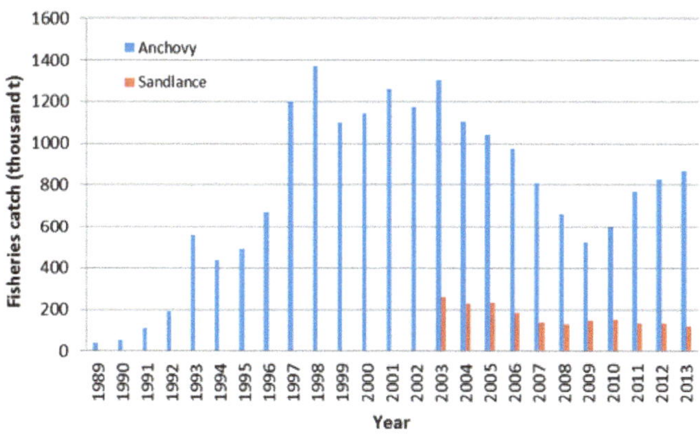

Figure 5: Fishery catch of anchovy and sandlance in China during 1989–2013 (data source: MOA (1990–2014)).

Jackson et al. (2001) have shown that overfishing of large fish fundamentally altered coastal marine ecosystems during each of the cultural periods examined. Forage fish is also an important part of the regional marine ecosystems. With pressure from both habitat loss/degradation and overfishing, it is hard to imagine how the regional coastal and nearshore ecosystems will evolve when both the large and forage fisheries are destroyed.

Proliferation of opportunistic species

Since 2007, there has been yearly outbreak of large-scale green tides over the Yellow Sea in summer. In 2013, at its peak, it covered a total area 790 km^2 distributed over a wide area of 29,700 km^2, while, in 2014, the numbers were, respectively, 540 and 50,000 km^2 (SOA 2014,2015). Figure 6 shows the distribution pattern of the green tides on 30 June 2013 and 14 July 2014. The green tides are caused by *Ulva prolifera* (*Enteromorpha prolifera*), a macro-algae species often found along the coasts of temperate Asia, including China. Proliferation of *U. prolifera* is associated with the large-scale mariculture system of *Porphyra yezoensis* over the extensive Su-Bei tidal shoals of China (Figs. 4 and 6) off the west boundary of the Yellow Sea (Liu et al. 2009).

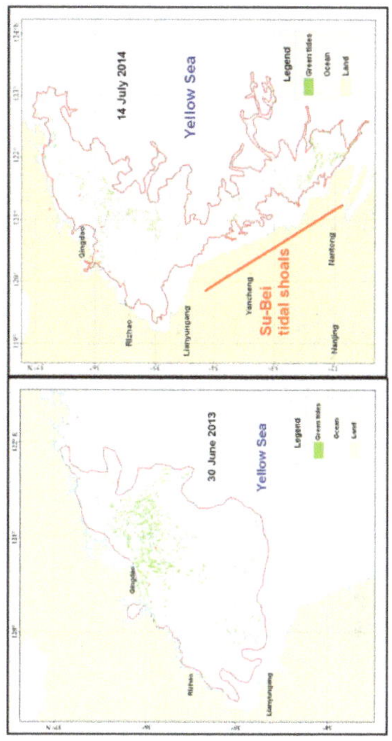

Figure 6: Distribution pattern of the green tides over Yellow Sea on 30 June 2013 and 14 July 2014 (figure source: SOA 2014, 2015).

Micropropagules of *U. prolifera* generally exist in the nearshore water of Yellow Sea. The micropropagule will grow only after attaching itself to a periodically exposed surface, including the rafts used for the *P. yezoensis* culture. A variety of macro-algae species would attach and grow on these rafts from late autumn, including *U. prolifera*, *Ulva linza*, *Ulva compressa*, *Ulva intestinalis*, and *Ulva clathrata*. *U. prolifera* became one of the most dominant species only in late March to mid-May (Fan et al. 2015). At the harvest of *P. yezoensis* in April, the attached macro-algae were stripped off and left on the shoals. Most of the macro-algae left on the shoals were drowned except *U. prolifera* which can float during photosynthesis. Once leaving the shoals, the floating *U. prolifera* would drift in an N-NE direction driven by the prevailing winds. The floating *U. prolifera* grew at an initial daily rate of 20–30 % and soon evolved to green tides (op cit). Some of the drifting green tides landed on the shores of Shangdong Province in June/July while the rest eventually sank offshore.

Figure 7 gives the yearly production, culture area, and productivity of the *P. yezoensis* culture in Jiangsu Province between 1988 and 2013 (most of this culture is over the Su-Bei Shoals). It is noted that the yearly production of *P. yezoensis* reached a plateau after 2007 even though the cultured area kept increasing. The productivity underwent a sharp drop then. The year 2007 was when the green tides over the Yellow Sea first became noticeable. From 2008 to the present, there is a yearly outbreak of green tides covering 250–2100 km^2 distributed over 20,000–60,000 km^2. It is possible that certain bloom-state threshold for the opportunistic species *U. prolifera* was reached after 2007, likely related to the size of the culture area of *P. yezoensis*(Fig. 7).

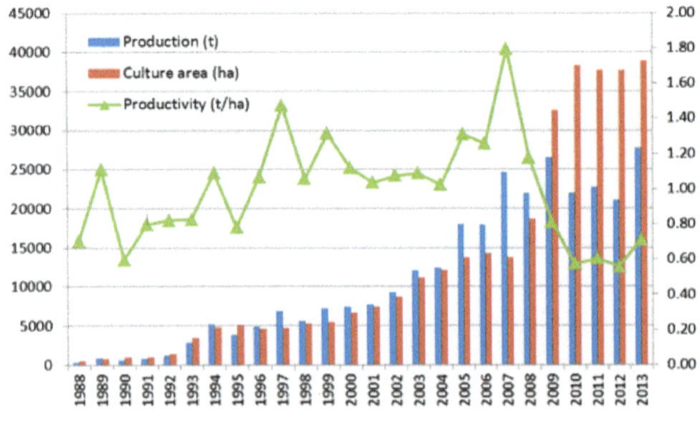

Figure 7: Yearly *P. yezoensis* culture production, culture area, and productivity in Jiangsu Province in 1988–2013 (source of data: MOA 1989–2014).

There is no reported estimate on the portion of the floating mass of *U. prolifera* that sink offshore, nor is there any study on the fate of these submerged macro-algae. It is well known that, in summer, the Yellow Sea Cold Water (YSCW), a water mass, occupies the central part of Yellow Sea deeper than 40 m below a sharp thermocline (e.g., Su 1998). Figures 7-8 and 7-6 in Hong (2012), respectively, showed that the bottom dissolved oxygen in YSCW is highest in spring and lowest in autumn, and in addition, over the past 30 years, it has a continuous decreasing trend with an average drop of about 0.77 mg $^{-1}$. A recent ocean acidification survey found that in late autumn of 2012, large area of YSCW had a saturation state of aragonite below 1.0 (SOA 2013), indicating a corrosive environment for many marine organisms (Fabry et al.2008). It is known that YSCW serves as important oversummer grounds for *Calanus sinicus*, an important copepod providing a wide size spectrum of food for fish stocks (Wang et al. 2003). How the annual outbreak of *U. prolifera* blooms will exert further pressure on the ecosystem services of YSCW is yet to be studied.

Alteration of biochemistry in water/sediment

Although mariculture ponds and enclosures and nearshore longline culture areas are usually counted as artificial wetlands, they are fundamentally different from natural coastal wetlands because their ecosystem services are significantly different. Mariculture is also a major driver on changes in the biogeochemistry of its neighboring water and sediment.

The deleterious environmental effects of the intensive and industrialized mariculture have been well recognized (e.g., National Research Council 2010). Finfish cage culture can cause significant rise in dissolved nitrogen in water column, particulate phosphorous in sediment, and chemical oxygen demand (COD) in both the water column and sediment (e.g., Holby and Hall1991; Wang 1998). Large-scale oyster (*Crassostrea gigas*) culture may alter the sediment geochemistry by raising NH_4^+ and HPO_4^{2-} by 7.7–11.5 and 1.8–8.0 times and sulfate reduction rates by 2.4–5.2 times (Hyun et al. 2013). Given the scale of bivalve culture in China (total annual production of 12.7 Mt over a total area of 1.6 Mha) (MOA 2014), the biochemical alteration of inshore sediments is considered significant enough to induce eutrophication, change in P/N ratio, and outbreaks of HABs (Bouwman et al. 2013). This was echoed by Hyun et al. (2013), who observed enhanced benthic microalgal biomass and primary production in the water column at the bivalve farms. As discussed in "China's approach to ecological mariculture" section, it is difficult even for an IMTA system to recycle all the nutrients and wastes within its system. Indeed, eutrophication effect of polyculture (with finfish, shellfish, and seaweed) can be significant in a sheltered bay (Zhang et al. 2003).

Nutrients in the reduced form in the ocean are of particular concern. For example, reduced N forms were preferentially taken up during the harmful microalgae blooms of *Karenia mikimotoi*and *Prorocentrum donghaiense* in nearshore waters of the East China Sea (Li et al. 2009). Both of these are common HAB species in coastal China, and the blooms of *K. mikimotoi* usually cause great losses in mariculture (SOA 2005–2015).

Among all forms of mariculture pollutions, direct release of nutrients in the reduced form into the ocean is of more concern (Bouwman et al. 2013). Laboratory experiments showed that ammonia and urea excreted by the cultured animals were readily taken up by phytoplankton and could stimulate their growth (Arzul et al. 1996; Yang et al. 2004). When the level of such nutrient is higher than the assimilative capacity of the marine area, it may fuel the production of algae on which mixotrophic HABs may feed (op cit). In turn, such mixotrophic HABs can kill or intoxicate the mariculture organisms and increase risks to human health. The scallop*Argopecten irradians* culture at the nearshore region of the northwestern Bohai Sea (Fig. 4) is number one in China by area scale (Zheng et al. 2011). Coincidentally, large-scale *Aureococcus anophagefferens* brown tides, a form of mixotrophic HABs, have occurred in the northwestern Bohai Sea in early summer for consecutive years since 2009 and have caused significant negative impacts on the scallop culture there (Zhang et al. 2012). China is the third country in the world hit by brown tides, but at a much greater scale. At its peak, the brown tides covered an area of over 3000 km^2 (op cit).

Particulate wastes from aquaculture usually accumulate in nearby sediments, driving an increase in COD value. As organic matter is decomposed by microbes through respiration, the redox potential will decrease, leading to bottom dissolved oxygen (DO) depletion (Wang 1998). Accumulation of wastes and microbial/phytoplankton biomass in proximity to mariculture sites may give rise to hypoxia conditions resulting in an anoxic layer of sediment and bottom waters (Yang et al. 2004), and this develops very often in stratified water. In fact, seasonal bottom oxygen depletion in western Bohai Sea, especially in the vicinity of large-scale bivalve culture sites, has now become normality and is getting worse (Zhai et al. 2012). The progressive depletion in the summer bottom dissolved oxygen in Bohai is rather alarming and raises further concern about Bohai Sea's ecosystems. Observation showed that its minimum bottom oxygen level in August steadily dropped over the years, being 6.3 mg $^{-1}$ in 1982, 5.1 mg $^{-1}$ in 1992, 4.5 mg $^{-1}$ in 2006, 3.5 mg $^{-1}$ in 2011, and below 3 mg $^{-1}$ in 2014 (Zhai Weidong, private communication). Analysis by Zhai et al. (2012) indicated that both the bottom DO depletion and acidification in summer were caused by remineralization of biogenic particles, which were supplied by either coastal HABs or regional nearshore mariculture.

CONCLUSION

As the world's population keeps on increasing with high aspiration for better life, rapid intensive reclamation and large-scale mariculture practices will no doubt be replicated in developing countries around the world's coasts. Such human activities are conducted over intertidal wetlands and nearshore waters, which lie in a narrow strip along the coast. These habitats are critical in the life history of many species of large and forage fish. At the same time, overfishing, either for food or for fishmeal, exerts tremendous pressure on the stock of both large and forage fish. During past cultural periods, overfishing of large fish was shown to have fundamentally altered coastal marine ecosystems (Jackson et al. 2001). If the large and forage fish are sharply reduced, both the coastal and shelf ecosystems will likely be significantly altered.

Mariculture replaces intertidal wetlands with artificial wetlands and brings artificial ecosystems into nearshore waters. These artificial ecosystems bring with them different but undesirable biological and biochemical characteristics. Spatial and temporal extents of the effects of these artificial ecosystems, i.e., the boundaries for the watershed/aquaculture zone level, are often hard to define. The speed of mariculture development far outpaces the progress of our knowledge about its impacts. Examples given above show that the coastal and nearshore ecosystems are rather vulnerable when facing intensive large-scale mariculture development.

The fast pace of the worldwide coastal development will likely cause the coastal and nearshore ecosystems to evolve into states unfriendly to either the coastal communities or the ocean ecosystems at large. As our understanding of such possible changes is lagging far behind the economic and social development, much research is urgently needed, including those related to large-scale intensive mariculture.

ACKNOWLEDGMENTS

This work arises from two task force studies, the Problems Facing China's Sea Enclosing and Land Reclamation (2008–2009) and the Scientific Questions and Policy Recommendations for Ecosystem Eveluation and Management of Coastal Waters of China (2014–2016), both funded by the Chinese Academy of Sciences.

REFERENCES

1. An S, Li H, Guan B, Zhou C, Wang Z, Deng Z, Zhi Y, Liu Y, Xu C, Fang S, Jiang J, Li H (2007) China's natural wetlands: past problems, current status, and future challenges. Ambio 4:335–342
2. Arzul G, Clément A, Pinier A (1996) Effects on phytoplankton growth of dissolved substances produced by fish farming. Aquat Living Resour 9:95–102
3. Barrington K, Chopin T, Robinson S (2009) Integrated multi-trophic aquaculture (IMTA) in marine temperate waters. In: Soto D (ed) Integrated mariculture: a global review, vol 529, FAO Fish Aquac Tech Paper., pp 7–46
4. Bouwman L, Beusen A, Glibert PM, Overbeek C, Pawlowski M, Herrera J, Mulsow S, Yu RC, Zhou MJ (2013) Mariculture: significant and expanding cause of coastal nutrient enrichment. Environ Res Lett 8. doi:10.1088/1748-9326/8/4/044026
5. Chen JH, Ouyang L (2014) Let the migrating birds fly freely. Ecol Econ 30(11):10–13
6. Chopin T (2012) Integrated multi-trophic aquaculture (IMTA). In: Meyers R (ed) Encyclopedia of sustainability science and technology. Springer Reference, N.Y., pp 184–205
7. Committee on China's Freshwater Aquaculture Experience (CoCFAE) (1961) China freshwater aquaculture. Science Press, Beijing
8. Costa-Pierce BA (2010) Sustainable ecological aquaculture systems: the need for a new social contract for aquaculture development. Mar Technol Soc J 44:88–112
9. Deng JY, Jin XS (2000) Study on fishery biodiversity and its conservation in Laizhou Bay and Yellow River Estuary. Zool Res 21:76–82
10. Dong SL (2011) History, principles, and classification of integrated aquaculture in China. J Fishery Sci China 18:1202–1209
11. Dong LX, Su JL, Deng JY, Chen Q (2007) The importance of estuarine gravitational circulation in the early life of the Bohai Penaeid Prawn. J Mar Syst 67:253–262
12. Fabry VJ, Seibel BA, Feely RA, Orr JC (2008) Impacts of ocean acidification on marine fauna and ecosystem processes. ICES J Mar Sci 65:414–432
13. Fan SL, Fu MZ, Wang ZL, Zhang XL, Song W, Li Y, Liu GX, Shi XY, Wang XN, Zhu MY (2015) Temporal variation of green macroalgal assemblage on Porphyra aquaculture rafts in the Subei Shoal, China. Estuar, Coastal Shelf Sci. doi:10.1016/j.ecss.2015.03.016
14. Fang JG, Sun HL, Yan JP, Kuang SH, Li F, Gary F, Grant J (1996) Polyculture of scallop*Chlamys farreri* and kelp *Laminria japonica* in Sungo Bay. Chin J Oceanol Limnol 14(4):322–329
15. FAO (2014) The state of world fisheries and aquaculture 2014. Rome. 223 pp
16. Gao HD, Zhang SR (2002) Discussions on sea bass seedling resource management measures. Hebei Fish 122:52
17. Gao AG, Yang JY, Chen QZ, Wang ZP, Zhang J, Dong YT, Ning XR (2003) Comparative studies on macrobenthos between cultured and non-cultured areas in Xiangshan Bay. J Fish China 27(1):25–31
18. Grant J, Curran KJ, Guyondet TL, Tita G, Bacher C, Koutitonsky V, Dowd M (2007) A box model of carrying capacity for suspended mussel aquaculture in Lagune de la Grande-Entree, Iles-de-la-Madeleine, Quebec. Ecol Model 200:193–206
19. Grigorakis K, Rigos G (2011) Aquaculture effects on environmental and public welfare—the case of Mediterranean mariculture. Chemosphere 855:899–919
20. Han QX, Liu DY (2014) Review on the effect of coastal bivalve raft mariculture on macrobenthic assemblages. Mar Sci Bull 33(3):352–359
21. Holby O, Hall PO (1991) Chemical fluxes and mass balances in a marine fish cage farm II. Phosphorus. Mar

Ecol Prog Ser 70:263–272

22. Hong HS (ed) (2012) Regional oceanography of china seas-chemical oceanography. Ocean Press, Beijing, 374pp

23. Hyun JH, Kim SH, Mok JS, Lee JS, An SU, Lee WC, Jung RH (2013) Impacts of long-line aquaculture of Pacific oysters (*Crassostrea gigas*) on sulfate reduction and diffusive nutrient flux in the coastal sediments of Jinhae–Tongyeong, Korea. Mar Pollut Bull 74:187–198

24. IMTA White Paper (2010) White paper integrated multi-trophic aquaculture: a workshop. September 14–15, 2010. Port Angeles, Washingtonhttp://www.nmfs.noaa.gov/aquaculture/docs/imta/imta_white_paper.pdf Accessed on February 27, 2015.

25. Jackson JBC, Kirby MX, Wolfgang HB, Bjorndal KA, Botsford LW, Bourque BJ, Bradbury RH, Cooke R, Erlandson J, Estes JA, Hughes TP, Kidwell S, Lange CB, Lenihan HS, Pandolfi JM, Peterson CH, Steneck RS, Tegner MJ, Warner RR (2001) Historical overfishing and the recent collapse of coastal ecosystems. Science 293:629–638

26. Ji RB, Mao XH, Zhu MY (1998) Impacts of coastal shellfish aquaculture on bay ecosystem. J Oceanogr Huanghai Bohai Seas 16(1):21–27

27. Li J, Glibert PM, Zhou MJ, Lu SH, Lu DD (2009) Relationships between nitrogen and phosphorus forms and ratios and the development of dinoflagellate blooms in the East China Sea. Mar Ecol Prog Ser 383:11–26

28. Liu H, Fang J, Zhu J, Dong S, Wang F, Liang X, Zhang J, Lian Y, Wang L, Jiang W (2004) Study on limiting nutrients and phytoplankton at long-line-culture areas in Laizhou Bay and Sanggou Bay, northeastern China. Aquat Conserv Mar Freshwat Ecosyst 14(6):551–574

29. Liu DY, Keesing JK, Xing QG, Shi P (2009) World's largest macroalgal bloom caused by expansion of seaweed aquaculture in China. Mar Pollut Bull 58:888–895

30. Liu H, Fang JG et al (2014) Environmental impact of large scale mariculture. In: Su JL (ed) Scientific issues and policy recommendations on ecosystem safety of ocean and coastal engineering in China. Science Press, Beijing,

31. Loucks RH, Smith RE, Fisher EB (2014) Interactions between finfish aquaculture and lobster catches in a sheltered bay. Mar Pollut Bull 88:255–259

32. Luo R (2012) Pay attention to disappearing coastal wetlands. China Nat 9:8–11

33. Mai KS (2010) Achievements and prospects for China aquaculture and feed industry. Sci Fish Aquac 11:1–2

34. Ministry of Agriculture (MOA) (1979–2014) China fishery statistical yearbook. China Agriculture Press, Beijing

35. Murray NJ, Clemens RS, Phinn SR, Possingham HP, Fuller RA (2014) Tracking the rapid loss of tidal wetlands in the Yellow Sea. Front Ecol Environ 12:267–272. doi:10.1890/130260

36. National Research Council (2010) Ecosystem concepts for sustainable bivalve mariculture. National Academies Press, Washington DC. http://www.nap.edu/catalog/12802.htmlAccessed on March 5, 2015

37. Naylor RL, Goldburg RJ, Primavera JH, Kautsky N, Beveridge MCM, Clay J, Folke C, Lubchenco J, Mooney H, Troel M (2000) Effect of aquaculture on world fish supplies. Nature 405:1017–1024

38. Niu ZZ (1982) Several technical issues on gathering and transporting shrimp broodstock. Hebei Fisheries Sci Technol 5:17–19

39. Nunes JP, Ferreira JG, Bricker SB, O'Loan B, Dabrowski T, Dallaghan B, Hawkins AJS, O'Connor B, O'Carroll T (2011) Towards an ecosystem approach to aquaculture: assessment of sustainable shellfish cultivation at different scales of space, time and complexity. Aquaculture 315:369–383

40. Reid GK, Chopin T, Robinson SMC, Azevedo P, Quinton M, Belyea E (2013) Weight ratios of the kelps, *Alaria esculenta* and *Saccharina latissima*, required to sequester dissolved inorganic nutrients and supply oxygen for Atlantic salmon, *Salmo salar*, in integrated multi-trophic aquaculture systems. Aquaculture 408–409:34–46

41. Ren L, Zhang J, Fang J, Tang Q, Zhang M, Du M (2014) Impact of shellfish biodeposits and rotten seaweed on the sediments of Ailian Bay, China. Aquacult Int 22:811–819

42. Shan XJ, Jin XS, Li ZY, Chen YL, Dai FQ (2012) Fish community structure and stock dynamics of main releasing species in the Bohai Sea. Progress Fish Sci 33(6):1–9

43. Shi J, Wei H, Zhao L, Yuan Y, Fang JG, Zhang JH (2011) A physical-biological coupled aquaculture model for a suspended aquaculture area of China. Aquaculture 318:412–424

44. Shi HH, Zheng W, Zhang XL, Zhu MY, Ding DW (2013) Ecological-economic assessment of monoculture and integrated multi-trophic aquaculture in Sanggou Bay of China. Aquaculture 410–411:172–178

45. Soto D, Aguilar-Manjarrez J, Hishamunda N (eds) (2008) Building an ecosystem approach to aquaculture. FAO/Universitat de les Illes Balears Expert Workshop. 7–11 May 2007, Palma de Mallorca, Spain. FAO Fisheries and Aquaculture Proceedings No. 14. FAO, Rome, p 221

46. State Forestry Administration (SFA) (2011) The basic situation of forestry in China in 2011.http://www.forestry.gov.cn/ Accessed on October 28, 2014

47. State Oceanic Administration (SOA) (2005–2015). China marine disaster bulletin, 2004–2014.http://www.soa.gov.cn/zwgk/hygb/zghyzhgb/ Accessed on March 16, 2015

48. State Oceanic Administration (SOA) (2013). China marine environment bulletin, 2012.http://www.soa.gov.cn/zwgk/hygb/zghyhjzlgb/ Accessed on June 10, 2015

49. Su J (1998) Circulation dynamics of the China Seas: north of 18°N, In: Robinson AR and Brink K [eds] "The sea," vol. 11, The global coastal ocean: regional studies and syntheses, Wiley, 483–506

50. Tang QS, Fang JG, Zhang JH, Jiang ZJ, Liu HM (2013) Impacts of multiple stressors on coastal ocean ecosystems and integrated multi-trophic aquaculture.

Progress Fish Sci 34(1):1–11

51. Task Force CCICED (2013) Ecosystem issues and policy options addressing the sustainable development of China's ocean and coasts. China Environmental Science Press, Beijing, p 493
52. Wang XP (1998) The nitrogen and phosphorus in sediment interstitial water of cage cultural waters in the Depth Bay of Niutou Island. J Fish Sci China 5(3):126–128
53. Wang R, Zuo T, Wang K (2003) The Yellow Sea Cold Bottom Water—an oversummering site for*Calanus sinicus* (Copepoda, Crustacea). J Plankton Res 25:169–183
54. Wang ZX, Sun PX, Liu CX, Liu XJ, Lian Y, Xu RS (2011) The species diversity of macrobenthic fauna in Sanggou Bay, China. Period Ocean Univ China 41(7/8):79–84
55. Wang W, Liu H, Li YQ, Su JL (2014) Development and management of land reclamation in China. Ocean Coastal Manag 102:415–425
56. Wu CD, Tang TD, Lin ZX, Li WJ (1980) Co-culture of shrimp and mullet. 1980 (3): 11
57. Xie CL (1981) High economic return generated by bivalve and seaweed co-culture. China Fish 1981(5):20
58. Xu ZL, Chen JJ (2010) Analysis to population division and migratory routine of populations and migratory routines of *Argyrosomus argentatus* in the north China waters. Acta Ecol Sin 30:6442–6450
59. Yang YF, Hou LC, Nie XP, Tang DL, Chung IK (2004) Development of mariculture and its impacts in Chinese coastal waters. Rev Fish Biol Fish 14:1–10
60. Yang JY, Gao AG, Ning XR, Zhang DS (2007) Characteristics on macrofauna and the responses on aquiculture in Yueqing Bay. Acta Ecolog Ica Sinica 1:34–41
61. Zhai WD, Zhao HD, Zheng N, Xu Y (2012) Coastal acidification in summer bottom oxygen-depleted waters in northwestern-northern Bohai Sea from June to August in 2011. Chin Sci Bull 57(9):1062–1068
62. Zhang J, Wu AY, Shi QS (2003) Mariculture and its environment effect in the Xiangshangang Bay. Donghai Mar Sci 21(4):54–59
63. Zhang QC, Qiu LM, Yu RC, Kong FZ, Wang YF, Yan T, Gobler CJ, Zhou MJ (2012) Emergence of brown tides caused by *Aureococcus anophagefferens* Hargraves et Sieburth in China. Harmful Algae 19:117–124
64. Zheng XR, Zhang FC, Wu XM, Zhang B, Wang LS (2011) Nutrients in bay scallop culture areas in Changli County. Hebei Fish 205:35–38
65. Zhi YL, Hou JL, Zhang T, Zhang LZ, Zhao F, Yang G, Zhuang P (2013) Spatiotemporal distribution patterns of elver (*Anguilla japonica*) in the Yangtze River Estuary of China. Chin J Ecol 32(10):2750–2755
66. Zhu YG (1981) Co-culture of clam *Meretrix meretrix* in a shrimp pond. Mar Fish 1981(2):25
67. Zuo YH, Lin GL (2008) Resource utilization and environment protection in the coastal zone. Science Press, Beijing, p 169

CHAPTER 6

ENVIRONMENTAL GOVERNANCE: PERSPECTIVES ON SECTOR MANAGEMENT OF GROUNDFISH FISHERY

Julia Olson and Patricia Pinto da Silva

Keywords: Fisheries governance, Co-management, Northeast United States (US) Multispecies Groundfish Fishery, Institutions, Social capital

INTRODUCTION

The groundfish fishery—a complex of bottom-dwelling species including Atlantic cod, haddock, yellowtail flounder, and Atlantic halibut—was one of the first colonial endeavors in New England, and remains its iconic fishery. Despite the implementation of the first Fishery Management Plan for groundfish in 1987, however, "most New England groundfish stocks have been heavily exploited and overfished for decades" (Brodziak et al. 2008: 358). The history of groundfish management in the Northeast US, which in addition to sixteen implemented amendments includes fifty regulatory adjustments (called frameworks) to the management plan, is also one of frequent industry opposition, resentment, criticism, and actual lawsuits (see overview in Acheson 2011). Although the reasons for overfishing may be complex, observers have pointed to a number of factors related to the management regime per se, including inconsistent policies and incentives that encouraged early fleet expansion (Acheson 1984), reliance on indirect controls, which ineffectively limited fishing effort (Anthony 1990), and the need for better integration "of science, management, and harvesting" (Hennessey and Healey2000: 210).

Groundfishermen are highly diverse, from size of operation to type of gear to choice of fishing grounds. Such diversity would seem to reduce the likelihood of fishermen coming together to form groups to manage their resources, as Ostrom's (1990) well known design elements would suggest and as some have specially argued for New England groundfishermen (Acheson 2006). Yet robust elements of social capital, often defined in terms of social networks, norms and relations of trust that enable cooperation and collective action (e.g. Pretty and Ward 2001), have been found in the groundfish industry (Holland et al. 2010), while community-based processes such as cooperation and mutual dependence have been found to characterize a diverse economy in fishing communities in New England (St. Martin 2005, 2007). While elements of participatory decision-making have been noted in previous regulatory actions of the groundfish industry (Hall-Arber 2005), since 2010 the allocation of the quota of New England groundfish has occurred primarily through the use of

Citation: Julia Olson and Patricia Pinto da Silva, Environmental Governance: Perspectives on Sector Management of Groundfish Fishery, DOI: 10.1186/2212-9790-13-3.

sectors, cooperative associations of fishermen (Figure 1). The use of sectors for the groundfish fishery was first proposed and used by a local, community-based user group, but the general strategy soon became part of a wider government promotion to consider "catch shares" as a management tool more generally, a complicated transition that we detail in the next section. Needless to say, this context of change and the possibility of further reductions[a] in fishing opportunities for fishermen created enormous challenges for fishermen seeking to preserve their fishing businesses and livelihoods. As such, both the management change with its ensuing need to create new structures and practices, coupled with concern over how the allocation of quota would be determined, provoked significant conflict, concern, and expectations (Holland et al. 2013). For some, sectors were the lesser of two evils. For others, they were a medium for more decentralized and community-oriented management. And for still others, especially those struggling under low allocations and those concerned about a consolidating industry, the management change was despised.

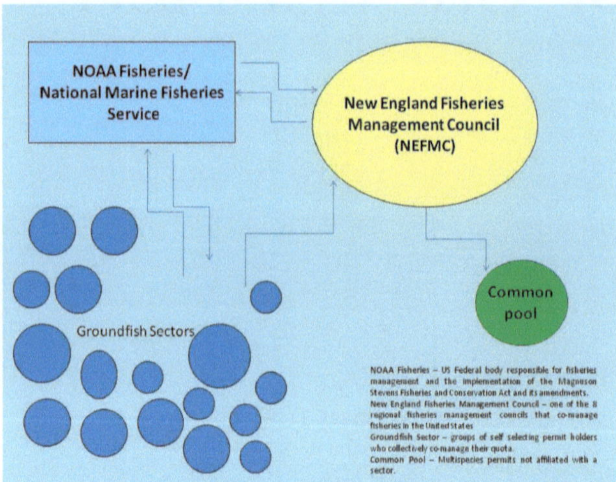

Figure 1: Institutional map of groundfish management.

But such forms of management also mark a more general emergence of hybrid institutional forms of environmental governance that involve different kinds of actors across spatial scales, such as private-public partnerships (Lemos and Agrawal 2006). As Bulkeley (2005) has argued, such hybrid efforts have involved not only a "rescaling of environmental governance" but may also create "new spheres" of authority as well, in which the very nature of activities undertaken may change. While the decentralization embodied by such hybrid forms has sparked hope for greater democratization of decision-making, issues of accountability and power imbalances mean that "contingent outcomes of contemporary shifts in governance, therefore, depend crucially on the ways local actors mobilize and establish alliances across sociopolitical and administrative scales of governance" (Lemos and Agrawal: 305). The sector system, though controversial, is a new direction in fisheries management in this region not only because it changed from an effort-based to a catch-based system, but because it represents an institutional change in governance that has necessitated such mobilizations and alliances. In the midst of crisis, fishermen have had to work together more formally, while at the same time sectors have also provided an opportunity for fishermen to organize around issues and practices far broader than quota allocation per se. This paper, based primarily on interviews with sector managers, seeks to explore some of the important dimensions of this institutional change: the background in which it evolved; the context in which it is has functioned; and the expectations for future management. In particular, we have sought to detail the processes that actors have had to adopt or to make in order to facilitate communication and cooperation, and the formal and informal relations and networks that have enabled them. Sector managers, as we explore in more detail in following sections, play an important new role as informal "boundary agents" in this new regime: similar to those described by Cash et al. (2003) who enable essential efforts at communication, translation and mediation; as well as Larner and Craig's (2005) notion of "strategic brokers" who mediate public-private partnerships and foster new spaces of governance. In addition to issues like consolidation that have impacted other similar experiments with sector management (see van Ginkel 2009 on the Netherlands), these interviews with managers revealed the importance of trust, connections and relations; the historical persistence of management and the misfit between expectations and reality; and the changing roles of managers and fishermen themselves.

As Ostrom (2010) has written, current theoretical efforts to understand collective action involve better understanding the role of trust. Indeed, past uses of social capital elements like trust have been critiqued for a circularity of reasoning that does not specify how social capital actually engenders collective action (Harriss and de Renzio 1997; Portes 1998; Ishihara and Pascual2009). Moreover, Çalışkan and Callon (2010: 21) argue that "trust" covers only a part of the "emotional, corporal, textual and technical elements that contribute to the maintenance of markets" and "black-boxes" their actual operations. As anthropologists and social scientists, we are keenly aware of how trust and other social processes are socially constructed through everyday forms of interaction that vary across and within groups. If adaptive governance depends on processes like trust that are built up through networked interactions (Folke et al.2005), then the historical, political and sociocultural context of such collaborative learning is also key. Thus it is surely significant that while ideas that fishermen may have about equity and fairness have mediated relations in many sectors, the development of sectors has occurred in a context in which there are no clear goals for the groundfishing industry (Singer 2011: 67) and in which a prevalent fear among fishermen is that future access will depend on maintaining significant catch histories (Brewer 2011). As initial observers noted (e.g. Holland and Wiersma2010), sectors represent the potential for a more collaborative form of management where different sectors could be designed to achieve different goals and ways of distributing fishing access. However, sectors, whatever their overall goals, have generally distributed allocations to sector members as individual quotas[b]. Why sectors have not more fully benefited from the collaborative aspects of management or embraced other ways of doing business—at the same time that negative impacts have been considerable on some segments of the industry—raises acute questions for those interested in groundfish management in particular, as well as for those interested in how other shifts in management, such ecosystem-based management, may continue to devolve the scale of governance.

The research was part of a wider effort to document and understand the impact of catch shares and sectors on fishermen and fishing families in the Northeast US (see, for example, Clay and Colburn 2013). We are among a handful of social scientists in the National Marine Fisheries Service(NMFS) [c] who are charged with, among other things, investigating the sociocultural and economic impacts from different regulatory measures. The project detailed in this paper sought to understand the governance and organizational differences between sectors, as they developed out of different backgrounds and histories. Data collection included semi-structured interviews with sector managers, a permit bank manager, and the program director of the Northeast Sector Service Network (NESSN). Sector managers came from a variety of backgrounds, though most had previous professional experience or interest in the fishing industry. For example, one interviewee was a former fisherman, another was a former observer on fishing vessels, and others came from a variety of backgrounds in fisheries business, policy and research. All regular sector managers were contacted, though not all were interviewed. A total of twelve managers in charge of sixteen sectors were interviewed, covering 80 percent of the groundfish sectors. Interviews began in the spring of 2011, and due to manager turnover, not all managers interviewed are current managers. Most interviews lasted around one hour. Interviews were generally recorded if permission was granted, then transcribed and initially coded and analyzed using ATLAS-ti software for textual analysis. Topics of discussion included sector background, governance and organization, and impacts on fishermen (see the question guide in Appendix 1). Interviews and analysis were also informed by a review of Council documents, letters, and newspaper articles.

BUILDING INSTITUTIONS FOR COLLABORATIVE MANAGEMENT

Sectors as a grassroots mobilization

Federal management of groundfish by the New England Fishery Management Council (NEFMC, or the Council) began in 1986 with the passage of the Northeast Multispecies (Groundfish) Fishery Management Plan. This plan manages the fishing of 12 groundfish species found primarily in New England waters, not including a separate amendment managing small-mesh multispecies. Until the passage of Amendment 13 in 2004, groundfish had been managed primarily through effort controls on allowable Days-at-Sea (DAS), coupled with additional trip limits, area closures, mesh regulations, and so on. Amendment 13 provided for the allocation of a percentage of the Atlantic cod quota to a fishermen's organization, paving the way for the introduction of more such "sector allocations" to be introduced in future amendments. The implementation of Amendment 16 in 2010 codified this new system of collaborative management into the management of groundfish in the Northeast, representing a major shift in how groundfish is managed in the region. In Amendment 16, fishermen were encouraged to form sectors, each of which would then be assigned an "Annual Catch Entitlement" (ACE) based on the landings history (resulting in a "Potential Sector Contribution", or PSC) associated with each sector member. Groundfish permit holders who opted not to join any sector were allowed to remain in the "common pool" where they would become subject to former regulatory controls such as DAS. Amendment 16 saw the approval of 17 new sectors. There are currently 20 sectors, two of which are permit banks and one that is a lease-only sector (see Figure 2). Most active permit holders, i.e. those vessels with recorded landings indicating that they have fished during the year, representing the vast majority of landings history, have joined a sector. In 2011, 772 vessels with groundfish permits joined sectors. Of these 446 were active and 301 landed at least one pound of groundfish during the year. The ACE of the majority of the non-active vessels was moved to other vessels through internal sector transfers or leasing. Most groundfish vessel owners have one permitted vessel, though some own multiple vessels, including fleets of more than ten vessels (Murphy et al. 2012: 11, 78).

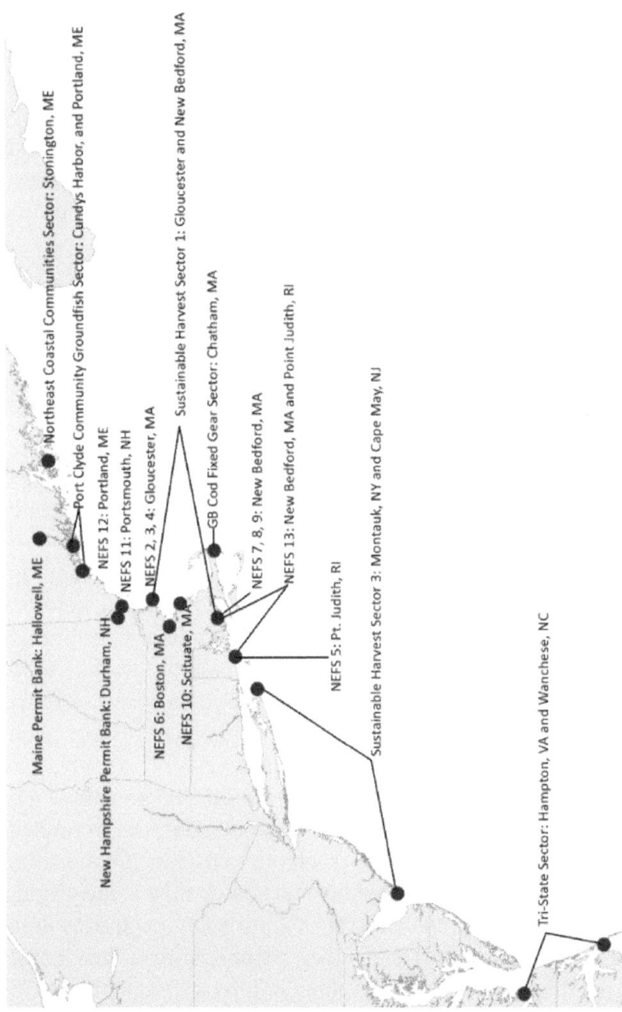

Figure 2: Most important landings ports for sectors in 2010 and 2011 [i].

Most of the current sectors began through initial connections with more politically oriented groups. The first sector, Georges Bank Hook Sector, was established in 2004 as a pilot case in Amendment 13 to the Groundfish Plan, through the efforts of the Cape Cod Commercial Hook Fishermen›s Association (CCCHFA, but now known as the Cape Cod Commercial Fishermen›s Alliance) in Chatham, Massachusetts. The impetus to form this sector, equivalent to a harvest cooperative, came from an increasing frustration felt by its fishermen with Days-at-Sea management restrictions. Hook Sector organizers drew from research on and national and international examples of other cooperatives and discussed the idea of forming the sector in association meetings for several years. Their preference for a group quota system over individual ones was based on recognition, as a spokesperson explained, that fishermen would have more voice at the table if they came together as a group. The CCCHFA founded a second sector, the Georges Bank Fixed Gear Sector, in 2008 (the two CCCHFA sectors merged in Amendment 16). While the history of sector formation has been detailed elsewhere (Singer 2011), it is important to note that the formation of sectors coincides with a broader trend of increasing organizational capacity among fishermen. Many non-governmental organizations (NGOs) and other groups—North Atlantic Marine Alliance, Downeast Groundfish Initiative, CCCHFA, NSC and the Gulf of Maine Research Institution among others[d] —came together during the 1990s and early 2000s to support alternatives modes of management (Pinto da Silva and Kitts 2006). As one manager put it, while members in his sector had known each other for years, they had never been organized formally before because "it is against their morals». That independently minded fishermen have begun to organize themselves more formally testifies to their acknowledgement of the challenges facing their fishing operations and heritage, and to the new importance of such connections.

The first sectors thus developed in part through community-based processes, but mainstreaming the concept required mobilizing a previously disorganized group of permit holders. Taking sectors to a system-wide scale also presented challenges to the Northeast Regional Office (NERO) and the Northeast Fisheries Science Center (NEFSC), who quickly had to develop new reporting systems and associated databases (e.g. dockside and at sea monitoring, sector weekly reports) and enact modifications to the existing systems to accommodate new uses. The new sector system also required the creation of new connections and relationships between fishermen and managers. For many sectors, this was through the efforts of the Northeast Seafood Coalition, though other organizations also came to play important organizational and financial roles. Twelve of the 20 current sectors were organized by NSC (one is a lease-only sector)[e]. As NSC has noted (NEFMC 2012), it began to plan for sectors when there seemed to be no other viable options emerging for groundfish management. NSC began to reach out to leaders in the fishing industry, both NSC membership and non-members, identifying fishermen by geographic region, fishing gear type, fishing area, vessel size, homeport and other commonalities, then began organizing meetings, and eventually providing a template for sector organization. As one sector manager described, one of the original leaders in his sector was a fisherman respected by his peers, who knew and personally approached others about joining the sector so as to preserve the right to fish in his state. Many managers similarly noted the importance of previously established relations that made fishermen comfortable doing business together, and of key individuals who could effectively lay out options for decision-making. Managers of non-NEF sectors also noted the importance of key NGOs, philanthropic groups, and industry groups for financial and organizational support—such as writing plans and by-laws, and helping with related programs.

Yet despite its grassroots beginnings in Amendment 13, many noted how the process became more top down during Amendment 16, as sectors were soon perceived not as a choice but as the only viable option. This was also found by Holland et al. (2010), where 46% of respondents felt that they were 'forced' into joining sectors and an additional 23% felt that it was the only viable alternative available to them. At the same time there was a push at a national level to create a catch shares policy to further their consideration in the fisheries management process[f]. As a recent review of experience in community based natural resource management (CBNRM) has described, there is a common tendency for universal models to usurp local specificity and meaning, often with negative results (Dressler et al. 2010). While many competing proposals were suggested by many different grassroots groups during the early stages of Amendment 16, including area management, transferable quotas, and so on, the council focused primarily on developing sectors when deadlines became pressing, such that "Suddenly, there were industry groups that were supporting very different approaches now gravitating to sectors as potentially the only opportunity for meaningful change in Amendment 16" (Singer 2011: 35). Yet this diversity—differences in goals and values that reflect "spaces of experimentation" (Healy 2009)—is to a certain degree reflected in the different sectors now operating in the groundfish industry, an issue which we address in the next section; we return to the issue of conflicting goals and objectives in groundfish as a whole in later sections of the paper.

Sector membership and goals

Sectors share many features, shaped as they are by the exigencies of groundfish management. In theory, each sector can tailor its internal design, catch distribution, goals, and other features to reflect the particular values and interests of its members. In addition to different goals and objectives, sectors can be shaped by the unique characteristics of their participants and the existing or evolving relationships between these individuals. Most sectors, as their original, full titles indicated, broadly share the same gear type and fishing groups, but many have also coalesced around shared qualities such as ethnicity, port location, or vision. Joint and several liability (in which all members are held liable if another sector member exceeds the sector allocation) was also a factor in sector formation and highlighted the importance of knowing and trusting the other members in one's sector. Style of fishing and vessel similarity were also important, in part because of discard rates, as we discuss later. According to Holland et al. (2013), on average, fishermen knew 56% of their sector's members very well, but didn't know 35% of their sector's members at all. The members of some sectors have very close relations: they may fish together, share a similar heritage, and even be related by kinship or marriage. In others, members may have known each other but not been particularly close, but sectors, as one manager put it, "provide an opportunity for fishermen to really know each another on a level they hadn't necessarily before". While some permit holders had been in business with others in the past, for most sector participants the new regulations represented major shifts in how they were connected to each other.

Most sector managers stated the primary goal of their sector was to help fishermen survive. Many sectors were brought together through the experience of similar problems, such as frustration with DAS, loss of fishing rights, or difficulties for small owner-operators. This took different emphases, from better using their full allocation, to sharing grounds more fairly, to improving profitability and "living within sustainable limits". Sectors that were primarily family-run businesses were more likely to be concerned about preserving fishing rights and fishing communities in their area. Many of the port-based sectors were characterized as very cohesive but certainly not all, and sectors not sharing a common geography could be cohesive if they shared something important, such as knowing each other well enough to feel comfortable and committed to working together or had a common and clear goal.

However, some sectors were formed by members who did not know each other, and who came together more to avoid other sectors whose members they did not know or trust or to avoid remaining in the common pool (Holland et al. 2013). One manager alluded to pressures some fishermen may have felt to join sectors in their primary port, and another explained how some fishermen had sought specifically to avoid others with whom they could not work easily. Some have found that through meetings (especially those related to Amendment 16) their personal networks have expanded to find more commonalities with other fishermen: as one manager said, "it has become less about having common gear type and more about common goals and mind-set". Yet those who do not share or cannot build up social and other capital, struggle more. Though many sectors are trying to make something out of a difficult situation, substantial concerns and challenges remain, some of which are detailed in the final section. And despite the different goals and objectives between sectors, all of the interviewed sectors return their member's PSC to them, effectively treating them as an individual allocation.

Governance and organization

Most sectors have a board of directors consisting of 3 to 9 officers elected by the sector members. Sector managers are responsible for day-to-day business, and in some sectors the board may also be strongly involved or consulted. Big decisions, such as electing the board or letting in new members, are generally voted on during full membership meetings, with one vote per owner, though at least one sector has weighted votes for some matters. Most have an annual meeting of all members and some meet more often, from several times a year to monthly, though attendance varies. In some sectors the boards meet several times a year, others monthly. Well attended sectors appear to be politically engaged, or heavily involved in groundfish (and perhaps also suffering more or with more at stake). Some managers have experimented with web-based meetings but found additional support necessary for such efforts to be effective. Learning how to communicate has been an unexpected challenge in creating the sector system so quickly. Since sector managers are an important boundary agent between fisheries managers and fishermen, regular communication in both directions is critical. Managers had to learn how to best communicate with their members, and solutions have varied between sectors. Some managers rely on e-mail, though as one manager put it, that was initially "a logistical nightmare" because not all members had or used e-mail. Others have found e-mail or texting ineffective when "it's still a phone driven business" and still others rely heavily on face-to-face encounters. Both members and managers have had to adapt, through setting up new communication means, learning how to use them, and finding the right mix to ensure quick and reliable linkages.

Most mangers voiced satisfaction with the use of a board of directors and the way it functions between the manager and the wider membership. Some sectors have specifically striven to ensure their boards are representative of different groups in their sector, such as different gear users or vessel sizes. Attention to structuring the board has been part of the learning process. One manager noted problems of mistrust created when membership, fearing consolidation, couldn't trust the board members who represented primarily larger operations; they have now made an effort to make the board more diverse. Another learning issue has been problems with attendance; sectors have expanded the number of individuals serving on their boards, as well as created alternates, to ensure enough for a quorum. Some sectors also have other internal committees, such as an enforcement committee, and one sector uses a "citizen's committee" of local non-fishermen residents who are available to mediate disputes between members. A few sector managers said that committees or groups formed around issues of interest, such as research projects or upcoming management issues, did arise spontaneously, and another noted that a number of members had expressed interest to him in their sector doing so. However, as he put it, such issues usually concerned "non sector items. People think that the sector encompasses everything [...] but we only deal with allocated species and sector issues [...] we don't have the manpower to do all that stuff". This represents one of many difficulties that sector managers face taking on these boundary agent roles in a complex and challenging regulatory environment.

Sectors thus display the potential for collaborative and decentralized management, but they also face many challenges and need more support, as we discuss later. Many fishermen have stated their lack of trust in the fishery management Council system and in the National Marine Fisheries Service, and have been little involved in management activities or public comment. Yet many fishermen have surely wanted effective ways to express and organize around their views, a need that newly forming interest groups within sectors now express. Yet for many of the sectors with close ties to other organizations, especially NEF sectors, the day to day work of managers and members tend to be operationally focused, not policy or vision focused, for the primary task for many managers is careful accounting of ACE usage. This means that few managers have any relationship with the Council. This re-routing of governance away from the New England Fishery Management Council to the NMFS Northeast Regional Office represents a key shift in governance relationships in the region. For example, NSC (on whose board of directors may sit both NSC-wide members and NEF sector board members) provides policy representation for NEF sector members, while NESSN (incorporated in 2011 after organizational development by NEF Sector and NSC leadership) provides technical support for its member sectors by facilitating discussion between sectors, alerting NEF sector managers to important issues (e.g. anti-trust), and providing a larger space within which NEF sectors can trade. NEF sectors have right of first offer to lease fish and right of first refusal for the sale of federal limited access Northeast multispecies permits, which means that if a sector member wants to sell either his fish or his permit outside the sector, the sector members are given the opportunity to supersede that sale, in order to keep fish within their existing fishing communities and/or the 12 sector network. The economies of scale created through this network also extend beyond trading to include weekly reports and checking quota usage and balance. This has been one major advantage of sectors forming within the NEF sector network. The network has also provided a forum for negotiating and coalition-building between NEF sectors on policy issues, extending in some cases to non-NEF sectors. One sector manager outside the NEF sectors, however, called its double right of refusal (double because the right of first refusal includes first the sector itself and then the 12 sectors in the NEF network) "a real problem" because "they have locked up two-thirds of total fish quota", making it harder for non-NEF sectors to lease fish; others saw it, however, as a way of keeping their fishing communities viable.

Whether or not sectors rely on a parent organization for policy work, there are differences among sectors and sector members in the extent of engagement or interest in management in this broader sense. Some boards receive input from membership through formal recommendations, consensus statements, or face-to-face interaction. In sectors with low attendance at membership or board meetings, managers noted a number of commonalities: members who didn't fully know each other had little sense of community cohesion, or sectors in which there was no unified vision. Indeed vision may be seen as a luxury when many are simply trying to adjust and make ends meet. Some managers noted a feeling among sector members that some of the larger organizations didn't represent many diverse interests in the industry, leading some fishermen to feel marginalized from decision-making processes. Another factor for some sectors was that membership was geographically spread out, or that its fishermen were primarily leasing groundfish rather than actively fishing it; in both cases, there could be little interest in active participation or policy representation. A number of sector managers noted the importance of key individuals in the sector who can link the manager to other sector members; others, however, noted problems if a sector becomes dominated by key individuals who don't represent, or reach out to, other members. Effectively identifying and resolving conflict or disagreements was cited as crucial. Lack of leadership, insufficient outreach, or ineffective communication between leaders and members was cited for lack of involvement; yet members in some relatively satisfied sectors were noted to have little interest in sector decision-making just as in sectors marked by infighting. Dissatisfaction with sector managers was also cited as a reason some fishermen move between sectors. And there is the perennial problem that fishermen are out fishing and can't go to meetings; as one manager said: there are simply "too many issues and not enough time".

Key relations also exist between sector managers. As many explained, they tend to work and trade with some managers more than others, in part due to needing particular fish but also through relations they build over time. Some managers are also more active and involved, and they may consult with each other about problems or questions, and, given a level of trust, conduct such business as trades informally through verbal agreements. Some felt that this collaboration might only be a temporary feature of the learning and adjustment period. Those who did not mention such camaraderie, on the other hand, likened their role to simply running their own business independently in that they felt little need to consult or organize with other managers. Likewise, many have found that most trading happens internally within their sector, with external trades channeled through already established relations; others found it to be "just a marketplace" that is not based on prior relations. Similarly, some leasing is arranged by fishermen in private deals, while others rely on sector managers to mediate. Managers innovated by forming an e-mail system to facilitate trading. Most noted their relations with the NERO, in contrast with the Council, as being frequent and improving. This too has been a learning process and has involved building relations over time, especially through monthly meetings. Communication with the NERO, sector managers noted, has improved as more personnel and ways to interact have developed; as one manager put it, "interaction with the NERO was difficult at first because so many infrastructural elements weren't working... but I've never seen the Service more helpful than they have been with sectors". Just as relations of trust had to be built between sector managers and fishermen, face-to-face connections have been built between managers and NMFS through repeated encounters, building trust but also sometimes distrust. Virtually all managers said they had excellent relations with those staff they directly interacted, but more than a few also noted that "something happens after it has gone beyond their offices". This bureaucratic element, as we discuss later, hinders reaching fuller advantages from collaborative and decentralized management, with implications for the ultimate effectiveness of sector management. Moreover, as Brewer (2011) has written, the sector system may also decrease public accountability in that internal decision-making and other functions are less transparent than public entities, and thus can undermine the movement to co-management.

Changing roles and opportunities

Some sector managers have taken advantage of the cooperative structure of sectors, and the ensuing ability to organize themselves, to work on different policy issues. A number of sectors have members involved in research and collaborative projects, such as gear technology and fuel efficiency, especially in conjunction with groups such as the Gulf of Maine Research Institute (GMRI). One manager explained how their projects arose through joint member meetings with Sea Grant and GMRI representatives, where they brainstormed research ideas and voted on which ones to undertake. This stakeholder-driven research, he said, differs from previous collaborative research, which was top-down. However, as another sector manager noted, though some outside groups had hoped that managers would be a liaison between fishermen and researchers, this has not always been possible given their heavy workload, especially when sectors were new and many operational elements were lacking. Others also mentioned legal issues that could prevent such projects, such as insurance liabilities. Fishermen, too, another explained, may be interested in research but suffer fatigue and frustration when their work is not used in policy or assessment.

Another development that a number of managers expressed interest in was in value-added activities such as local markets and branding. As one manager put it, "sectors offer flexibility to cater to local markets and understand when the vessels are landing to coordinate". Some were engaged in feasibility studies on local seafood demand and developing the necessary infrastructure (such as local processing contacts or facilities) and communication media (such as websites to facilitate education and outreach with local consumers). One sector had conducted a pilot Community Supported Fishery (CSF). Yet, for the most part, local branding or marketing was happening by individual fishermen, not through sectors. Again, one of the most important issues was the limited time, capacity, and capital to undertake such new initiatives. The sector with a pilot CSF, for example, relied on an outside NGO to run it. While managers were keenly aware such efforts could provide greater financial return for their members, it would be risky for it would require learning how to be a fish dealer, making the necessary business relations, and acquiring the necessary licenses and capital investments in infrastructure. It would also be risky because such efforts would require commitment from sector members, more difficult for small sectors or less cohesive ones, and for those whose members might be inclined to lease their allocation or who wished to retain more independence. Additionally, a number of managers noted deep concern about anti-trust laws. One sector had anti-trust monopoly language written into its operating agreement and was concerned that even individual fishermen marketing their catch might be in violation of this agreement. Another manager noted that without knowing more about anti-trust laws, his sector didn't want to risk violations, and even a report written by lawyer hired by the NSC and GMRI to provide guidance was difficult to understand. Again, this speaks to the need for better meta-support from government agencies or other groups to better foster potentially innovative ways to sustain sector members.

These elements of governance, within and between sectors, speak to the changing roles of fishermen and their intermediaries. For fishermen, the voting structure within sectors and the new positions of leadership in which increasing numbers serve, imply the potential for a greater voice in management. Some mangers also noted changing opinion of some of their members regarding sector management, from skepticism to greater involvement in its decision-making to even becoming community leaders. Most noted this was out of a feeling of necessity rather than commitment to sectors *per se*; as one manager put it, "individually they have no voice[... now] you speak united [...] at least they are listening to us". Yet another noted, with an example of monitoring marine mammal takes with the sector's own data, that sector members were realizing the benefits of being proactive, that "they can take a little bit of control, as a sector, of their fate". Other managers noted how independently-minded fishermen, even if they knew each other, might not have shared information or worked on shared issues to the same extent as they do now, which speaks to how new institutional arrangements may engender changing identities and subjectivities with respect to resource use (Agrawal 2003). In contrast, the provision for joint and several liabilities has influenced who wanted to join which sectors, but at the same time it has strained some existing relations, demanding self-policing and feelings of tattling that go against deeply held values. One sector, for example, had to expel a member who had been careless about his quota, but this action upset some and produced feelings of mistrust.

In all this, sector managers play a key role, effectively becoming the boundary agents between sector members and government, as well as between sector members and other support organizations. Through managers, the fishing industry is working directly with the NERO to manage most of groundfish, changing the role of the Council and, to a certain extent, decentralizing some aspects of management. Moving from nearly 1,500 separately permitted vessels to 16 individuals is a significant change in how information is potentially shared, and feedback received, from industry. Managers have also had to find ways to add value to their members allocation and to help pay for their operating costs to ensure the long term sustainability of both their members and their sector. For some, this has entailed creating permit banks and increasing membership. Others have also focused on policy issues, such as working to create quota market restrictions to insure their members aren't disadvantaged. Managers and members are writing letters to the Council or directly to its committees (such as the Statistical and Scientific Committee), to the NERO, as well as to their congressmen, through understanding the science and taking a proactive role in things before they become regulation. These are all further elements of co-management that have been galvanized through the creation of networks and the linking at different organizational levels (Rosen and Olsson 2013). Additionally, one manager believed that friction between different industry groups may be diminishing due to the role played by managers, since they can interact on their behalf, both with other managers and with government, but aren't themselves direct stakeholders. Sector managers work for the sectors and thus can bring sector issues to the various forums in which they participate. Acting as representatives then allows for policy suggestions to be more organized and more powerful, yet, as this same manager cautioned, it is still vital that fishermen be brought to the table as well. There are many crucial challenges that fishermen, and managers, continue to face, to which we now turn.

CHALLENGES

Many managers noted the initial challenges for everyone caused by the compressed timeline for sector implementation: sudden changes (e.g. reporting, setting up trading), obtaining PSC information, deciding which sector to join, technical difficulties, communication problems, and troubles interpreting the new rules and regulations. Many of these challenges were detailed by sector managers in a Council-held "Lessons Learned" workshop held in 2011[g]. As one manager noted, NMFS didn't have the infrastructural support or the "people power" in the beginning to support the new system; another described the steep learning curve involved in becoming a sector manager. Most, however, noted the good rapport and communication networks that developed with NERO staffers as they all "stumbled their way through it" together. Yet many managers noted the frustration caused by persistent infrastructural issues, most especially with data redundancies and quality issues. While streamlining data processes was a priority identified at early workshops, one manager felt data issues had only gotten worse. Another manager felt that sectors from the beginning had been held to such a high standard that NMFS itself couldn't meet. Another detailed the redundancy caused by poor NMFS data, requiring them to maintain separate books and hire staff to manage the data. Instead of multiple reporting requirements (such as vessel trip reports, dockside reports, dealer reports), one manager suggested finding ways to let sectors focus on their comparative advantage in getting fisherman-based data, while removing some of the other reporting burdens that often just repackage already existing NMFS data. Others suggested having NMFS focus on data quality and data standards, while sectors could focus on data collection to which they were better suited. On a related note, some managers noted that they filled out NMFS data requests but were unsure of the purpose for these data, demonstrating the need for better outreach and involvement. Finally, one manager said their biggest challenge was "answering to two masters", with the agency increasing their workload and using them as the liaison to sector members, when managers "work for the sector, not the agency".

The new system has posed many challenges for fishermen. Given that the sector system started in a period of declining total allowable catches (TACs) and the introduction of restrictive annual catch limits, many fishermen have had to deal with low and decreasing allocations. With the abundance of some stocks in the groundfish complex expected to remain low in the near future, these difficulties are likely to continue. While sectors have many potential benefits, such as the ability to target a market or spread out landings, according to sector managers many fishermen have not been in a position to realize such benefits. Although some have argued that continued management through DAS would have been more onerous, and indeed most managers believed their members were also unhappy under the DAS system, nonetheless the low allocations combined with transferability have had many predictable effects common when resources become privatized (see Olson 2011). One of the most important is the issue of consolidation (Kitts et al. 2011). As one manager described, some groups of fishermen were better able to take advantage of the sector system: those with relatively larger and more diverse allocations, and who had a stronger capitalized base, could more easily buy fish from those with fewer options, driving consolidation. Many managers noted that smaller vessels or operations in particular have had problems affording higher prices and buying quota. Smaller vessels are also disadvantaged by the lack of any baseline restriction on the leasing market, as in DAS leasing. This, as one manager stated, "doesn't recognize the difference in purchasing power and access to capital, [which] puts smaller boats at a disadvantage". Moreover, many smaller vessels were also disadvantaged by allocations based on catch history during 1996–2006, when measures such as trip limits and rolling closures had greater effects on certain segments of the industry. There is a widespread sentiment that allocations have favored the larger, more rapacious fishermen. As van Ginkel (2009: 284) has written of the Dutch experiment with sectors, while many fishermen have been pleased with the more participatory nature of management, it has come with "considerable ecological and social costs", including such concerns as highgrading of catch and concentration of ownership.

The combination of an allocation too low to fish profitably and/or too little cash flow to finance leasing creates a downward spiral, as decreasing profits increasingly cut away at available funds. The response in the first year in particular was characterized by managers as one of shock and of paralysis. But, as another explained, those who dared to go fishing amidst the paralysis got better prices and did well. "If you don't have the finances, you can't play this game", as one manager put it. It is also relatively more risky for those with less adequate finances, for there is no guarantee that leased fish will be profitable or even caught. More than one could tell of fishermen who leased high and landed when prices were low, going out of business. Leasing, as one manager described, was "out of control" with prices too high for smaller operations to afford. Managers noted the increasing stress and family problems that have resulted from financial difficulties and worry over overfishing. Some fishermen have tried to adapt by selling their vessels and then working for others. One manager noted how impacts also extend beyond many of these smaller businesses, often family-run, who have had to let go crew they have fished with for years. Another manager related how some fishermen, out of a deep sense of personal responsibility, have tried to keep their crews by leasing more, but with increased leasing costs still a negative factor on both owner and crew.

Yet it is not only small-vessel fishermen who have had to adapt. Fishermen who had designed their businesses around DAS, for example, may have had to sell some vessels at a loss to adapt to the sector system. Many voice worry about shouldering the costs of at-sea monitoring and the continued uncertainty over who will pay. A fear of what the future will bring, in terms of a changing policy environment as well as fluctuating allocations, creates uncertainty and difficulty in making business decisions. As another manager described, the new fishing regulations require a new way of thinking: those who are able and willing to adapt and take a long-term view will benefit, but those who don't will lose out; but as this manager continued, many fishermen are no longer willing to adapt anymore. Conversely, some managers noted that the sector leasing system has made it more profitable for some fishermen to treat their allocation as an "annual annuity" rather than fish it. Though sector allocations are not officially individual permanent allocations, more than one manager described the "ITQ-think" (for Individual Transferable Quotas, or privatization of resources) that developed during Amendment 16 proceedings, in which some fishermen were encouraged to think about sectors as individual quotas. In contrast, the initial interest in sectors had been much broader and was thought that individual sectors could tailor their allocations to their own goals and objectives, such as profit-sharing (Holland and Wiersma 2010). Indeed one manager felt it would cost less and be easier to manage community quotas than individual ones. But others described how leasing itself encouraged thinking about sectors individually, since it helps create a race to buy and sell before prices change, and encourages a sense of entitlement to a particular allocation[h].

The spatial distribution of fish and the mix of species within an allocation have also been important. Day-boat fishermen, for example, with smaller vessels might receive allocations for species that can't target, whereas the price of species they do target can easily be driven up by other fishermen to cover their bycatch. As one manager explained, one "driver of consolidation" is that larger vessels can afford to lose money on secondary species, even when it's not priced at its shadow value, but which they buy to go fishing on their primary interests. It's "quite dramatic how this has separated the ability for these small vessels to participate in the quota market and compete with these larger vessels". As another manager said, "They've created this commodity market for people who basically did nothing. It was never that type of market when it was DAS…some people are reaping the benefit and some people are paying through the nose". Moreover, smaller vessels are more sensitive to finer scale issues such as local depletion in their traditional fishing grounds; larger vessels can benefit more if and when total abundance rises, whereas smaller ones need a fuller recovery of the stocks in the local areas in which they fish. Indeed, as Acheson (2011: 70) has explicitly written, from "the perspective of small boat operators in Maine", the disappearance of the option of area management from Amendment 16, such that fishermen can fish anywhere they want, "means that the large boat owners from southern New England have been able to get the rules they wanted again". Additionally, different fishermen (and different sectors) may not fish groundfish exclusively, and the seasonality of different fisheries may make following strict timelines for groundfish sectors more difficult. For example, those who are done early in the season are disadvantaged by not knowing how much quota they can carry over, whereas those who fish later in the fishing year may scramble to catch their full quota. Many fishermen are also concerned about being locked out of fisheries when access is based on recent history; they can't move into them when different species are doing well and groundfish aren't. Sector management seems to amplify the effects of single-species management on fishermen, for as Brewer (2011: 15) has written, fishermen must "negotiate for access to paper fish, suppressing their own observations of fish life histories, spatial patterns, and inter-species and habitat interactions" because "Many fear that not catching the maximum limit could jeopardize future individual allocations because days-at-sea are based on catch histories, and individual quotas would likely be allocated similarly".

As Berkes (2004) has argued, while "mutual learning" and "trust building" are both key to adaptive co-management, they take time to develop and cannot be imposed. Managers provided many ideas about ways to improve the sector system, but more than one noted how sectors are embedded in a large bureaucracy, i.e. NMFS and the Council, that cannot collect or respond quickly to new information, so that the purported benefits of the sector system are harder to materialize: "We have this sector system designed to react to things happening on the water and it's being hamstrung by a management and monitoring system that is years behind". Many noted with frustration that suggestions on ways to improve or incorporate new information and studies were not acted upon. A number of managers pointed to inequities from assumed discard rates, and suggested better observer coverage to base discard rates on actual practices. Generic rules neither encourage accountability nor reward stewardship, creating lost opportunities to encourage collaborative management. An example that another manager provided is hand gear, which has an assumed discard rate of 50%; this manager believes the penalty assessed to that gear will eventually eliminate the gear from the Northeast, despite studies showing high survivability of released juvenile fish. Others questioned the need for any observers given the design of sectors to encourage joint liability. And one pointed deeper to what he believed were flaws in the Magnuson-Stevens Fishery Conservation and Management Act, the primary legislation governing federal management of fisheries in the US, and that a new commitment to participatory governance and ecosystem-based management was needed, not micro-management of individual species. One avenue was thought to be better capitalizing on bottom-up data that sectors and fishermen could provide, as a first step in bringing stakeholders and scientists together. As this manager put it, "We can't base this all on scarcity and fear; we have to base it on abundance and hope".

CONCLUSIONS

The sector system was introduced to manage groundfish in the Northeast US with many promised benefits. Yet in the first few years of sector management, as interviews with sector managers have revealed, the accrual of these benefits has been uneven. Certainly, some groups are well on their way to capitalizing on value-added pursuits that sector management is supposed to better enable. Others were overwhelmed and didn't know where to begin. Given the importance of formal and informal relations, those with the least, in terms of both financial and social capital, seem to have fared the worst. Yet many factors have been important, such as the amount of allocation, the strength of support networks, and the need for business relations and acumen. Fishermen of different ages have faced differing challenges, with younger fishermen discouraged from entry and older fishermen concerned about being able to retire. Moreover, sectors have had different levels of outside support and funding. Such asymmetries are a barrier to the full realization of sector potential. It seems that many have assumed that sector managers could be tasked to do things without full consideration for the support they might need to be successful.

Yet sectors are also a new form of environmental governance in Northeast fisheries. Both formal and informal relations and networks have been critical to creating new institutional forms of communication and cooperation, and in which the scope of management activities has broadened. Sector managers as boundary agents have played a new and crucial role in creating new alliances and networks of environmental governance in the groundfish fishery. We have noted in this paper the need for further support and greater flexibility to sectors and to sector managers. Yet success means different things to different people. Sectors have been saddled with conflicting goals and objectives, with expectations to be both economically efficient vehicles of quota allocations, and institutions to promote cooperative and devolved management. Given the expectation that maintaining catch history is critical to resource access, it is not surprising that only one of these expectations has gained the most ground.

ENDNOTES

[a]The implementation of this program was complicated by the assessment of overfishing on most species in the groundfish complex, expected reductions in available quota, and the further need by law to set annual catch limits (ACLs) and accountability measures (AMs) for fisheries subject to overfishing.

[b]Wilson and Jacobsen (2013) have argued that the ability to communicate and to reach resilient management solutions depends on not "reduc[ing] discussions of complex systems to simple categories", and suggest further that sector management in New England may represent a flexible "boundary object" since the actual content of sector organization is not imposed. However, the similar adoption by sectors to distribute their allocation on an individual basis, despite no requirement to do so, speaks to how the flexibility of a boundary object also depends on the context in which such communication occurs,

a more general point to which we will return throughout the paper.

[c]The National Marine Fisheries Service (NMFS), part of the National Oceanic and Atmospheric Administration (NOAA), is the federal regulatory agency that oversees fisheries management in US federal waters (the 200-mile Exclusive Economic Zone, excluding state waters from shore to 3 miles). NMFS has six regional offices and six regional science centers across the country that provide fisheries policy assessment and advice, and which also participate in analyzing fishery management plans formulated by the eight regional fishery management councils. In the Northeast US, this includes the Northeast Regional Office (NERO), the Northeast Fisheries Science Center (NEFSC) and the New England Fishery Management Council (NEFMC).

[d]See organization websites, including: http://www.namanet.org, http://www.ccchfa.org,http://www.northeastseafoodcoalition.org, http://www.penobscoteast.org, andhttp://www.gmri.org

[e]These 12 sectors are referred to as Northeast Fishery sectors (NEF) II through XIII, which are distinct non-profit 501(c)(6) organizations. Members of all NEF sectors are represented by the NSC on policy matters concerning the groundfish fishery.

[f]See http://www.nmfs.noaa.gov/sfa/management/catch_shares/index.html (accessed 7 November, 2013).

[g]See http://www.nefmc.org/nemulti/cte_mtg_docs/111025-26_sector_ws/111025-26.htmlfor documents provided by all sector managers. Concerns expressed ranged from the difficulties facing fishermen to data issues and communication problems, echoing issues raised in the interviews we conducted.

[h]NOAA has been at pains to emphasize that sectors are not ITQs since they don't represent a permanent allocation to a vessel (see for examplehttp://www.nero.noaa.gov/sfd/Amend16Docs/Reply%20to%20NB%20reply.as%20filed.3.2.11.pdf, accessed December 5, 2013). What is significant in this case is not only the concurrent efforts at a national level to articulate a policy that promoted the consideration of catch shares more generally for fisheries management, but the history of mistrust characterizing fishermen-federal interactions in New England, which only intensified a desire to maintain one's catch history (see also Brewer 2011).

[i]Note: Landings include all species, not limited to groundfish. Permit bank location is based on office location.

APPENDIX 1. INTERVIEW GUIDE FOR SEMI-STRUCTURED INTERVIEWS

Background

- How was your sector formed?
- How would you describe your membership?
- Are there other groups that have been important for forming or running your sector?

Governance

- How are decisions made in this sector (e.g. one owner/one vote)?
- What different roles or structures have you created for running your sector?
- How are these decision-making and other activities working out for you?
- Have there been key individuals instrumental in the organization of your sector?
- Have your members expressed interest in taking on more decision-making in the sector?
- What are the particular goals of your sector?
- Have there been particular problems you have encountered in trying to meet your goals?
- What solutions would you offer for these problems?

Membership

- Have there been, or does it look like there will be, any major changes in your membership?
- Are there members of your sector that have benefitted more than others from being in a sector?
- What do you think is the ideal composition of a sector?
- What are the costs of sector membership? What services to members get from sectors? Can everyone afford the costs?

Other

- Does your sector interact with/trade quota/exchange information/cooperate in some other way with other sectors? Which ones, and why?
- Relationship with Council/Regional Office - is this changing over time?
- Does your sector have any marketing activities?
- Has your sector worked on creating ties between fishermen and consumers?
- Has your sector worked on creating or changing ties between fishermen and dealers or processors?
- What other activities does your sector engage in?

ACKNOWLEDGEMENTS

We would like to thank all the sector managers who graciously took time out of their busy schedules to meet and talk with us, and Matt McPherson, who participated in a number of the interviews and helped to craft the interview guide. Thanks to the editor and the two anonymous reviewers for helping us to clarify and deepen our argument. Any misunderstandings are of course our own. The information in this paper does not reflect, represent, or form any part of the support of the policies of NOAA or the Department of Commerce, nor imply that NOAA or the Department of Commerce agree with the information contained herein.

AUTHORS' CONTRIBUTIONS

JO and PPdS both carried out interviewing, compiled and analyzed data, wrote the manuscript, and read and approved the final version.

REFERENCES

1. Acheson JM: Government regulation and exploitive capacity: the case of the New England groundfishery. *Human Organization* 1984, 43(4):319–329.
2. Acheson JM: Lobster and groundfish management in the gulf of maine: a rational choice perspective. *Human Organization* 2006, 65(3):240–252.
3. Acheson JM: Coming up empty: management failure of the New England groundfishery. *MAST*2011, 10(1):57–86.
4. Agrawal A: Sustainable governance of common-pool resources: context, methods, politics.*Annual Review of Anthropology* 2003, 32: 243–262. 10.1146/annurev.anthro.32.061002.093112
5. Anthony VC: The New England groundfish fishery after 10 years under the magnuson fishery conservation and management act. *North American Journal of Fisheries Management* 1990, 10(2):175–184. 10.1577/1548-8675(1990)010<0175:TNEGFA>2.3.CO;2
6. Berkes F: Rethinking community-based conservation. *Conservation Biology* 2004, 18(3):621–630. 10.1111/j.1523-1739.2004.00077.x
7. Brewer JF: Paper fish and policy conflict: catch shares and ecosystem-based management in Maine's groundfishery. *Ecology and Society* 2011, 16(1):15.
8. Brodziak J, Cadrin SX, Legault CM, Murawski SA: Goals and strategies for rebuilding New England groundfish stocks. *Fisheries Research* 2008, 94(3):355–366. 10.1016/j.fishres.2008.03.008
9. Bulkeley H: Reconfiguring environmental governance: towards a politics of scales and networks.*Political Geography* 2005, 24(8):875–902. 10.1016/j.polgeo.2005.07.002
10. Cash D, Clark W, Alcock F, Dickson N, Eckley N, Guston D, Jager J, Mitchell R: Knowledge systems for sustainable development. *Proceedings of the National Academy of Sciences* 2003, 100(14):8086–8091. 10.1073/pnas.1231332100
11. Çalışkan K, Callon M: Economization, part 2: a research programme for the study of markets.*Economy and Society* 2010, 39(1):1–32.
12. Clay PM, Colburn LL Paper presented at the National Working Waterfronts and Waterways Symposium, March 25–28. In *Understanding Impacts on Fishermen and their Families from New England Catch Shares*. Tacoma, Washington; 2013.
13. Dressler W, Büscher B, Schoon M, Brockington D, Hayes T, Kull CA, McCarthy J, Shrestha K: From hope to crisis and back again? A critical history of the global CBNRM narrative.*Environmental Conservation* 2010, 37(1):5–15. 10.1017/S0376892910000044
14. Folke C, Hahn T, Olsson P, Norberg J: Adaptive governance of social-ecological systems. *Annual Review of Environment and Resources* 2005, 30: 441–473. 10.1146/annurev.energy.30.050504.144511
15. Hall-Arber M: Co-management at the eleventh hour? Participation in the governance of the New England groundfish fishery. In *Participation in fisheries governance*. Edited by: Gray TS. Dordrecht, The Netherlands: Springer; 2005:141–162.
16. Harriss J, de Renzio P: 'Missing link' or analytically missing?: The concept of social capital.*Journal of International Development* 1997, 9(7):919–937. 10.1002/(SICI)1099-1328(199711)9:7<919::AID-JID496>3.0.CO;2-9
17. Healy S: Economies, Alternative. In *International Encyclopedia of Human Geography*. Edited by: Rob K, Nigel T. Oxford: Elsevier; 2009:338–344.
18. Hennessey T, Healey M: Ludwig's Ratchet and the Collapse of New England Groundfish Stocks. *Coastal Management* 2000, 28(3):187–213. 10.1080/089207500408629
19. Holland DS, Wiersma J: Free form property rights for fisheries: the decentralized design of rights-based management through groundfish "sectors" in New England. *Marine Policy* 2010, 34(5):1076–1081. 10.1016/j.marpol.2010.03.007
20. Holland DS, Kitts AW, Pinto da Silva P, Wiersma J: Social capital and the success of harvest cooperatives in the New England groundfish fishery. *Marine Resource Economics* 2013, 28(2):133–153. 10.5950/0738-1360-28.2.133
21. Holland DS, Pinto da Silva P, Wiersma J: *A survey of social capital and attitudes toward management in the New England groundfish fishery*. Woods Hole, MA: US Dept Commerce, NEFSC Ref Doc; 2010:10–12.
22. Ishihara H, Pascual U: Social capital in community level environmental governance: a critique.*Ecological Economics* 2009, 68(5):1549–1562. 10.1016/j.ecolecon.2008.11.003
23. Kitts A, Bing-Sawyer E, Walden J, Demarest C, McPherson M, Christman P, Steinback S, Olson J, Clay PM: *2010 final report on the performance of the northeast multispecies (groundfish) fishery (May 2010-April 2011)*. Woods Hole, MA: US Dept Commerce, NEFSC Ref Doc; 2011:11–19.
24. Larner W, Craig D: After neoliberalism? Community activism and local partnerships in Aotearoa New Zealand. *Antipode* 2005, 37(3):402–424. 10.1111/j.0066-4812.2005.00504.x
25. Lemos MC, Agrawal A: Environmental governance. *Annual Review of Environment and Resources* 2006, 31: 297–325. 10.1146/annurev.energy.31.042605.135621
26. Murphy T, Kitts A, Records D, Demarest C, McPherson M, Walden J, Daniel Caless D, Bing-Sawyer E, Steinback S, Olson J: *2011 Final Report on the Performance of the Northeast Multispecies (Groundfish) Fishery (May 2011-April 2012)*. Woods Hole, MA: US Dept Commerce, NEFSC Ref Doc; 2012:12–30.
27. NEFMC: *Amendment 18 to the NE Multispecies FMP Gloucester, MA Scoping Meeting – comments by Jackie Odell, Northeast Seafood Coalition. Minute 56–61.* 2012. . Accessed 23 July 2013 http://www.nefmc.org/nemulti/cte_audio/Gloucester%20A18%20Scoping_r.

mp3 . Accessed 23 July 2013

28. Olson J: Understanding and contextualizing social impacts from the privatization of fisheries: an overview. *Ocean and Coastal Management* 2011, 54(5):353–363. 10.1016/j.ocecoaman.2011.02.002

29. Ostrom E: *Governing the Commons: The Evolution of Institutions for Collective Action*. Cambridge: Cambridge University Press; 1990.

30. Ostrom E: Beyond markets and states: polycentric governance of complex economic systems.*American Economic Review* 2010, 100(3):641–672. 10.1257/aer.100.3.641

31. Pinto da Silva P, Kitts A: Collaborative fisheries management in the Northeast US: emerging initiatives and future directions. *Marine Policy* 2006, 30: 832–841. 10.1016/j.marpol.2006.04.003

32. Portes A: Social capital: its origins and applications in modern sociology. *Annual Review of Sociology* 1998, 24: 1–24. 10.1146/annurev.soc.24.1.1

33. Pretty J, Ward H: Social capital and the environment. *World Development* 2001, 29(2):209–227. 10.1016/S0305-750X(00)00098-X

34. Rosen F, Olsson P: Institutional entrepreneurs, global networks, and the emergence of international institutions for ecosystem-based management: the Coral Triangle Initiative.*Marine Policy* 2013, 38: 195–204.

35. Singer LT: *The Development of Catch Shares: Lessons Learned from New England*. Portland, Maine: GMRI; 2011.

36. St. Martin K: Mapping economic diversity in the First World: the case of Fisheries. *Environment and Planning A* 2005, 37(6):959–979. 10.1068/a36296

37. St. Martin K: The difference that class makes: neoliberalization and non-capitalism in the fishing industry of New England. *Antipode* 2007, 39(3):527–549. 10.1111/j.1467-8330.2007.00538.x

38. van Ginkel R: *Braving Troubled Waters: Sea Change in a Dutch Fishing Community*. Amsterdam: MARE Publication Series No. 4, Amsterdam University Press; 2009.

39. Wilson DCK, Jacobsen RB: Analyzing resilience with communicative systems theory an example from European fisheries. *Maritime Studies* 2013, 12(1):1–18. 10.1186/2212-9790-12-1

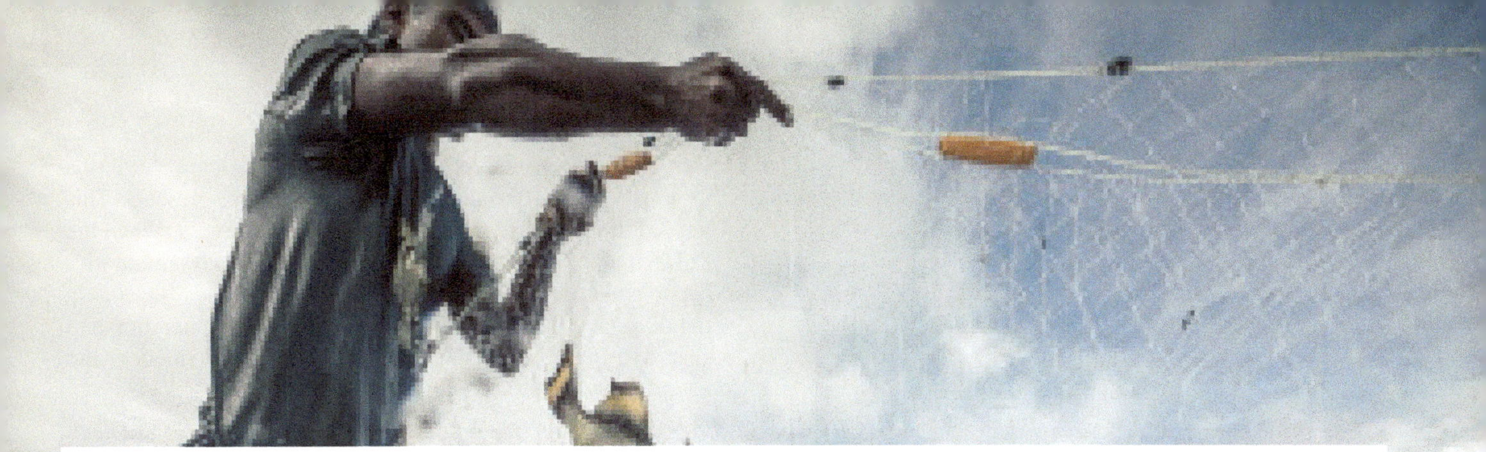

CHAPTER 7

MARINE SPATIAL PLANNING

Svein Jentoft and Maaike Knol

Keywords: Marine spatial planning; Marine protected areas; Fisheries and coastal governance; Stakeholder conflict; Subsidiarity principle North Sea

INTRODUCTION

The North Sea is one of the most crowded marine areas in the world (Halpern et al. 2008; Jay2012a). It is also a major fishing ground. Seven countries border the North Sea, each with their own spatial claims, cross-cutting ecosystem boundaries, an array of user groups, different governing systems, a shortage of cross-boundary mechanisms for settling disputes, and pressing demands for environmental protection. Fisheries, ocean and coastal governance are a daunting challenge beyond anyone's imagination.

It could still be argued that the North Sea is in many ways in a better situation than marine regions are in many other parts of the world. To begin with, North Sea countries share a common democratic culture, a factor that is not the case for all of Europe's regional seas. The North Sea countries also have a high score on the World Bank's "good governance index". Save for Norway, all of them are within the European Union (EU) and thus have an opportunity and the commitment to work together within common institutional frameworks, such as the OSPAR Convention, the Marine Strategy Framework Directive, Natura 2000 and the Common Fisheries Policy (CFP), to note but a few. If Marine Spatial Planning (MSP), which is the topic of this paper, needs cannot be realized and/or met under such cooperative frameworks, one would be hard-pressed to envision a situation in which it could.

From a MSP perspective, the situation in the North Sea nevertheless remains complex. Governments are not cooperating as well as they could; legislation is not as developed and synchronized as one could hope for (Drankier 2012); the interaction between different sectors and user-groups could be more constructive (Ritchie and Ellis 2010); and the status of the marine ecosystems and fisheries in the North Sea are hardly positive (ICES 2011). All of these aforementioned issues are problems that marine spatial planning MSP would need to relate to.

In this paper we argue that MSP cannot solely be thought of as a technical instrument. Rather, it should perceived as a multi-scalar and multi-sectoral governance mechanism (Kidd and Shaw2013). MSP must also be about creating order and negotiating conflict between user-groups who do not always

Citation: Svein Jentoft, Maaike Knol, Marine spatial planning, DOI: 10.1186/2212-9790-13-1.

share the same goals, and whose powers, interests and worldviews differ sharply. MSP is therefore about much more than setting goals and implementing them.

To put MSP in a governance context where it belongs, we shall commence by defining what governance means, followed by a similar conceptual clarification of MSP. Coming from a fisheries research background we are particularly interested in what MSP represents as a challenge for this particular sector, which in the case of the North Sea has ancient roots. For that a micro-perspective is needed. The focus must also be on what MSP may mean to fishers and their local communities. Here, one should both have an eye for the risks as well as the opportunity which this new tool provides. From a fisheries sector and community perspective, attention must be paid to the competition for space and resources between fisheries and other stakeholders. Who are these stakeholders and what characterizes their spatial usage? Finally, the paper addresses some key governance principles as they apply to both fisheries governance and MSP in the North Sea, in particular those that relate to scale, subsidiarity, and participation in decision-making.

MARINE GOVERNANCE DEFINED

There is currently a huge and ever-growing body of literature on governance, in and of itself an indicator of a general interest in governing systems and mechanisms that are more inclusive and less state-centered than they used to be (Kjær 2004). Increasingly, societal governing is organized through partnership arrangements that involve industries and civil society. This is also occurring for many industries that operate in the North Sea (van Hoof and van Tatenhove2009; van Leeuwen and van Tatenhove 2010). The move towards "governing as governance" (Kooiman 2003) is mostly perceived as a step in the direction of a more user-friendly, democratic, responsive, interactive and effective form of governing. Hence, such a move is associated with "good governance" as defined by the World Bank, the EU and others. Notably, governing as governance is partly an observation of what is actually going, as well as an instruction and norm-i.e. something that we should have more of. Not only does it promise a better working governing system, it also abides by some old governance principles that have intrinsic value.

Many definitions of *marine* governance in particular figure in the academic literature. In line with the general governance concept, a common denominator is the emphasis on the participatory approach involving market and civil society stakeholders. For instance, van Leeuwen and van Tatenhove (2010:591) define marine governance as, "the sharing of policy making competences in a system of negotiation between nested governmental institutions at several levels (international, supranational, national, regional and local) on the one hand, and state actors, market parties and civil society organizations on the other in order to govern activities at seas and their consequences." Below are two examples, one ancient and one modern, of such governance institutions for the marine environment that are relevant in a MSP and fisheries context.

Polder model

Despite its modern applications, the Polder model is not a recent invention. The particular historical context of this model in the Netherlands is rooted in the need to arrive at some consensus on how to tackle the constant challenge water presents. This has led to innovative thinking about how to govern. When the Polder model was initiated in the Middle Ages, communities had to put aside their differences and work together to maintain the dikes and pumping stations. Without this cooperation and shared responsibility, the polders would have flooded and everyone would have suffered.

Interestingly, the polder model still plays a role in Dutch politics. Glasbergen (2002), for example, talks about a "green polder model" as a tool for environmental governance. There are, however, mixed experiences with this model (Schreuder 2001; Woldendorp and Keman 2007), as is the case with most governance systems. Nevertheless, the fact that it is still relevant suggests that marine governance in the modern age is not so much about introducing entirely new and untested principles and governance mechanism, but rather about invoking solutions that have a deep-seated history.

Regional advisory councils

The set up and design of the regional advisory councils (RAC) as a form of "trans-national governance" (van Hoof and van Tatenhove 2009) of fisheries by the EU between 2002 and 2004, is largely structured as an interactive governance mechanism, based on similar principles as the Polder model, albeit on a much larger scale.

The first RAC to be established was for the North Sea (cf. Degnbol and Wilson 2008). Stakeholders of various sorts, be they the fishing industry, environmental organizations, consumer groups, women's networks or others, are represented. The RACs do not, however, have the same authority to make autonomous decisions as the Polder communities had. They are instead limited to providing advice to the Commission on fisheries regulatory matters. Broad participation is also meant to give legitimacy to the management process. Notably, the design and basic operating principles and experiences of the North Sea RAC (NSRAC) are relevant to MSP also because it has several sub-units, one of them devoted to MSP.

Not surprisingly, some of the seven RACs throughout Europe seem to work better than others. As for the NSRAC, according to Degnbol and Wilson (2008:193) "[it is] the relationship between the conservation non-governmental organizations (NGOs) and the industry that is the basic story about the NSRAC." According to these researchers, the latter tends to be the stronger partner-to the extent that the NGOs have been questioning their role and contribution. The frustration is not only directed at other stakeholder representatives as much as at the EU fisheries authorities. Fishing industry representatives often feel that their advice is not taken seriously into account and that they are therefore wasting time by participating (Ounanian et al.2012).

Rather than concluding decisively as to their potential, however, RACs could be seen as a lesson for MSP. Although the founding governance principles are relevant, one cannot expect organizations like the RACs to work well if the design does not match their mandate and if the external support is not there. That may well also become the fate of MSP. As the RAC experience demonstrate, to plan and govern overlapping activities in seas that are shared by several countries is demanding. Brussels is too far away to be able to feel the pulse of what is going on at local scales. Marine governance in the form of multi-scale MSP is about trying to reconcile conflicting demands. Although some distance may be needed in order to see the broader ecosystem picture, such reconciliation cannot easily be achieved from a distance.

MARINE SPATIAL PLANNING DEFINED

MSP is defined as "a process of analyzing and allocating parts of three-dimensional marine spaces to specific uses, to achieve ecological, economic, and social objectives that are usually specified through the political process" (Douvere and Ehler 2009:78) The idea of MSP is to foster rational use of a shared marine space that is scarce, and to secure the sustenance of fragile ecosystems. Additionally, MSP endeavors to create more conducive interactions among marine user groups in a way that balances the demands for economic development, human wellbeing and environmental conservation.

While some are more advanced than others, all of the countries that border the North Sea are now engaged in MSP in one way or other. Jay and colleagues (2012b) refer to MSP as "the new frontier." Although MSP holds great promise and widely adopted, it is hardly a panacea. It is a process that is socially, institutionally, and legally, and therefore also politically, intertwined, which sets the stage for political ambition that may cause conflict as interests and worldviews differ among the involved parties.

Marine spatial planning as a wicked problem

MSP is applied to social and ecological systems (SESs) that are inherently diverse, complex and dynamic, and work at multiple scales. Accordingly, MSP deals with problems that Rittel and Webber (1973) termed "wicked" (cf. also Jentoft and Chuenpagdee 2009). In such situations, one should not expect quick fixes; the problems are not expected to go away anytime soon. As wicked problems they are not solved once and for all but need constant attention, and it is hard to determine if and when they are solved, and for how long. They can at best be "tamed." Solutions are less than ideal and never given. Instead they are negotiated among stakeholders who tend to have different ideas about what the problem is. Decisions would therefore require compromise if agreement cannot be made.

In fact, the latter makes MSP a wicked problem in itself. As such it makes sense in theory, but is onerous in practice. As marine space gets more and more congested by the day, users find themselves in a situation where more space for one means less for others. Agreeing on how to allocate space among different user groups requires negotiation and cooperation. A complicating factor is that cross-border interaction in these spaces is unavoidable. For instance, fishing fleets are mobile and cannot easily be fenced in. While the fishing grounds may be stationary, the fishers themselves must move between varying areas depending on the fishing season and the species targeted. Likewise, oilrigs and marine aquaculture installations situated on fishing grounds or near shipping lanes may interfere with each other trigger conflict. This also frequently happens when marine protected areas (MPAs) interfere with fishing patterns.

NATURE CONSERVATION AND MARINE PROTECTED AREAS

As Jay and colleagues (2012c) observe, MSP has its origin in marine nature conservation and is a logical extension of marine protected areas (MPAs). The two share the ecosystem perspective and a holistic and integrated approach (Kidd et al. 2011). Experiences from the implementation of MPAs should therefore be useful to MSP.

In the EU, Natura 2000 creates a network of protected areas throughout Europe based on the criteria outlined in the EU Habitat and Bird Directives (see Fock 2011). In the case of the North Sea, nature conservation is integral to MSP. From a spatial perspective, it is not the number of MPAs *per se* that is of importance, but instead their size and regulatory content. In Germany, about 45 percent of the waters in the North Sea and the Baltic Seas are marine protected areas (MPA) under the Natura 2000 network[a]. This is impressive when compared to Scotland, where just 4 percent of the territorial waters are currently covered by Special Areas of Conservation (SAC) designation. There are no Special Protection Areas (SPAs) designated entirely in the Scottish marine environment[b]. The Dogger Bank, shared by Germany, the Netherlands, the UK and Denmark is renowned as an important fishing area and nursing ground, but it is also the focal point of a new wind energy development program. Falling under the category "*sandbanks slightly covered by seawater all the time*", areas of the Dogger Bank are fit to fall under the Habitats Directive to become a Special Area of Conservation (Degnbol 2012).

MPAs (and MSP) are meant to be proactive mechanisms of area-based ecosystem management (Douvere and Ehler 2008). According to the precautionary principle lack of information is no excuse for inaction. All seven North Sea countries have ratified the Convention on Biological Diversity (CBD) and therefore this principle governs by implication. Yet, things are more difficult to implement in reality than they initially seem on the surface. MPAs cannot be easily imposed from the top down. Its development requires interactive processes at the regional scale. In Belgium this lesson was learned the hard way. As noted by Bogaert et al. (2009), the initial MPA designation process was top-down and received local resistance, and therefore failed. It had to be terminated, and could not be resumed before a more interactive process involving stakeholders was initiated six years later. This would also be a lesson. There is little reason to assume that MSP would be any different.

MPAs are most needed when numerous and cumulative activities expose risks to the environment. This is usually the situation near the coast and in limited ocean spaces such as the North Sea. It is not the same as when a MPA is established in the middle of the Atlantic where human presence is considerably limited. The rule seems to be that the easier their implementation due to minimal local resistance, the less is the need for them (Gilliland and Laffoley 2008). Marine ecosystems do not need MPAs if humans do not tamper with them, but that seems to be the case to a diminishing extent.

User-conflicts are likely to magnify and intensify as more and more MPAs are created. Uses for the same space are often incompatible and, if ways for different uses to co-exist cannot be found, it remains logical they must be kept apart. No wonder therefore that MPAs have received mixed support by fishers who find them as an intrusion into their freedom of access and movement (Jentoft et al. 2007). Fishers have their own precautionary principle in place (Degnbol and Wilson 2008). In the face of uncertainty, they do not easily put their livelihood or community at risk. That is also one reason why MPAs globally have received less support in the fishing population than many would have expected, and why they have been much slower in their implementation than originally anticipated (Chuenpagdee et al. 2013).

When fishers talk about MPAs and MSP, issues of exclusion are of primary concern. They are often told to fish somewhere else, while the recreational sector enjoys continued access. Such discrepancies in conservation policy need not be necessary, however. In fact, nature conservation instruments can allow multiple uses, including fishing under some form of regulation. Fishers are therefore not always opposed to nature conservation, and might find that it may actually work in their best interests. It is argued, thus, that nature conservation should be treated "as a means to achieve good environmental status of the ecosystem, based on its spatial, economic and conflict resolution dimensions" (Zacharoula et al. 2013).

Those affected are therefore better consulted or directly involved–as in the Polder model. The stakes are high for those who have much to lose. Fearing displacement, the fisher may invariably be skeptical, even when the plans for change promise a brighter future (Agardy et al.2011). One tends to be less apprehensive when part of the process and has the opportunity to learn what is happening and why. Conversely, one is also inclined to be more positive when heard and respected.

FISH AND FISHERIES

With more than 230 fish species, the North Sea is a rich fishing area. Norway, the UK, Denmark, Germany and The Netherlands each have a high presence in the North Sea fisheries, harvesting hundreds of thousands of tons every year, and in some instances with the consequence of severely reducing the biomass to a level where the International Council for the Exploration of the Sea (ICES) have advised against continued fishing. The North Sea cod was at an historical low in 2006. Despite quota regulation, by-catch and the discard resultant from mixed fisheries continues to be a problem. The same is the case with "ghost fishing", when gear are lost in the sea but continues to fish. The good news according to the Norwegian Institute of Marine Research in Bergen, however, is that herring shows a more positive development lately and the spawning stock is now in a better shape, well above the precautionary reference point. The same is true for the haddock, but less so with saith (ICES 2011).

The fishing industry is highly diversified with different kinds of users. As in all countries with a coastline, North Sea fishers make a living from harvesting marine resources using technology at various scales. In the UK, for instance, the commercial fishing fleet is largely made up of small boats. As of 2010, 78 percent of licensed vessels are smaller than 10 meters (cf. Cardwell 2012). In comparison, the Dutch fleet is of considerably larger scale[c]. In most instances, the smaller the vessels, the closer to shore they fish, the more simple technology they employ, and the smaller their ecological footprint. There is, however, an extensive large scale fishery on the North Sea, which is responsible for a major part of the total harvested volume of fish. Bottom trawls, typically employed by large scale fishing vessels, are harmful to the sea floor benthos, such as coral reefs and other marine life on the seabed. So-called beam trawls, a popular fishing method in Germany, Belgium and the Netherlands, have therefore been highly controversial, pitting biologists and ecologists up against the fishing industry (see van Ginkel 2009).

Fisheries have an ancient presence in the North Sea, and therefore a cultural heritage to maintain. In recent decades new marine industries have appeared, one of them aquaculture. Salmon raised in pens in the sea is now a major industry in Norway and the UK. For Scotland, for instance, (which today accounts for 90 percent of the UK production), the sky seems to be the limit. Scotland's National Marine Plan aims at an increase of finfish farmed at a rate of 4 percent per annum, which should lead to a 50 percent increase by 2020 (Scottish Government2011). The expected rise of farmed shell fish, especially mussels, will likely experience a similar expansion. According to the most recent available statistics, the Netherlands is farming oyster mussels, eel, catfish and tilapia at over 4000 tons per year[d]. This is about one tenth of what Denmark produces[e].

OTHER STAKEHOLDERS AND USE(R) S OF MARINE SPACE

Fishers find themselves in an increasingly congested area where space is getting increasingly scarce. This occurs both on land, on the coast and offshore due the expansion of existing usages as well as the introduction of new ones. As economic growth by exploiting marine resources, there is need to carefully plan for the management and conservation marine ecosystems. For fisheries this is no easy challenge, but they cannot anymore be looking inwards; they must also have an eye for what is happening in their environment, and consider who the other stakeholders are, what their particular stakes are, and how they operate.

Oil industry

In the 1950s, few believed that the North Sea contained a large petroleum potential. This changed when the Dutch Slochteren field was discovered, and the North Sea became what Yergin (1991) calls, one of the greatest investment projects in the world and a "technological marvel of the first order." Norway, the UK and the Netherlands are the major players in North Sea oil and gas, with the first topping in volume of oil produced. Apart from the potentially harmful environmental effects, oil and gas activities have direct spatial implications[f]. In addition to the large number of actual oilrigs, there are security zones around the platforms (500 meters around the platform where marine traffic such as fishing is banned) and thousands of kilometers of pipelines and cables over the sea bed, which in many instances have caused problems for trawl fishing[g].

Shipping and military operations

To the list of stakeholders in the North Sea we must add shipping and military operations. Every year, roughly 34,000 sea-going vessels call at the port of Rotterdam. Many of the North Sea countries, namely France, the UK, Belgium, and, leading them all, the Netherlands, have pronounced sand mining industries. With sea-levels rising, sand mining will likely increase considerably. Estimates for the Netherlands suggest that sand mining will soon reach approximately 85 million cubic meters per year[h].

Coastal communities and recreation

As noted in the Dutch Water Management Plan, the coast is an attractive place to live, work and to enjoy leisure time[i]. The "lure of the sea" (Corbin 1994) is only getting stronger and it happens everywhere, giving rise to a booming recreational industry. Numerous coastal communities reside on and around the shores of the North Sea, and are where people have lived for ages, drawing their livelihoods from the exploitation of marine resources such as fish. It is to these communities one would go to experience our common cultural marine heritage. People in these communities have their own interests to defend, and they may therefore regard the new developments happening along the coast as a threat. They do not, for instance, always like the new wind farms distorting their sunset view of the horizon.

Wind farms

The move towards a greener economy based on renewable energy has its own spatial demands. Wind parks at sea are now being introduced at an increasing tempo in all countries surrounding the North Sea. The Directive on Renewable Energy of 2009 established a binding target to supply 20 percent of the total EU energy consumption from green energy by the same year. Thus offshore wind parks in the North Sea have the potential to meet a large share of Europe's energy demand. With consistent wind conditions at sea and the fact that the North Sea is pretty shallow, the costs would be relatively low (Schillings et al. 2012). Wind parks are also discussed in the context of economic development. For the United Kingdom, for instance, the estimate is that they will create 30 000 jobs by 2020[j].

Despite promise of a greener energy source, wind parks raise important concerns. One of these concerns pertains to the safety (Jay 2012). Due to their porosity, wind farms are also more dangerous to pass by than oilrigs. Fishing or recreational vessels that trespass these farms may suddenly appear from among the turbines. In bad weather they are not easily spotted, as they may be difficult to distinguish on the radar.

MSP FROM A FISHERIES PERSPECTIVE

Pioneered by European countries, most of which are currently implementing it in one way or another, MSP offers an approach to better marine governance, also within a fisheries context as it builds on the ecosystem perspective now central to fisheries governance. MSP is as challenging within and between sectors domestically as it is at regional scales[k]. Accordingly, North Sea MSP and fisheries governance retain both a vertical and horizontal dimension. Although MSP in the North Sea brings forward many challenges, there are obviously lessons to be learned that have wider implications. The same is true for fisheries governance in the context of MSP.

When areas of common usage are congested, frictions between different user-groups are bound to occur. It should therefore come as no surprise that competition for marine space is often described by invoking the language of war. Different groups are exploiting different resources and, accordingly, have different spatial needs that are often incompatible. However, peaceful co-existence is often possible. For instance, the Dutch MSP approach has been undergoing revision (IDON 2011) towards multiple uses of the same area. In the North Sea, exclusion through zoning is now primarily seen as a MSP measure of the last resort, after the multiple use option is exhausted. However, the cumulative effect of multiple uses on the environment remains a central concern. Resource depletion, pollution, habitat loss, biodiversity decline have increasingly become pressing issues. Protection of marine areas that are vulnerable to human interference is still a relevant measure in many instances.

Essentially, ocean space is not a *tabula rasa* absent of pre-existing norms, rights and regulations, and "Mare Liberum" does not guarantee full freedom. The usage of marine space and fisheries resources is already heavily influenced by management interventions. Thus, when stakeholders negotiate on how to divide or share space, they do not start from scratch and negotiate outside context. They are ready to defend their interests with all the powers they can muster, including the existing institutional mechanisms if they work to their favor.

Marine legislation and management traditionally follow a sectoral approach, such as for fisheries specifically. They are not designed to fit the whole system, nor do they always facilitate cross-sector interactions. As a consequence, when aggregated to a regional level, i.e. the North Sea, the entire governance system may be less than optimal and prone to create conflict. With such fragmentation, fishers being a weaker party relative to other stakeholders may easily be unable to influence the process and protect themselves from being pushed aside. MSP is no guarantee of a fair and just process unless it is able to break this pattern and create avenues for a more holistic and equitable approach.

For instance, with the "spatial turn" (St. Martin and Hall-Arbor 2008) in marine governance, fishers find themselves in situations where they need to negotiate spatial claims with other user-groups, many of whom only recently arrived in their traditional area. Sometimes fishers are forced to yield, simply because they do not have the clout they need to set the stage. In other case, diplomacy may work. Interestingly, fishers from Urk, a community in the Netherlands, who were cited by Toonen and Mol (2013: 298), expressed:

"Of course, closing fishing grounds is hurtful. We are fishermen! For generations, our family has been fishing at the North Sea. But today's world is different, fishermen are no longer kings in a sole domain. We have to deal with all kinds of other marine activities demanding space. And we have to face that eNGOs, consumers and the whole public is watching us. Collaboration and finding compromises are the only way forward."

Here, Urk fishers realize that they have much to gain by engaging constructively in the process together with other stakeholders. It is this sense of diplomacy that the MSP probably requires. Even if, as Degnbol and Wilson (2008) argue, MSP is a new approach, one that is essential if institutions for ecosystem-based marine and fisheries management is to be established, fishers are likely to be hesitant if not vociferously antagonistic. They would put MSP in a larger context and see it as another step towards enclosure of the fishery commons.

As stated in a pamphlet (*Fishing on a Square Inch*) produced by the Dutch Product Board: "If developments continue at this rate, and everything points that way, our fishermen will have only a tiny fraction of the sea left at their disposal" (cf. Degnbol and Wilson 2008; 196-7).

However, there is also room for ambivalence. MSP involves risks but may also provide opportunities to reduce vulnerability. MSP can help to protect fishers against the intrusion of other user-groups. MSP is legitimized by the ecosystem approach to natural resource governance (Gilliland and Laffoley 2008; Douvere 2008), which is what fisheries also need. This approach makes scientists stakeholders in the cause (Knol 2010; Degnbol 2012; van Haastrecht and Toonen 2011), but also a potential ally. How much weight science will carry in the process is, however, an open question. Negotiating conflicting spatial claims is a political game where power counts and where other interests may have more influence than fishers and scientists at the end of the day. MSP is also a political issue, as Kidd and Ellis (2012:62) argue: "There is clearly a politics of MSP at local, national and international scales, which is currently largely hidden."

Use of maps as tool in conflict resolution

As Toonen and Mol (2013) illustrate in their article about the Dutch North Sea plaice fishery, being able to table good maps is crucial in many cases. Stakeholders who are unable to do so are more likely to be pushed aside by those who can and are more powerful. If fishers cannot provide a credible map that exemplifies their fishing grounds and spatial needs, they are more likely to be displaced when, for instance, a location for wind parks is being researched (Smith and Brennan 2012). A problem for fishers is that their geographical focus is less static than the focus of most other users. This also explains some of their resistance toward mapping in MSP. They fear that their much needed mobility will be reduced as they might become bound by their own maps. In the same way that maps might fence other user-groups out, they may also fence user groups such as fishers in. Thus a map is not just neutral representation of land-or seascape features and uses. They have economic, social and political consequences that raise issue of social justice. As observed by Nietschmann (1997): "Maps are power. Either you will map or you will be mapped."

The role of mapping in MSP is in itself a research issue (Hersoug and Johnson 2012; Knol 2011; Smith and Brennan 2012). New geo-technology, such as GIS (Geographical Information Systems) that affords users the ability to merge and analyze different datasets that exist within different sectors, is an important facilitator of MSP. In fact, such new geo-technologies make MSP both possible and desirable. However, it is important to recall that maps can be biased; some things are highlighted while others are left out. Based on experiences on the east coast fisheries of the US, St. Martin and Hall-Arber (2008: 779) talk about "the missing layer" in marine mapping using GIS. In particular, they refer to an absent social and human dimension:

"Neither the complexity of human communities nor their relationships to locations and resources at sea are represented in current data collection initiatives despite the insistence that marine ecosystems (and ecosystem-based science and management) include human processes, impacts, knowledge, and needs." It rests upon the government, as the representative of the common interest, to ensure that the maps are equally empowering to all stakeholders and that inherent biases are minimized. This would require a more systematic effort in generating economic and socio-cultural data about the missing layer. That may however not be as easy as one would think. As Degnbol and Wilson (2008) observe, in the case of the North Sea RAC, fishers often reluctantly provide the requisite information about their fishing practices and spatial movements. Yet, fishers can hardly have it both ways: like it or not, fishers have no other option than to be proactive in MSP. As a fisheries industry representative quoted by Degnbol and Wilson (same page) explains: "It is better to be on the train than being hit by it."

The subsidiarity principle and the issue of scale

As spatial conflicts are about interests, norms and values, they do not dissipate easily. More scientific knowledge about the SES does not necessarily help. Conversely, it tends to influence the balance of power, as in the case of the MPA initiative in Belgium where "the authorities used scientific knowledge as an argument of power" (Bogaert et al. 2009:881). For this reason, MSP must facilitate a governance process that is interactive and involves stakeholders in the planning process (Kooiman et al. 2005; Ritchie and Ellis 2010; Pomeroy and Douvere 2008; Gopnik et al.2012). This is not to say that government does not have an important role to play. In fact, certain actions, like legislation, can only be taken by the government. Moreover, spatial conflicts that cannot be resolved by stakeholders require a higher level of authority for appeal.

The "subsidiarity principle" (van Kersbergen and Verbeek 2004 as stressed by the EU Commission in 2008 in a road map for MSP (EU 2008). Subsidiarity is a guiding principle in the Dutch Water Management Plan: "Decentralized whenever possible, centralized whenever needed". The same cannot apparently be said of the CFP, which Degnbol and Wilson (2008:189) characterize as "perhaps the most top-down fisheries management system on the planet." The challenge is to find agreement on what is "possible" and when things are "needed". Neither the higher nor the lower level can be expected to easily let their power go. What appears to be impossible, however, can sometimes be made possible, for instance through institutional innovation. Thus, the formation of the RACs for fisheries management has provided opportunities for decentralization and stakeholder involvement, including the fishing industry, that did not exist before. Subsidiarity is therefore also about creating opportunities (Kooiman2003), which in turn is fundamentally about governability–i.e. the capacity to govern (Kooiman2008)–a facet that can be enhanced through MSP given its particular institutional designs that because of the multi-scalar issues involved requires governance mechanisms that for the North Sea cut across national borders and sector boundaries.

Today, MSP is mostly carried out at the national level, which leaves issues at local and supra-national levels in an untouched void. MSP also needs to happen at the level of the local community to avoid conflict and to ensure that stakeholders are involved in the process. MSP cannot just deal with issues that are broad; it also must pay attention to detail specific to a particular place. Henry Mintzberg suggests a governance principle that would also apply to MSP; "the more complex the environment, the more decentralized the structure (Mintzberg1983: 138). This has also been applied for the co-management of fisheries (Jentoft 2007). The reason for this is further elaborated by Pressman and Wildavsky (1984: 205), who argue that "the closer one is to the source of the problem, the greater is one's ability to influence it, and the problem solving ability to complex systems depends not on the hierarchical control but on maximizing discretion at the point where the problem is most immediate."

Depending on what the problem is, this may call for decentralization of decision-making authority. In some instances the problem is mostly felt at higher scales at which the decision-making capacity may be low. Such is the case when several countries share a management problem and there is no institution that binds them together and facilitates collective action. Marine ecosystems are not always aligned with national boundaries, and fish stocks pass from one country's EEZ to another. Therefore there is a need for both MSP and fisheries governance at a regional scale. With the establishment of the RACs, there may be opportunities for decentralization of fisheries management functions that are not currently put to use, Lack of institutions such as RACs may make fisheries ineffective and therefore vulnerable to MSP initiatives, given the anticipated play of power that such a planning initiative may invite to.

Coordination and cooperation at regional scales is supported by the Marine Strategy Framework Directive (adopted in June 2008), which aims at improving the environmental status of Europe's seas. Dogger Bank is a case in point. Countries evaluated (Germany, Belgium and the Netherlands) by the MASPNOSE project do not readily share information with each other, and not all are equally accommodating to stakeholder involvement[1]. According to this project's findings, the Netherlands seems to be more prepared to do so than Belgium and Germany, perhaps a testament to the culture of the Polder model that again has influence here, but efforts on behalf of all the involved countries are essential.

CONCLUDING REMARKS

With a congested North Sea and the developments that can be seen on the horizon, the stakes are high, not at least for those whose livelihoods depend on the harvest and use of natural resources such as fish. The North Sea needs to improve its environmental status and the health of its fish stocks as some of experiencing major difficulties due to overfishing and environmental degradation. From this perspective, MSP and its ecosystem-based approach seems to be "an idea whose time has come" (Gilliland and Laffoley 2008: 787). It may not be the only answer to these challenges, but is likely to an essential part of it.

Still there is a need to be cautious. MSP may not appear to be a win-win. On the contrary, for the least powerful stakeholders, those who are poorly represented and who have few resources to back up their claims, they risk being ignored in the planning process. For them MSP may add further pressure and result in a loss. As illustrated in this paper, there is reason to assume that fishers and their communities may be within this group of stakeholders. MSP is a governance mechanism that may provide a more secure tenure but it also involves the risk of losing it, for instance as a result of spatial mapping. They have therefore reason to be ambivalent regarding this new tool and how it works from a fisheries governance perspective.

Leaving stakeholders to fight for space on their own is a recipe for social and ecological failure. It would also have implications regarding social justice as some user-groups do not only have more urgent needs and stakes than others, but their powers to protect their interests, define the problems, and shape the planning process are not equally distributed. Fisheries have large stakes but relative to other users such as the oil or shipping industry, they are a small player risking to be displaced. Therefore, spatial ordering through zoning is necessary also as a protective measure, but territorial boundaries cannot always be impenetrable, especially for fishers who need to remain mobile.

Yet, as Douvere (2008) has argued, MSP cannot just be about zoning, it must also be about negotiating interactions among user groups within complex SESs that are highly diverse, multi-scalar and dynamic, and with targets that are mobile and fluid due to natural as well social characteristics. This is what makes the marine environment different from its terrestrial counterpart. In dealing with dynamic systems that are complex, diverse and multiscalar, governance systems must assume similar traits to be up to the task (Duck 2012; Kidd and Ellis2012 Kooiman et al. 2005). MSP is no exception from this rule.

As far as the North Sea is concerned, with the institutional and cooperative frameworks backing up governance at both the national and EU levels and for particular sectors such as fisheries, there is reason to believe that MSP has a greater chance of succeeding here than in many other regions of the world. Still, the need for improved coordination of MSP efforts at all scales and across sectors remains a priority for the North Sea. The social and economic dimensions need attention, and stakeholder participation must be facilitated and playing fields levelled not only in the interest of justice and democracy, which are important principles in themselves.

As Mintzberg (1983) and Pressman and Wildavsky (1984) observe, this is also fundamentally about effective governance and the need to situate decision-making authority at most appropriate level. Some issues must certainly be dealt with at higher scales, such as for the North Sea as a whole, but not every issue can be handled from far away. This is now recognized within the EU, where the subsidiarity principle is adopted albeit not systematically-and certainly not at the local level as compared with the national level (van Kersbergen and Verbeek 2004).

These elements in turn require novel institutional designs, somewhat along the lines of the Regional Advisory Council system in fisheries and the Polder model originally for water management and an emphasis on the criteria of "good governance" across the board. Neither of these institutions is perfect, as no governance system is, but they do build on sound and well-proven governance principles that continue to be relevant and effective and have broad application. It is these principles, the values and norms that underpin them and the concrete organizational solutions they have led to, that are worth considering in MSP and fisheries governance for the North Sea, for other marine regions in Europe, and elsewhere.

ENDNOTES

[a] http://www.birdlife.org/eubiodiversityreport2012/?page_id=1888.

[b] http://www.scotlink.org/files/policy/ParliamentaryBriefings/LINKfilenoteMPA1Oct2011.pdf.

[c]Exact comparison is difficult as fishing vessels in the Netherlands are registered by their engine horsepower. In 2009 there were 308 cutters, 14 trawlers, and 193 miscellaneous. A third of the cutters are between 24-60 meters. Others are smaller but most of them larger than 10 meters presumably.

[d] http://www.compendiumvoordeleefomgeving.nl/indicatoren/nl1538-Viskweek.html?i=20-110.

[e] http://webfd.fd.dk/stat/Akvakultur_tab/prod_art_12_eng.html.

[f]For example, in 2007, there were 143 oil and gas platforms in the Dutch part of the North Sea. These include drilling rigs, service platforms and platforms that are no longer in use, but have not been dismantled yet. Around 130 production locations are in use, of which the major part is related to gas winning. Most of the platforms are located in the central part of the Dutch Continental Shelf. (http://www.noordzeeloket.nl).

[g]http://www.statoil.com/no/EnvironmentSociety/Environment/impactassessments/fielddevelopments/Downloads/Utbygging%20av%20Valemon%20-%20Konsekvenser%20for%20fiskeri.pdf.

[h] http://www.noordzeeloket.nl.

[i] http://www.government.nl/issues/water-management/water-quality/towards-better-water-quality.

[j] https://www.gov.uk/government/news/multi-million-pound-investment-in-offshore-wind-industry-to-unlock-billions-in-uk-economy.

[k]In this paper, we talk about marine spatial planning instead of maritime spatial planning (as used by the European Commission.) Despite slightly different emphases, both concepts center around the spatial regulation of *human activities* in a complex and dynamic marine environment. We use *marine* spatial planning, which is most widely used in the scholarly debate."

[l]The multi-national MASPNOSE project is looking at the entire North Sea and funded by the DG MARE, including the Dogger Bank issue. http://www.wageningenur.nl/en/show/Maspnose-Maritime-spatial-planning-in-the-North-Sea.htm.

ACKNOWLEDGEMENTS

This paper was first prepared for the symposium "The North Sea; the Past, the Present and the Future", held at the Royal Palace in Amsterdam on November 2, 2012. We are grateful for constructive comments by Maria Victoria Solstrand, Glen Smith, and two anonymous reviewers.

AUTHORS' CONTRIBUTIONS

Both authors carried out research for this paper and contributed with written text. SJ had the responsibility for the overall structure. Both authors read and approved the final manuscript.

REFERENCES

1. Agardy T, Di Sciara GN, Christie P: Mind the gap: addressing the shortcomings of marine protected areas through large scale marina spatial planning. *Marine Policy* 2011, 35: 326–232.
2. Bogaert D, Cliquet A, Maes F: Designation of marine protected areas in Belgium: a legal and ecological success? *Marine Policy* 2009, 33: 878–886.
3. Cardwell E: Invisible fishermen: the rise and fall of the British small boat fleet. In *European fisheries at a tipping-point*. Edited by: Højrup P, Schriewer K. Murcia: Universidad de Murcia; 2012:367–411.
4. Chuenpagdee R, Pascual-Fernandéz J, Szelianszky E, Alegret JL, Fraga J, Jentoft S: Marine protected areas: re-thinking their inception. *Marine Policy* 2013, 39: 234–240.
5. Corbin A: *The lure of the sea: the discovery of the seaside in the western world, 1750–1840*. Univ of California Press; 1994.
6. Degnbol D: *Slightly covered all the time: the science of sandbanks on the Dogger Bank. PhD thesis*. Aalborg University: Department of Development and Planning; 2012.
7. Degnbol D, Wilson DC: Spatial planning on the North Sea: a case of cross-scale linkages.*Marine Policy* 2008, 32: 189–200.
8. Douvere F: The importance of marine spatial planning in advancing ecosystem-based sea use management. *Marine Policy* 2008, 32(5):762–771.
9. Douvere F, Ehler C: Introduction. *Marine Policy* 2008, 32: 759–761.
10. Douvere F, Ehler C: New perspectives on sea use management: initial findings from European experience with marine spatial planning. *Journal of Environmental Management* 2009, 90: 77–78.
11. Drankier P: Embedding maritime spatial planning in national legal frameworks. *Journal of Environmental Policy and Planning* 2012, 14(1):7–27.
12. Duck RW: Marine spatial planning: managing a dynamic environment. *Journal of Environmental Policy and Planning* 2012, 14(1):67–79.
13. EU: *Roadmap for Maritime Spatial Planning: Achieving common principles in the EU*. Brussels: Commission of the European Communities; 2008.
14. Fock HO: Natura 2000 and the European common fisheries policy. *Marine Policy* 2011, 35: 181–188.
15. Gilliland PM, Laffoley D: Key elements and steps in the process of developing ecosystem-based marine spatial planning. *Marine Policy* 2008, 32: 787–796.
16. Glasbergen P: The green Polder model: institutionalizing multi-stakeholder processes in strategic environmental decision-making. *European Environment* 2002, 12: 303–315.
17. Gopnik M, Fieseler C, Cantral L, McClellan K: Coming to the table: early stakeholder engagement in marine spatial planning. *Marine Policy* 2012, 36: 1139–1149.
18. Halpern BS, Walbridge S, Selkoe KA, Kappel CV,

Michelli F, D'Agrosa JF, Bruno KS, Casey C, Ebert HE, Fox R, Fujita D, Heinemann HS, Lenihan EMP, Madin MT, Perry ER, Selig M, Spalding RS, Watson R: A global map of human impact on marine ecosystems. *Science* 2008, 319: 948–952.

19. Hersoug B, Johnson JP (Eds): *Kampen om plass på kysten: interesser og utviklingstrekk i kystsoneplanleggingen*. Oslo: Universitetsforlaget; 2012.
20. ICES: *Report of the ICES Advisory Committee: Book 6*. Copenhagen: North Sea; 2011.
21. IDON: *Integrated Management Plan for the North Sea 2015 (Revision)*. Rijswijk, the Netherlands: Interdepartmental Directors' Consultative Committee North Sea (Interdepartementaal Directeurenoverleg Noordzee-IDON); 2011.
22. Jay S: Marine space: manoeuvring towards a relational understanding. *Journal of Environmental Policy and Planning* 2012a, 14(1):81–96.
23. Jay S, Ellis G, Kidd S: Marine spatial planning: a new frontier? *Journal of Environmental Policy and Planning* 2012b, 14(1):1–5.
24. Jay S, Flannery W, Vince J, Liu W-H, Guifang Xue J, Matczak M, Zaucha J, Janssen H, Van Tatenhove J, Toonen H, Morf A, Olsen E, Suárez de Vivero JL, Rodríguez Mateos JC, Calado H, Duff J, Dean H: Coastal and Marine Spatial Planning: International Progress in Marine Spatial Planning. *Ocean Yearbook 27* 2012c, 171–212.
25. Jentoft S: Limits of governability? Institutional implications for fisheries and coastal governance. *Marine Policy* 2007, 31: 360–370.
26. Jentoft S, Chuenpagdee R: Fisheries and coastal governance as a wicked problem. *Marine Policy* 2009, 33: 553–560.
27. Jentoft S, Van Son T, Bjørkan M: Marine protected areas: a governance system analysis.*Human Ecology* 2007, 35: 611–622.
28. Kidd S, Ellis G: From the land to the sea and back again? Using terrestrial planning to understand the process of marine spatial planning. *Journal of Environmental Policy and Planning* 2012, 14(1):49–66.
29. Kidd S, Shaw D: Reconceptualising territoriality and spatial planning: insights from the sea.*Planning Theory & Practice*. 2013, 14(2):180–197.
30. Kidd S, Plater A, Fried C (Eds): *The ecosystem approach to marine planning and management*. London: Earthscan; 2011.
31. Kjær AM: *Governance*. Cambridge: Polity Press; 2004.
32. Knol M: Scientific advice in integrated ocean management: the process towards the Barents Sea plan. *Marine Policy* 2010, 34(2):252–260.
33. Knol M: Mapping ocean governance: from ecological values to policy instrumentation.*Journal of environmental management and planning* 2011, 54(7):979–995.
34. Kooiman J: *Governing and governance*. London: Sage Publication; 2003.
35. Kooiman J: Exploring the concept of governability.*Journal of Comparative policy analysis* 2008, 10(2):171–190.
36. Kooiman J, Bavinck M, Jentoft S, Pullin R (Eds): *Fish for life: interactive governance for fisheries*. Amsterdam: Amsterdam University Press; 2005.
37. Mintzberg H: *Structures in fives: designing effective organizations*. Englewoods Cliffs: Prentice-Hall International, Inc.; 1983.
38. Nietschmann BQ: "The Making of the Maya Atlas". In *Maya Atlas*. Berkeley, CA: North Atlantic Books; 1997:136–149.
39. Ounanian K, Delaney A, Raakjær J, Ramirez-Monsalve P: On equal footing: stakeholder perspectives on the marine strategy framework directive as a mechanism of the ecosystem-based approach to marine management. *Marine Policy* 2012, 36: 658–666.
40. Pomeroy R, Douvere F: The engagement of stakeholders in the marine spatial planning process.*Marine Policy* 2008, 32(5):816–822.
41. Pressman JL, Wildavsky A: *Implementation*. Berkeley: University of California Press; 1984.
42. Ritchie H, Ellis G: A system that works for the sea? Exploring stakeholder engagement in marine spatial planning. *Journal of Environmental Planning and Management* 2010, 53(6):701–723.
43. Rittel H, Webber M: Dilemmas in a general theory of planning. *Policy Sciences* 1973, 4: 155–169.
44. Schillings C, Wanderer T, Cameron L, van der Wal JT, Jacquemin J, Veum K: A decision support system for assessing offshore wind energy potential in the North Sea. *Energy Policy*2012, 49: 541–551.
45. Schreuder Y: The Polder model in Dutch economic and environmental planning. *Bulletin of Science, Technology & Society* 2001, 21(4):237–245.
46. Scottish Government: *Scotland's National Marine Plan–Pre-Consultation Draft*. 2011. Available from: http://www.scotland.gov.uk/Publications/2011/03/21114728/0 Available from:
47. Smith G, Brennan R: Losing our way with mapping: thinking critically about marine spatial planning in Scotland. *Ocean and Coastal Management* 2012, 69: 210–216.
48. St. Martin K, Hall-Arber M: The missing layer: geo-technologies, communities, and implications for marine spatial planning. *Marine Policy* 2008, 32: 779–786.
49. Toonen H, Mol AP: Putting sustainable fisheries on the map? Establishing no-take zones for North-Sea plaice fisheries through MSC certification. *Marine Policy* 2013, 37: 294–304.
50. Van Ginkel R: *Braving troubled waters: sea change in a Dutch fishing community*. Amsterdam: Amsterdam University Press; 2009.
51. Van Haastrecht EK, Toonen HM: Science-policy interactions in MPA site selection in the Dutch part of the North Sea. *Environmental Management* 2011, 47(4):656–670.
52. Van Hoof L, Van Tatenhove J: EU marine policy on

the move: the tension between fisheries and maritime policy. *Marine Policy* 2009, 33: 726–733.

53. Van Kersbergen K, Verbeek B: Subsidiarity as a principle of governance in the European Union.*Comparative European Politics* 2004, 2: 142–162.

54. Van Leeuwen J, Van Tatenhove J: The triangle of marine governance in the environment governance of Dutch offshore platforms. *Marine Policy* 2010, 34: 590–597.

55. Woldendorp J, Keman H: The Polder model reviewed: Dutch corporatism 1965–2000.*Economic and Industrial Democracy* 2007, 28(3):317–347.

56. Yergin D: *The prize: the epic quest for oil, money, and power*. New York: Simon & Schuster; 1991.

57. Zacharoula K, Maes F, Rabaut M, Vincx M, Degraer S: The integration of nature conservation into the marine spatial planning process. *Marine Policy* 2013, 38: 133–139.

CHAPTER 8

FISHERS' POLITICS AND MARINE CONSERVATION

Vasudha Chhotray

Keywords: Fishers rights, Sanctuary, Environmentalism, Justice, Indigeneity, Territory, Refugees, Immigrants, India

INTRODUCTION

This paper critically examines the politics of a protracted conflict involving fishing communities in Odisha on the eastern coast of India. At the centre of the story is a tussle over fishing rights in the Gahirmatha marine sanctuary, notified by the Odisha state government in 1997. A key objective of the sanctuary was the protection of the endangered Olive Ridley turtle, which visits this coastline en masse for breeding from December until March. Created following little or no popular consultation with local fishers, the sanctuary has imposed a total ban on fishing. Those affected include Odiya and Bengali fishers, of different castes, who variously pursue both motorized and non-motorized fishing. The mainly Odiya owners of mechanized trawlers that come from other parts of Odisha also resent the sanctuary.

Conservation, for its advocates, is 'self-evidently an unimpeachable political value' (Jayal 2010: 69). Political values like rights and justice are extended in relation to the natural environment, making it *just* to demarcate protected areas and sanctuaries, and fence these away from human use. This viewpoint may disregard the question of justice for the people that are affected by conservation. If, as is often the case, the local communities in question are poor, marginalised tribals, or immigrants and refugees or other minorities, then the ethic of conservation is placed under even greater strain (Jayal 2010: 69).

The politics of justice in the context of conservation is fairly complex. Conservation is not simply a debate about nature versus people, but involves arguments between certain groups of people versus other groups of people (Low and Gleeson 1998; Saberwal and Rangarajan 2003). Who is speaking for which cause, which constituency do they support and what are the consequences for framing a particular environmental conflict or struggle in a particular manner are all pertinent. These questions may matter especially when there are conflicting claims to space and there is disagreement over who has priority in resource use, like in Gahirmatha.

Odiyas comprise the local population of the state, whereas the Bengalis- being referred to here- are immigrants. These comprise both those who claim to be from West Bengal and others who have travelled from East Pakistan or former Bangladesh at various points since 1947. Bengali settlers are

Citation: Vasudha Chhotray, Fishers' politics and marine conservation, DOI 10.1186/s40152-016-0043-3.

scattered in numerous villages and hamlets of Rajnagar and Mahakalpada blocks of Kendrapara district of Odisha. Many of these are adjacent to the sanctuary and neighbouring estuarine areas. Odiyas and Bengali speak related languages, they share many cultural similarities and importantly, the Bengali immigrants here are principally Hindus, like the majority Odiya population. Many, if not all, have acquired citizenship documents over the decades.

The paper centres on a detailed case study of Matsyapalli,[1] a large fishing village and gram panchayat[2] in Mahakalpada block in Kendrapara district, which is wholly Bengali with about 2000 households. Matsyapalli has been purposively selected as a case due to its special status in the Gahirmatha conflict for three reasons. One, it is located at a key point at the entrance to the sanctuary. It therefore serves as a jetty and is a nucleus of fishing activity in the area. Two, it figures prominently in the political organisation of fishers as the President of the Odisha Traditional Fishworkers' Union is from this village. Three, its residents are Bengalis, many of whom are immigrants from East Bengal (now Bangladesh). For all of these reasons, Matsyapalli provides an important window not only into the wider conflict between fishers' and the sanctuary, but also the social landscape of fishers' groups involved in this problem. In addition, the paper draws on additional case work done in two smaller inland neighbouring villages: Ambapalli, of approximately 500 households, and Narayanpur with about 300 households, also largely comprising Bengali immigrants but with some Odiya households too. The fishers here mainly practise river-based and estuarine fishing, which brings them into direct conflict with Odiya caste fishers from adjacent areas. The three villages together provide a unique perspective into fishing conflicts around the Gahirmatha sanctuary. These coastal villages are also flanked by Bhitarkanika Wildlife Sanctuary that was notified in 1975, with the Bhitarkanika National Park being designated from within its core area in 1998[3].

A purposive case study approach has allowed us to investigate the subject critically, using the case study villages (Matsyapalli, Ambapalli and Narayanpur) as the empirical centre of the problem while tracing other actors 'outwards' through an inductive interview process. We have not been interested in a survey approach across various fishing villages as we are not trying to obtain generalizable knowledge. Our chosen methods therefore support the principal research questions: 'How does a conflict between fishers and a sanctuary generate the question of legitimate resource rights? How is a regional Odiya environmentalism constructed, and how does it relate to the everyday social relations amongst coastal fishers living next to a sanctuary?'

Our primary research methods accordingly comprised focus group discussions (FGDs) and detailed key informant interviews with small fishers from Matsyapalli, Ambapalli and Narayanpur, who were both male and female, and a mix of Bengali and Odiya. In addition, we conducted key informant interviews with trawler operators (all of whom are Odiya), heads of fishers' cooperatives (both Bengali and Odiya), Odiya fishing caste leaders and fish godown operators (all Odiya). We interviewed retired and current government officials, NGO persons and environmentalists. Approximately ten FGDs and forty long interviews were carried out between November 2011 and May 2012. In considering the injustice of this particular conservation initiative at Gahirmatha through an analytical perspective of representation and framing, the paper hopes to engage directly with environmentalism, as a way of thinking about politics and society based upon an understanding of environmental problems (Sharma 2012), and the question of justice as recognition (Fraser1997, 2000; Schlosberg 2004; Young 1990). Recognition or misrecognition depends on meanings and values, albeit partial and fragile, prevailing within social and environmental discourse (Li 2004). Taking this as a starting point, the paper contains a detailed account of diverging representations of fishers' rights in relation to the conflict at Gahirmatha, and through it, the dynamic positions and stakes of powerful stakeholders.

The paper is organized in four sections. Section 2 relates justice as recognition to 'environmentalism'. It discusses eco-nationalism and eco-naturalism as increasingly popular strains of environmentalism in India. It also describes the significance of regional environmentalisms within the construction of eco-nationalism in a country as diverse as India. Section 3 sets out the state's view of fisheries development and the tensions in reconciling these with its conservation agenda. It also introduces the historical role of the state in enabling the particular regional environmentalism being described in this paper. Section 4 characterises life for the fishers living next to the Gahirmatha sanctuary, as evident from the case studies. It juxtaposes their everyday social relations against the complex Odiya environmentalism that seeks to blame Bengali immigrants for the decline of marine fisheries in the state, and seeks to expose the contradictions in the narrative being constructed. Section 5 focuses on the sanctuary itself, policing the borders of which allows a more abstract state to appear in material form. It is here, at sea, that some visible effects of such environmental thinking- however fragile and contradictory- can be observed; even as small fishers, both Odiya and Bengali, encounter state authority from within their disadvantaged position in the marine class hierarchy.

CONCEPTUAL FRAMEWORK

The conceptual framework to this paper is in two parts. The first briefly recapitulates the significance of justice as recognition, and discusses its relevance for assessing the effects of environmentalism in society. The second part describes environmentalism as eco-nationalism and eco-naturalism, in the Indian context, while focusing on the significance of regional imaginations of eco-nationalism. It also shows how territory, nature and people are brought together within an effective set of representations that justify particular modes of state intervention through conservation.

Justice as recognition, environmentalism and frames of meaning

Environmental justice scholarship has been marked by a long running debate between justice as distribution versus justice as recognition. Justice theorists Fraser (1997, 2000), Schlosberg (2004) and Young (1990) show us that while distributional issues—as highlighted by John Rawls- are crucial to a satisfactory conclusion, it is a mistake to reduce social justice to distribution. They argue that the politics of recognition inheres in all societies and contexts, and that justice demands substantive recognition within the political community. The concern here is beyond formal rights alone, though these matter significantly, for effective recognition within society is necessary in order to penetrate and contest marginalising discourses (Holston 2008). Moreover, the literature suggests there is a very strong relationship between justice as distribution and justice as recognition. When particular storylines get the backing of more vocal/dominant groups in society, and they enrol powerful institutions, then the outcomes may be substantively unjust (Hajer 1995 in Forsyth 2006).

Within the burgeoning literature on environmental justice, there is a keen sense of which meanings and values prevail within social and environmental discourse (Walker 2012). 'Whose justice' is a question that intensely engages scholars as they detect the pervasive power of particular discourses in marginalising groups. Baviskar et al. (2006) and Mawdsley (2004) have persuasively demonstrated how a 'middle class environmentalism' in Indian cities has adversely affected working class populations engaged in 'polluting' activities. Middle class environmentalism is based around a 'storyline' (to use a term by Hajer 1995 cited in Forsyth2006), which constructs a peculiar narrative about environmental degradation, with predefined notions of blame to certain social groups.

Such representation or framing inevitably involve simplification, even falsification, of history and the stereotyping and classification of populations, while serving to entrench powerful interests in society. And yet, we also know that the domination of social discourses is contingent upon dynamic power relations and fluid coalitions of interests (Roseberry 1994). Going further, there are continuous attempts at producing dominant frames of meanings, but there is also a constant possibility of their fracture, contestation and re-articulation (Li 2004).

Territory, nature, people, state: Eco-nationalism as environmentalism

The 'nature-for-nationalism' or 'eco-nationalism' breed of environmentalism is about the imagination of particular territorial landscapes through the invocation of specific national cultural symbols and historical continuities. It has deep roots in India, going back to colonial times when environmental imagery became an important means of developing a national consciousness against the British. The 'new traditionalist discourse' has continued to propagate a nostalgic vision of pre-colonial India where 'traditional ecology' was balanced by self-contained communities endowed with a special conservationist ethic (Sinha, Gururani and Greenberg 1997). Other strains of environmentalism also derived from particular uses of national environmental imageries, often in strikingly different ways, like Gandhian environmental thinking which challenged the new modernist projects of development.

In a plural society like that of India, there are many different national imaginings at work. Sivaramakrishnan and Cederlof (2012) have proposed the concept of 'ecological nationalism' to express how both 'cosmopolitan' as well as 'nativist' forms of 'nature devotion' may be articulated in terms of nationalistic pride. This may be of two types: cosmopolitan or 'metropolitan-secular' which claims to work in the interests of nation-building projects, and seeks to appropriate nature for the wider 'public' good, and nativist or 'indigenist-regionalist' which generally includes a reaction to development interventions by the state or global forces for their marginalising effects on the lives of local communities (Sivaramakrishnan and Cederlof2012: 7). Examples of the former include Jawaharlal Nehru, India's first Prime Minister, who expressed his vision of the imagined landscape of the Indian nation of secular modernism in relation to physical imagery like of the Himalayas or of various 'mighty rivers' (Sivaramakrishnan and Cederlof 2012: 30).

The latter however is distinguished for its assertions of indigeneity and claims to authenticity with respect to particular places, and various historical memories may be espoused to convey affinity with those environments. Adivasi assertions of indigeneity in parts of India (like in Odisha, in contestation against international mining projects; see Sahu 2008) may be an example. However, there is also much evidence of the misappropriation of nature, both symbolically and materially, by regionally dominant communities or particular social and cultural groups to promote their own agendas of regional control. Sivaramakrishnan and Cederlof give the example of the Roman Catholic Mukkuvar fisherfolk in Kanyakumari district in South India for their attempts to claim fishing rights through an assertion of belonging to the locality and the nation, and their support of the Hindu nationalist Bharatiya Janata Party in this regard (2012: 9).

In other quarters too, the popularisation of the concept of indigeneity amongst local groups has evoked concern, especially amongst anthropologists, who worry that the search for 'cultural authenticity' is quite pointless, and it is preferable to focus on the political strategies through which indigeneity is being articulated and claimed (Li 2000 as cited in Dove 2006). Scholars have also attacked the proponents of indigenous knowledge for over- emphasising cultural purity and continuity, and in the process, marginalising other poorer groups who may not be able to claim indigeneity defined in such ways (Dove 2006). Li has persuasively argued that 'indigenous identity is in any case a narrow target, which is easily over-shot or undershot' (2000 cited in Dove 2006: 194). This is not the place for an extended consideration of the contested nature of indigeneity, our limited point being that there is a seriously performative aspect to this notion, which has and can be deployed instrumentally by dominant elites for limited ends.

In India moreover, the complicating factor is also the sheer plurality of regional identities that mediate the host of possible ecological nationalisms. To borrow from Aloysius' provocative contention, 'there is in India, no nation' (1999 cited in Sivaramakrishnan and Cederlof 2012). There are examples galore of articulations of regional identities as mediated by geography, but simultaneously underpinned by some reference to their place as citizens of the Indian state (like in Jharkhand and Uttaranchal, both of which experienced ethno-regional movements for separate statehood). Regional ecological nationalisms refer to a pristine aesthetic of nature that emanates from a particular space, even as a larger construct of India as a sacred and inviolable territory is endorsed. In this regard, it is worth highlighting the convergence between eco-nationalism and Hindu religious chauvinisms. The imagining of national landscapes as unique is central to Hindu political thought, 'making the *natural* contours of the nation seem supreme' (Sharma 2012: 30, italics added for emphasis). Hindu nationalism is increasingly shaping mainstream environmental thinking in India, with the use of various popular discourses to censure foreign domination, attack immigrants and challenge Muslims and Christians (Sharma2012).

In fact, it may be argued that conservation itself represents a form of metropolitan-secular eco-nationalism, wherein a particular imagination of the national territory is translated into concrete state policy. Indeed, state creation itself draws on a longer historical narrative of modernity and progress, which in turn rested on a 'fundamental reordering of society in space' (Scott 1998 cited in Neumann 2004: 201). Colonial states across Asia and Africa legitimised sole and exclusive territorial control over all 'vacant lands' exclusive of those on recognized private land (Peluso and Vandergeest 2001). Indeed, 'states come into being through these claims and the assertion of control over territory, resources and people' (Neumann 2004: 202).

The Indian state has followed an approach to conservation precisely interpreted as 'bounding nature' away from human use that was popularised in the United States. This involved the creation of national parks, often involving the heavy use of state fiat and the deployment of particularly strategic ideas around science and expertise (Rangarajan 2001). India enacted the Wildlife Protection Act in 1972, and established 92 national parks and 500 wildlife sanctuaries, as well as 33 marine sanctuaries and national parks, designated together as marine protected areas (Singh 2002). Through conservation, states were reconstituting peoples' relationships with nature under the garb of a civilising mission, and India was no exception. The setting out of strict norms of citizenry, of how people ought to behave with respect to the modern state, was also a part of this project.

In fact, conservation produced a peculiar construct of a national citizen with responsibilities but no rights. People were expected to live responsibly around parks, foregoing or greatly curtailing their resource use in the interests of nature, which the conservation paradigm artificially divorces of human beings. Yet they did not have commensurate rights, like the right to be informed following the notification of a national park or sanctuary, and the right to represent their claims (Jayal 2010). There have also been more overt cases of coercion, like through the infamous patrolling approach of 'guns and guards' followed by the forest department officials to keep people out of parks. Yet, very often in practice, either the forest department has limited capacity to patrol effectively or forest guards are complicit in villagers' entering the park area in return for money, which may be benignly obtained or extorted (Baviskar 2003).

However, it should be stated that while conservation practices may reflect a dominant metropolitan-secular eco-nationalism, to relate all forms of eco-nationalism to concrete state policies is not always very straightforward. Especially with multiple plural competing regional eco-nationalisms, some of which may be overtly 'indigenist-regionalist' in flavour, the connections either with official state practices or dominant social norms cannot be easily assumed. In this paper, we show how the creation of a sanctuary sparked off a series of livelihood constraints amongst small fishers, raised questions about legitimate traditional fishing rights, and elicited a reaction involving the constructing of a uniquely regional Odiya eco-nationalism. The rest of the paper explores various aspects of this social discourse, its contradictions and fragility, as well as the role of the state in both facilitating as well as mediating such a discourse.

CONTEXTUALISING THE ROLE OF THE STATE

In this section, the paper moves on to a somewhat more detailed consideration of two aspects of the state, and the role it plays for the case at hand. In the first part, we consider the principal aspects of the state's policy for fisheries regulation in Odisha, and the particular regional resonance of wider national tensions between fishing and marine conservation. In the second part, we comment briefly on the historical stance of the Indian state towards the entry of refugees and immigrants following the great partition of 1947. We also refer to some particular controversies that have erupted in these coastal parts of Odisha regarding immigration from Bangladesh, and the role assumed by the state in Odisha in this regard.[4] It is hoped that this discussion will contextualise the role of the state in enabling the particular regional environmentalism being described in this paper.

State policy, turtle conservation and marine fishing

The Olive Ridley turtle is a critically endangered sea turtle species that visits Odisha between December and March each year. Turtle conservation has engaged both official policy as well as non-governmental advocacy since the 1970s. However, as this section shows, turtle conservation is tied in with the larger problem of regulation of marine fishing, which in turn, is an extremely contested terrain given the diversity of interests within the marine fishing sector. Moreover, while the state has been criticised by conservation groups for not doing enough to protect the Olive Ridley, it has equally come under attack by fishers groups and fisheries officials for not dealing adequately with issues of fishing productivity.

Widespread reports of turtle mortalities in the 1970s triggered off a massive response of state protection at the national level. In 1976, the endangered status of this species was sealed with its inclusion into Schedule I of the Indian Wildlife (Protection) Act or WLPA, 1972, and the Forest Department stopped issuing licenses for the collection of turtle eggs in Odisha. National and international scientific research played a critical part in this process, as did environmental campaigns that led to the deployment of the coast guard to Gahirmatha by Prime Minister Gandhi in 1982. Heavy policing contributed to a marked decline in illegal turtle trade in Odisha by the 1980s (Sridhar et al. 2005).

Besides WLPA, the Odisha state government also enacted the Odisha Marine Fisheries Regulation Act (OMFRA) in 1981, and introduced other subsequent regulations to curb destructive fishing practices that hurt turtles. Under OMFRA, the state government tried to set standards for 'sustainable fishing' by introducing different fishing rules for the different types of craft, as well as limiting the number of mechanized fishing boats. Only 'non-mechanized traditional fishing boats' were to be allowed within 5 km of the shoreline throughout the coast of Odisha, and all larger mechanized vessels could only operate beyond 10 km from the shoreline (Sridhar et al. 2005). For the purposes of this paper, we use the widely popular usage meaning of a mechanized boat to refer to that 'where fishing is done mechanically (like trawlers), as opposed to a non-mechanized boat, where the actual fishing is not done mechanically (these include motors in motorized boats, which are used only for locomotion)'.[5]

In 1994, the state government issued biennial orders prohibiting fishing in Gahirmatha and these were reissued periodically. By 1997, Gahirmatha had been declared a marine sanctuary under Section 26 (1) of the WLPA, with a core area measuring 725.5 sq. km and a buffer zone of 709.5 sq. km. This meant that absolutely no fishing was permitted in the core area at any time. Few or no consultations with local people preceded the sanctuary's notification on September 27, 1997, sadly echoing other experiences of state led conservation elsewhere in the country (Saberwal and Rangarajan 2003).

The government proactively adopted measures to protect turtles at other key nesting sites (mouth of Devi and Rushikulya rivers further south of Gahirmatha), and prohibited mechanized fishing within 20 km of the high tide line between 1 November and 31 May each year. There is a blanket ban on fishing anywhere along the Odisha coast from April 15 to May 31 in view of the fish breeding season. The government has made the use of turtle excluder devices (TEDs) compulsory for trawlers under Section 29 (B) of OMFRA (Wright and Mohanty 2006).

Despite these steps, turtle deaths continued to escalate through the 1990s, and were widely attributed to 'a surge in mechanized fishing, unsustainable fishing practices and deliberate violations of the law' (Wright and Mohanty 2006: 290). Prominent environment groups criticised the state for this situation. Operation Kachhapa (OpK), the prime turtle conservation initiative in Odisha (formed under the aegis of the Wildlife Protection Society of India in 1988), was especially concerned about 'overfishing' being sponsored by the state. In March, 2002, OpK filed an affidavit to the Odisha High Court against the government for having issued 'about 6000 fishing licenses - to about 900 trawlers and 5000 gill netters - against the official OMFRA quota of 1080' (Wright and Mohanty 2006: 292). OpK estimates that 16,000 fishing boats operate along the Odisha coast, of which about 8000 are large mechanized fishing trawlers and gill netters.

Greenpeace (2008) also campaigned to highlight what it describes as an 'exponential' increase in mechanized marine fishing: 250 % in 25 years and a fall in the numbers of non-mechanized and non-motorized boats. This drastic reduction in the numbers of wooden dinghies ('traditional' craft) directly affected the poorest fishers, who neither had the means to modernize nor could they find work easily within the mechanized fishing sector. This was given the level of demand for employment, or indeed in other sectors, where they had to compete with workers with more appropriate skills (Salagrama 2006). Marine fish production rose from 53,581 metric tonnes (MT) in 1985/86 to 156,081 MT in 1997/98 before declining through till 2003/04, reaching 116,880 MT.[6] For environmentalist campaigners, falling fish catch is a sign that 'marine fisheries in Odisha have reached a stage of stagnation and possible decline' (Greenpeace 2008: 8). In later years, marine fish production has experienced some relative growth from 121,928 MT in 2004–5, reaching a high point of 133,479 MT in 2010–11,[7] and then 133,211 MT in 2014–15.

In contrast with the narrative pursued by environmentalist groups is that of an unlikely alliance of state officials and many local fishers, who take the view that fisheries policy in Odisha has not been 'productivist' enough. A retired fisheries official viewed that adoption of multiple restrictions (sanctuary, seasonal fishing bans and trawler excluder devices) was an 'overreaction' to exaggerated environmentalist agendas.[8] Unlike other coastal states like Kerala (Sinha 2012), the Government of Odisha is generally criticized for neglecting marine fisheries (Sampath 2005), and with little or no investment in off-shore infrastructure, promotion of fishing cooperatives and the 'modernisation' of indigenous craft and gear.[9] The government has not invested adequately in deep-sea fishing (with no deep sea trawlers till date) or exploited the full potential of the Exclusive Economic Zone. Many fisheries officials consider that the overfishing alleged by environmentalists stems from overcrowding of near shore waters by vessels of all types, which is a deeper problem requiring a systematic response. In particular, small fishers and trawlers are equally indignant about the transgression of foreign vessels into Indian territorial waters. However, more recently, there is a recognition in the state of the stagnation in marine fish production, and of the need to adopt measures for modernisation and shore based infrastructure development besides the exploitation of deep sea fisheries.[10]

The Indian state, refugees and the 2005 controversy in odisha

The subject of refugees is an extremely contentious one in post-partition South Asia. Both the great partition of India and Pakistan in 1947 as well as the war of liberation of East Pakistan that saw the creation of Bangladesh in 1971 produced large scale movements of people. These have repeatedly raised difficult questions around citizenship and the rights of refugees. Interestingly, India has neither ratified the 1951 Convention on Status of Refugees nor the Protocol of 1967. Yet, right since the days of the Constituent Assembly—tasked with formulating the Indian Constitution- the debate on who should be legitimately regarded as a refugee and accorded protection has been repeatedly had (Jayal 2013).

Historian Joya Chatterji writes that in the two decades after the partition of India and Pakistan, millions of Hindus crossed over into India, especially in the 'turbulent wake of the Noakhali and Tippera riots in 1946 and the Khulna riots of 1960'; nearly another two million left 'after the theft of holy Muslim relics from the Hazratbal shrine in Kashmir in 1964' (Chatterji 2007, 111). The so-called refugee crisis intensified in 1971 when Bangladesh was formed, with a steady influx of both Muslim and Hindu refugees. This has continued steadily thereafter, triggering off profound unrest in especially affected states like Assam. A powerful student agitation against the 'swamping of Assam by foreigners' from 1979 to 1985 led to the signing of the Assam Accord with the central government. The government prescribed that 'a) all those who had migrated before 1966 would be treated as citizens, b) those who had migrated between 1966 and 1971 could stay provided they put themselves through an official process of registration as foreigners, and c) all those who had migrated thereafter were simply illegal immigrants' (Jayal 2013: 64). The provisions of the Assam Accord led to the 1985 amendment to the Citizenship Act of 1955. The amendment allowed:

'for every person of Indian origin (if either of his parents or grandparents was born in undivided India) that came into Assam on or after the 1st day of January 1966 and the 25th day of March 1971, and has ordinarily been resident in Assam, and has been 'detected as a foreigner' by a Tribunal constituted under the Foreigners Tribunal Order of 1964, would need to register himself or herself in accordance with the rules made by the Central Government on this behalf. His or her name would need to be deleted from the electoral rolls, and such a person would have 10 years from the date on which he/she has been detected as a foreigner the same rights and obligations as any other citizen of India, including the right to obtain an Indian passport'.[11]

No such provision was made for anyone entering after the cut-off date in March 1971. At a human and personal level, such a politics of citizenship has been extremely debilitating for the thousands of immigrants, forever exposing them to expedient questions around authenticity.

The subject of 'illegal immigrants' has been a deeply emotive one for many Indians, who feel bitterly towards politicians of all political parties for using immigrants as easy vote banks. The Illegal Migrants (Determination by Tribunal) Act or IMTD Act of 1983 created an Assam-specific exception to India's law on foreigners, which normally requires individuals in question to prove their own citizenship status. Under the IMTD Act, neighbours could simply file a complaint against suspected illegal immigrants, and the matter would then be decided by a tribunal. The Act had was partly meant to prevent a witch hunt against illegal migrants, but also had the professed aim of making it easier to detect and deport illegal migrants (Kapur 2012: 152).

However, the IMTD Act was not received well in Assam, and a writ petition was filed in the Supreme Court claiming that the Act had 'made it "impossible for citizens who are resident in Assam to secure the detection and deportation of foreigners from Indian soil"' (Sarbananda Sonowal 2005 cited in Jayal 2013: 64). Though the IMTD Act was eventually struck down, the law on citizenship continues to reflect the issue of illegal immigrants from Bangladesh. In particular, a 2003 amendment to the 1955 Citizenship Act modified the provision of citizenship by birth to exclude from it such persons born in India as have one parent who is an illegal immigrant at the time of their birth (Article 3 c ii).[12]

The state of Odisha received a steady stream of Bengali refugees from former East Pakistan in the 1960s, as well as subsequent decades even after the formation of Bangladesh. Interviews in our case villages revealed that several people arrived here from the government refugee camps that had been established in various places, such as the hot and dry parts of central India (Kudaisya 1997). They came here because they were attracted by the familiar coastal habitat to which they were accustomed back in Bangladesh. Others came in course of their journeys through India looking for work, having arrived in West Bengal, and then gone on to Paradeep, a port town in Odisha.

The areas immediately by the sea were covered with dense forest and were unpopulated. We heard a number of accounts about how villages and settlements around here grew organically, with the gradual clearing of forest land. In those early years, Bengali refugees were ostensibly welcomed by absentee Odiya landlords to cultivate their lands. We were told that the revenue department had initially awarded *pattas* or land titles to a small group of refugees, but most others had negotiated small homesteads through informal payments and bribes to revenue and forest officials over time. Not everyone has a land *patta*. A number of respondents openly admitted to arriving here in the 1980s and even later. They shared painful memories of Muslim atrocities meted out to the Hindu population long after the formation of Bangladesh. Those who came later on merged with previous settlers, and like them, went on to acquire a range of identity documents associated with citizenship. These ranged from certificates of landlessness, voter cards, ration cards, BPL (below poverty line) cards and so on.

Such grassroots processes of acquiring documents are typically associated with the vote bank politics concerning groups of immigrants. In coastal Odisha, it is well known that Bengalis have been patronised by a now deceased state legislator (or MLA). Much loved by the Bengali population, this politician was routinely criticised by his opponents. Some even argued that it was he that had influenced the illegal construction of a jetty at Matsyapalli, in violation of OMFRA, nearly 40 km away from the sea. It is also these micro-level interactions that run in parallel to national level restrictive pronouncements; the chasm between the ostensibly neat categorisations in law of 'legitimate refugees' and 'illegitimate migrants', and the far messier reality is plain for anyone to see.

There had been no other specific interventions or policies by the state government in Odisha with regard to immigrant Bengalis in these coastal parts. In 2005 however, with little warning, the state government produced an order, following a high court directive, for the identification of 1551 illegal immigrants, who have ostensibly entered India illegally after 1971, or born to parents that were born illegally. 'Quit India' notices were issued to this effect by the Kendrapara District Collectorate to persons on the list, scattered across several coastal villages with Bengali settlers, included Narayanpur and Ambapalli. The immediate context to this appears to have been the strict 2003 amendment to the Citizenship Act that excludes children born of 'illegal immigrants' from Indian citizenship. We interviewed the District Collector of Kendrapara for more insight into this, but, in response, we were merely told that 'infiltrators' had entered India without the prior permission of 'competent authorities'. We were also not given any further details about the precise enquiry that led to the production of such a list. The wider politics of the case have been examined elsewhere (Chhotray 2016); and it suffices to mention here that the episode and subsequent reactions revealed various fascinating processes at work, which we will refer to in the next section.

FISHERS' SOCIAL RELATIONS AND ODIYA ENVIRONMENTALISM CLOSE TO THE GAHIRMATHA SANCTUARY

In this section, we present the substantive part of our field research, first characterising the transformation of livelihoods and fishing practice with the arrival of Bengali immigrants and the later creation of Gahirmatha sanctuary, and then critically examining Odiya-Bengali fishers' social relations. The section will also show how a complex, contradictory and fragile regional Odiya environmental discourse is constructed.

Livelihoods next to the sanctuary, the arrival of Bengalis and marine class relations

Our case study villages are amongst the 70 odd villages in the Rajnagar and Mahakalpada blocks of Kendrapara district by the Bay of Bengal, that were directly affected by the total ban on fishing following the creation of Gahirmatha marine sanctuary in 1997. These are predominantly occupied by Bengalis, who have arrived here in waves since the 1960s. The curtailment of fishing in the sea has put an unbearable pressure on their fishing-oriented livelihoods. Flanked by the Bhitarkanika National Park and Wildlife sanctuary on the other side, people here have also faced restrictions on fishing in rivers and estuaries within the park area or entering the forest to collect forest products, ever since the sanctuary's notification in 1975. Other Oriya dominated villages a little away from the coast also face the effects of these restrictions.

Some people practise one-crop subsistence agriculture by cultivating paddy in the rainy season. However, farming is hampered by a continuous problem of salinization. There are no irrigation facilities and locals complained that they have not been assisted by the government, such as through the provision of special seeds that can cope with salinity.[13] People do not have access to credit either. These problems have been worsened since the super-cyclone struck the coast of Odisha in October 1999, with wind speeds of 260 km/h and a storm surge that travelled nearly 20 km inland entirely submerging low-lying coastal villages (Chhotray and Few 2012).

Government provision for alternative livelihoods has been thin and is a sore point with the local communities. Odisha (along with two other coastal states, West Bengal and Gujarat) implements the Integrated Coastal Zone Management Plan (ICZMP), an integrated approach coordinated by the central Ministry of Environment and Forests, the World Bank and the state government to 'coordinate the activities of various government agencies and departments for the sustainable management and usages of coastal resources maintaining the natural environment'.[14] ICZMP has a livelihood security component aimed at fostering support to fishers in the coastal stretch in southern Odisha (Gopalpur to Chilika) as well as northern Odisha (Paradeep to Dhamra) (the latter covers our study sites), but has had limited impact till date. The main model of help is through SHG based financing, but this is not understood to constitute a viable alternative to earnings from fishing.[15] Moreover, local fishing communities widely observed declining fish catch, which they attribute to industrial pollution and the dumping of effluents into the river Mahanadi and its tributaries. The irony of the situation is that women and men from these sea side villages regularly seek daily wage in these very factories.

This wider economic context is extremely important even as we try to reconstruct the story of how marine fishing began in these northern parts of the Odisha coastline. As Bengalis from both West Bengal and former East Pakistan (and later Bangladesh) arrived into these parts, they cleared densely forested and unoccupied lands and began fishing in rivers and estuaries. But as the restrictions against entering the Bhitarkanika sanctuary area became enforced more strictly and even more people arrived, they gradually inched towards the sea and tried small-scale fishing. This was quite an important development, as traditionally, the principal Odiya fishing castes (*kaibartas, gokhas and rajabanshis*) have fished in rivers and estuaries, but not in the sea. The only sea fishing in Odisha therefore, until the Bengalis began, was restricted to the*noliyas,* a community originally from Andhra Pradesh, but with long roots in Ganjam and Puri districts of Odisha state.

At first, the Bengali fishers used very small non-motorized dinghies, gradually moving on to motorized boats and more advanced fishing nets than Odiya fishers had traditionally employed. These were colloquially called *Hilsa* nets, designed to catch this prized fish, whereas Odiya fishers had generally used *patia* nets for their river fishing. The movement of Odiya fishers towards sea fishing probably took place around the 1980s, prompted also by the rising demand for marine fish all over India. This has also been reflected in changing diets in Odisha with the greater consumption of sea fish.

Quite understandably, the arrival of Bengalis into these parts and their initiation of sea fishing is a subject that evokes complex reactions all around. On the one hand, Odiya fishers we interviewed spoke admirably of Bengali fishers for having introduced them to new modes of fishing by sea. On the other, a number of local Odiya respondents mentioned that Bengalis owned gill net boats and had become greatly prosperous in comparison with other Odiya fishers from these parts. For example, the Odiya President of the Utkal Kesari Primary Marine Society from Rajnagar block in Kendrapara spoke of the society's efforts to promote the interests of fishers practising 'traditional' techniques, 'unlike the Bengalis' who were 'prospering' through gill net fishing (though he did admit that several Bengalis were also members of the society).

Moreover, an Odiya environmentalist from Operation Kachhapa, who also takes a keen interest in local fishers' matters, observed that the new Bengali settlers 'got sufficient income from fishing in estuarine areas and they also did paddy cultivation in the mangroves', and this enabled their transition to motorized fishing.[16] He added that it was the rising population of Bengalis (from Bangladesh, after 1971) that in fact made it impossible for the local Odiya fishers to continue fishing in rivers and estuaries, 'forcing them to go to the sea'. This narrative clearly downplays the contribution of the Bhitarkanika National Park and its restrictions in the transition towards sea fishing.

The Bengali sarpanch of Matsyapalli gram panchayat explained the story a little differently. He emphasised the hardship by Bengali settlers in the early years, when they came to find a new life in lands that were of little interest to Odiya people (who had 'occupied all the nice lands and were scared to live in these areas'). He also talked about the conflicts that Bengalis faced from Odiya caste fishers when they initially tried to fish in rivers and estuaries. It was only later when they carved a niche with sea fishing and earned profits (in the years before the sanctuary came into effect), that Odiyas became more interested in sea fishing. In more than one conversation with Bengali fishers, we detected a strong sense of discovery and pride when they recalled their entrepreneurial journeys, albeit these are now tinged with frustration at the restrictions imposed by the sanctuary. While such sentiments are important, they must not distract us from the larger class context of marine fishing where we see poor small fishers- both Odiya and Bengali- engaged in a conflict with the state over fishing rights due to the imposition of the sanctuary. Moreover, as this section has emphasised, wider economic pressures on the fishing and farming livelihoods of the coastal people have been exacerbated through the conservation initiatives of the state. And it is in this context, that we find the genesis of a new indigenist assertion that questions the presence of Bengalis, by labelling them as outsiders.

The bigger economic and demographic context of the marine fishing population in Odisha reiterates the common challenges being faced by the Odiya and Bengali small fishers of interest to this study. According to DoF (2002 cited in Salagrama 2006), there are just over a million fishers in the state, and a third of a million are marine fishers. Only 86,500 are active marine fishers and the total number of fishing boats is about 12,500, which suggests that the large majority of active fishers work as fishing crews and are extremely poor (ibid.). Besides, as previously mentioned (in section 2), many fishers who work as crew members now do so because of the wider reduction in the numbers of traditional craft, and many more still are underemployed.

Motorization of fishing craft began in 1956 but both motorization and mechanization gathered momentum only after 1980, with the availability of increased government support. Motorized traditional craft increased from 5 in 1985/86 to 3643 in 2000/1 (DOF 2002 cited in Salagrama2006: 20). The contribution of non-motorized boats to total fish landings concomitantly declined, from 52.5 % in 1985/86 to 33.4 % in 1996/97 and only 24 % in 2000/01 (ibid.). Over the same period, the contribution of the mechanized sector (including small, large and deep sea trawlers) to total landings increased from 25,000 to 89,000 tonnes (ibid.). While in 1987, there were 745 mechanized boats operating along the Odisha coast (of which 470 were trawlers and the rest were gillnetters),[17] in 2015, this number had increased to 1754.[18] 85 % of these mechanized vessels are trawlers.

Data on income earnings disaggregated according to these different categories of fishers are not available due to the largely unorganised nature of the sector, the dispersed nature of landings and various seasonal differences in income and migration (Salagrama 2006). However, the vast majority of marine fishers involved in non-mechanized (including artisanal) fishing are poor. Other links in the fishing chain like trader financiers, commission agents and boat builders are regarded as relatively better off compared to boat owners and crew members, but 'most livelihoods at best provide subsistence earnings that are subject to seasonal swings and shocks' (Salagrama 2006: 32). Even those in the mechanized sector are typically not very rich, and their livelihoods are tenuous on the whole.

An example from Matsyapalli village will illustrate the difficult economic situation of small fishers living next to the Gahirmatha marine sanctuary. There are 2000 households approximately; a quick estimate gathered during a focus group discussion revealed that roughly 115 have motorized boats with gill nets, another 200 have non-motorized boats only and the rest simply work as fishing crew on other boats or trawlers and practise subsistence cultivation. A typical new FRP (fibre reinforced plastic) boat costs nearly 3,00,000 rupees at current prices, and a gill net may cost between 70,000 and 2,50,000 rupees depending on its size. Besides, there are annual costs of maintenance (2000–30,000 annually on an average) and operation (a typical 4–5 day fishing trip costs nearly 8000 rupees towards fuel, ice and food). Our sources in Matsyapalli explained that even in an optimal scenario of a quintal of Hilsa caught and sold at 100 rupees per kilo in the wholesale market, it would not yield a clear profit once operational and maintenance costs were taken into account. Compounded with the fishing ban during the peak hilsa season, plus ever falling fish catch and fish seizures by forest department guards, there was no guarantee of making ends meet. The condition of assetless crew members is even more abysmal.

Boat owners are also dependent on middlemen (*gaddiwallas*) for finances, and typically receive advances to finance their fishing trips in return for selling their catch. They also pay a percentage of money (4 % according to the middlemen, 7 % as per Matsyapalli fishers) earned at the 'auction' in the wholesale fish market at Paradip, as commission to these middlemen. There are reported malpractices of middlemen 'fixing' prices with selected fish traders at below market prices and splitting the difference amongst them, depriving fishermen of their full share. Boat owners are unable to repay their advances if they suffer repeated losses. There have been a number of distress suicides in Matsyapalli. Crew members are inevitably affected too, as their earnings are restricted to sharing only 50 % of the total value earned by the boat owner. Many of them have turned to daily wage work.

It is clear that small fishers engaged in gill netting are positioned at the bottom of the marine class hierarchy.[19] In Odisha, most gill netters are non-mechanized, with simple motorized boats, though a small proportion of gill netters are also mechanized.[20] Historically, gill netters have opposed the introduction of trawlers along the Odisha coast, and smaller trawlers in turn have resisted the entry of larger capacity trawlers. The Paradip Marine Primary Fish Production and Marketing Cooperation Society Limited comprises of 175 members, that are all owners of Sona boats (small sized trawlers), and it has resisted the entry of large trawlers into the Odisha coast from neighbouring Andhra Pradesh. They are all opposed to deep sea trawlers and foreign fishing vessels especially as these allegedly foray into near sea waters illegally. It is hardly surprising then that *all gill netters*, Bengali, Odiya, and Telugu, are united in their view that the largest turtle deaths are in fact caused by trawl nets.

Social relations amongst fishers and the construction of Odiya environmentalism

This is an appropriate point in our narrative to take a critical look at the unfolding social relations amongst Odiya and Bengali fishers by the Gahirmatha sanctuary over time, referring in the process to the construction of a uniquely regional environmental discourse. This discourse relates to the indigenist-regionalist eco-nationalisms characterised by Sivaramakrishnan and Cederlof (2012), which as we have discussed earlier, are distinguished by assertions of indigeneity and claims to authenticity with respect to particular places. We are equally interested in describing the problems with this narrative as revealed by the concrete reality of everyday fishers' relations.

The *kaibartas* are the Odiya fishing caste group that ran into early troubles with the incoming Bengalis. In a historical sense, river fishing has been a matter of great pride and custom for the kaibartas, generally regarded to be the most enterprising and proudest of the fishing castes (Tietze, 1985). In previous decades, they were a relatively cohesive caste group, under the aegis of an old inland fishing cooperative founded in 1962, by the name of the 'Kalinga Karnadhar Fishermen's Society (KKFS). In fact, the society claims exclusive easementary rights over 57 river mouths within Jagatsinghpur and Kendrapara, two large districts along the northern coast.

In a personal interview, Behera, the principal *kaibarta* leader displayed a proud identity through a historical narrative of *kaibartas* as the 'sons of Odisha' who fought against a succession of invaders from pre-British through to colonial times.[21] He claimed that in 1861, the king (of Kujang, Jagatsinghpur district) had rewarded their loyalty by granting them sole responsibility over these 57 rivers (renamed as *sahar pentha*), in exchange for a yearly tax of 561 rupees. The modern society campaigned for state recognition of its customary privileges, which was eventually granted by the state government in 1972. As the *kaibarta* society pays an annual 'tax' to the government, it required that any non-*kaibartas* that fish in these rivers must in turn pay a fee to the society.

As a result of these restrictions and the proprietary influence claimed by KKFS, there were severe clashes between the Odiya *kaibarta* caste fishers and the incoming Bengalis who began fishing in the rivers and estuaries of these parts in the 1970s. We heard that Bengali fishers also formed their own cooperative society ('Jay Jagannath Society) in 1986, in order to organise themselves given these circumstances. On the 21st of August 1987, there was a physical scuffle between Bengali fishers from Matsyapalli and other villages and Odiya *kaibartas*. The police got involved and some officials from the state revenue department apparently reiterated the 'customary right' of *kaibartas*. This then led to a court petition by Bengalis in the Revenue Court and later the High Court. There was a huge mobilisation from several Bengali dominated villages in the area.

We did not independently investigate the outcome of this petition, but we received very different accounts of what followed from the two sides. Behera said that the courts upheld the traditional rights of the Kalinga Karnadhar Society, and he had been magnanimous later and 'allowed' Bengalis to fish in these rivers of Kendrapara for free after 'they apologised for their misdeeds'. In contrast, the Bengali Sarpanch of Matsyapalli told us that Bengalis had 'won' the right to fish in the rivers on the Kendrapara side. Even at the time of fieldwork, many Odiyas in Ambapalli believed that the Bengalis 'lost' their case because the High Court demanded caste certificates which the Bengalis could not produce.

The case is firmly in the past, though memories of it still produced strident positions by those whom we spoke to. In the present day, our research in Narayanpur and Ambapalli villages confirms that enforcing the old strictures has become quite difficult; multiple smaller sub-groups of *kaibartas* engage in separate acts of vigilantism producing a rather diffused network of *kaibarta* authority in the rivers of coastal Odisha. There is also no clear acknowledgement anymore of Behera's leadership, though he is well known in this parts. As for the KKFS, Behera himself admitted that 'politics has divided the society, and that members are supposed to pay the tax at regular basis, which is not happening'. There is a real question of the society's contemporary relevance, especially in dealing with new challenges facing fishers.

Interestingly, the general economic pressures described earlier are compounded by a feeling that the government is not doing anything for the 'caste fishermen', and the ensuing anger that outsiders- mainly Bengalis- have spoiled everything. The old *kaibarta* leader ruminated that in fact, Bengali immigrants were responsible for despoiling a quintessentially Odiya fishing culture, which was based around river mouths and estuaries. He said that they had exhausted rivers and estuaries of fish and strayed into the sea, disturbing breeding fish even as they journeyed through the river mouths. As the conversation progressed, so did his claims, as he constructed a discourse of difference between Odiyas and Bengalis with respect to their fishing practices: 'Bengalis simply do not use the kinds of nets that we *kaibartas* traditionally do. We use large threaded nets, whereas they use very fine nets.' He continued, 'They are responsible for the death of the Olive Ridley, and are exporting these to Bangladesh'.

Another key proponent of this kind of narrative is the environmentalist from OpK, a well-regarded body that has worked to highlight the grave and composite risks being faced here, including illegal fishing by mechanized boats and foreign vessels (see Wright and Mohanty2006). But referring especially to Matsyapalli, the principal OpK spokesperson said: 'They have lots of gill nets. Some of the people have 4–5 boats. And because most of them are Bangladeshi refugees- as it is, there is a huge proportion of illegal immigrants. They do not have refugee status actually.....I had filed a court case in 1998 for seeking an eviction of all illegal immigrants. But there was political pressure so these people are still hanging on'. He continued: 'the entire fisheries in the state has collapsed because of *these Bangladeshis* only...they have used every kind of fishing method to wipe clean the rivers and oceans of fish, and even put nets at the mouths of rivers.'

Our research showed that some of these arguments can be challenged. First, it is increasingly difficult to claim that only Bengalis do gill netting. 'Traditional' stick and net fishing practised by Odiyas in rivers and estuaries is on the decline, and some Odiya fishers have made the transition to motorized gill net fishing, and many more are keen to do so, though they have understandably been impeded by their general lack of resources and discouraged by the ongoing sanctuary restrictions. And second, it is the type of net used that is crucial and gill nets may be either 'monofilament' or 'multifilament'.[22] The latter is considerably thicker and said to be safer for turtles as they can escape more easily. We also heard from a number of Bengali gill net fishers that they used the *hilsa* and *kani* nets that allowed larger fish and turtles to escape. They also claimed to observe a strict restraint against using the *vecti* and *telia* gill nets where turtles get easily stuck. However, perhaps this is not the most interesting point to be made.

Following our previous discussion of the rise of regional environmentalisms that are indigenist-nativist in orientation, it is fascinating how a distinctive Odiya environmental discourse about the destructive effects of Bengali immigrants has taken shape. We also note the repeated references to illegal immigration from Bangladesh. The historical developments concerning the Indian state's treatment of refugees, post 1971, are significant here, especially the most recent 2003 amendment which creates new barriers to citizenship for those whose parents may have entered India illegally. This environmental narrative or environmentalism which emerges as a response to the debate on fishing rights in and around the sanctuary itself is specifically enabled by this history.

And yet, as the wider scholarship has shown, the plethora of indigenist assertions and regional eco-nationalisms that potentially exist across the Indian sub-continent are essentially claims, often by dominant elites, who resort to these narratives as a mode of exercising power. In this case too, these are only claims by disparate actors who have no unified position on the larger debate about fishing and marine conservation. More importantly perhaps, the environmentalism we describe has a rather limited hold on everyday interactions between Bengalis and Odiyas in these parts.

Some reactions are worth reproducing. When we discussed in a focus group discussion at Matsyapalli that some people were of the opinion that not only were many Bengali fishers of illegal standing, with 'fradulent Indian citizenship papers', they were also 'sheltering their friends' and 'relatives' from Bangladesh, the reaction was of total outrage. Our respondents in Matsyapalli posed the logical counter question that if there was indeed infiltration by sea, then 'what was the coastguard doing'? More importantly, they identified strongly as citizens of India, with 'proper' documents and a very clear sense of affinity to these lands and seas. One respondent summed it up for most when he said, 'We will never think of doing something against India. We are born here and maybe we have an ancestral history of Bangladesh, but we consider ourselves as Indians. We have been staying here with our cattle, coconut plants and family members'. Some even questioned the rights of fishers from neighbouring states like Andhra Pradesh to fish near this coast. An interview with officials (who were all Odiya) at the marine police station in Matsyapalli also did not provide any evidence to corroborate this 'infiltration' thesis.

We discussed the same issue amongst Odiya fishers in Narayanpur and Ambapalli and received a very similar reaction. People did say that Bengalis had come into these areas from Bangladesh after the country was formed in 1971 (which is a well-known fact nationally), but also that these processes of immigration had also slowed down over the years. The 2005 episode had come as a massive shock to these local communities, Odiya and Bengali, who have been coexisting in close proximity over the years. When the 'quit India' orders had been issued, then there had been a general outpouring of solidarity with those affected, barring a few isolated incidents. Eventually a 'united' Bengali-Odiya front formalised into a body called the Utkal Banga Suraksha Samiti (translated as 'Odisha-Bengal Security Committee) or UBSS that organised demonstrations, threatened road blockades in the event of police action and facilitated political attention into this issue. As the years have passed, nothing more has been done by way of 'deporting' those identified, though many people who find themselves on this list have had their identity documents 'nullified', which is producing other consequences (Chhotray 2016).

On this basis, we can conclude that there is limited traction of such Odiya environmentalism in the everyday social life of the fishers living close to the Gahirmatha sanctuary. The basic struggle being waged by these fishers is a class one, where they compete with larger trawling vessels for fishing in the context of severe restrictions imposed by the sanctuary. This is also revealed in the struggles of the Odisha Traditional Fishworkers Union (OTFWU), a registered trade union of fishers along the coast that claims to represent the class interests of all*traditional* fishworkers from Balasore in the north to Ganjam in the south. It came into existence in the late 1990s with the 'merger' of local fisher associations protesting against the illegal entry of trawlers from Andhra Pradesh into the Odisha coastal zone. Over the years, the union has taken up issues on behalf of marine fishermen all along the Odisha coastline; be it trawlers in Ganjam, prawn aquaculture in Balasore and Bhadrak, and the marine sanctuary in Kendrapara. It organized very energetic protests after the imposition of a blanket fishing ban in Gahirmatha, but has struggled to be effective in the face of continued restrictions and patrolling of the sanctuary.

However, the Union's claim to represent *traditional* fishers has led to some acrimony in recent years. Some Odiyas, including Behera, the environmentalist from OpK and a few other fishers, have raised questions about how *outsiders* like Telugus and Bengalis should legitimately be a part of the Odisha 'traditional' fishworkers union. One argued that 'traditional' in Indian law refers to those 'who have been living in these areas for at least 2–3 generations, or 25 years', and he cited the example of the recently enacted Forest Rights Act 2006 to identify indigenous forest dwellers. These arguments are framed in terms of a particular reading of Odiya indigeneity, but were vehemently challenged by many of our Bengali respondents as well as the Telugu Secretary of the OTFWU. They asserted that such viewpoints were merely divisive strategies that deliberately denied their long standing involvement with fishing. For the latter group, it was their class position as small fishers that was the basis for the pan-coastal union of interests. On the whole however, such arguments that favour 'Odiya indigeneity' as a prior basis for claiming a 'traditional' position do not translate into anything particularly consequential.

JUSTICE AT SEA

In this final section, the paper returns to the ever pressing matter of the sanctuary and what its restrictions actually mean for the fishers that negotiate its boundaries daily. The paper contends that it is at sea, where fishers, both Bengali and Odiya, and of all sizes, encounter authority. The state passes into material form, in the form of 'guns and guards', and there is an opportunity to witness the injustice of conservation that is already well documented.

In Gahirmatha, the injustice of curtailed access is magnified as fishers, both Bengali and Odiya, consider that the state is cynically disinterested in finding a rational solution to the 'impasse' of seemingly irreconcilable interests between fishers and conservation. They make two points: one, a yearly ban to fishing is not necessary at Gahirmatha because turtles nest between January and March, and there is no need to keep fishers out in November and December, which are the key months for *hilsa* fishing, and two, the only passage to the open sea (beyond 20 km of the restricted sanctuary area) is through the sanctuary (at the Barunei river mouth at Hukitola). They want the state to renegotiate the rules of access through the sanctuary, and especially smaller fishers want to be allowed to fish near shore during the peak season.

The predictable response of state officials is that the yearly ban is not only for Olive Ridleys but for the health of the marine ecosystem on the whole, and fishers have other avenues to do fishing. Forest department officials also talk of an alternative route to the sea that is not through the sanctuary. While this suggests a disjuncture of knowledge between the state and fishers, there are other factors that have eroded mutual trust. Typically, fishers complain that they are stopped as they are returning through this passage, often with fish caught *outside* the sanctuary, and their goods seized by lower level forest guards, even as they have no other route. With this motive, forest officials generally issued no receipts for fines charged. Senior forest officials dismiss this story by squarely blaming fishers for cynically stealing from the sanctuary out of greed. Fishers of all sizes insinuate that the state has a vested interest in allowing the status quo to continue. Trawler operators too reported that forest officials tried to extract a much higher penalty than the value of seized fish. Allegedly, forest officials did not follow the recommendation of the Central Empowered Committee, which was constituted upon the orders of the Supreme Court in 2004, to 'auction' all seized fish from the sanctuary in the presence of trade union representatives.[23]

Complying with as well as enforcing the law is difficult at sea. While denying any intention of transgressing sanctuary boundaries, fishers of all sizes resort to the familiar argument about technological lapses. Small fishers say they do not have GPS systems, while trawlers claim that their GPS systems often fail. And yet, trawlers are more effective at fishing inside the sanctuary and getting away quickly. There is a very strong sense of injustice amongst smaller fishers that do get caught. The marine class system can thus have very palpable consequences at sea, especially with respect to the enforcement of sanctuary boundaries by the state.

Enforcement of the law is difficult and risky for the state. Trawler crews have been known to use intimidation tactics at forest guards (Wright and Mohanty 2006), even throwing dynamite on occasion. State officials complain that the police are of little help during these 'encounters' as they are unfamiliar with the sea. There are regrettable consequences for both sides. In March 2003, three forest guards were reportedly 'abducted' by two gill net boats and one guard was pushed aboard and died. In 2005, a small fisher from Matsyapalli was shot dead by a forest guard, and a public controversy erupted here over how the government had 'tried its best' to project the dead man as a 'Bangladeshi pirate'. The Sarpanch told us how the villagers refused to cremate his body until the government had withdrawn the charges, which it did, and also paid a substantial amount of ex-gratia payment to the family of the deceased.

This tragic incident contains a powerful illustration of how an abstract and nebulous environmentalism can sometimes produce extremely serious effects. This is in fact a very sensitive subject, and one that the Odisha state government generally refrains from publicly escalating. As one forest official said, "Forest people and coast guards check papers at sea, registration papers, proof of ID, fishing licenses, and they all show that the fishermen are residents of Odisha. No doubt they are not Odiyas, but it is not the case that only Odiya people will live in Odisha. It cannot be'.[24] The same issue elicited a very different reaction from the environmentalist from the OpK: 'You see, I have got experience of working in that area, since 2005–6, personally going along with the forest department and patrolling boats in the sea. So many times during seizure of boats, they will say that they are from local Odisha villages, and yet, they will not be able to produce a single document and all of them will be speaking in Bangladeshi dialect, which is different from that of Midnapore (in west Bengal).'

And once again, the 'politics of authenticity' that so often constitutes indigenist-regionalist imaginations of eco-nationalism, as the one described in this paper, appears. The discourse itself is facilitated by a particular kind of historical role adopted by the state, but it may also mediate the various contingent acts of the state's multiple actors.

CONCLUSION

In this paper, the notification of a marine sanctuary accompanied by wider economic pressures create the context in which questions are raised about legitimate resource rights. The setting is the arrival of Bengali immigrants into the Odisha coast and the transformation of the fishing landscape here through the onset of marine fishing. The paper substantively describes how a historical context of resistance to the incoming Bengalis by Odiya caste fishers is magnified through contemporary economic constraints and the restrictions imposed by the Gahirmatha marine sanctuary, giving way to the construction of a distinctive regional Odiya eco-nationalism. This particular environmentalism emphasises the prior rights of indigenous Odiya traditional fishers and highlights the destructive influence of illegal Bengali immigrants. The paper also draws attention to the particular role of the Indian state, especially in its laws and policies towards immigrants and refugees, in enabling the making of such an environmental discourse. Importantly, the paper shows that this environmentalism is constituted by various different disparate voices that lack any unified position or even politics.

While the paper has been interested in revealing the construction of such a discourse, it has been equally keen to analyse its uneasy juxtaposition with the reality of fishers' lives and social relations. It explores the limits to Odiya environmentalism, as described, through fieldwork where Bengali fishers widely reject the accusations levied, and most Odiya fishers do not lend credence to these either. It draws from the important demonstration of Odiya-Bengali solidarity after the terrible 'quit India' directives of the 2005 state order. These discussions further underpin the central point of the paper that, in fact, the main line of struggle being witnessed is between small fishers and the state, within the wider economic context and the pressures of the sanctuary. The paper develops this point at some length, drawing out the common economic constraints faced by small fishers, both Odiya and Bengali, and their low position within the marine class hierarchy. But equally, even as these basic conflicts are re-enacted at sea, the paper shows that there are moments when such environmentalism can have profoundly serious and tangible effects.

Finally, the injustice that conservation policies may

contain for local communities is a well-known theme within the wider scholarship. In this paper, we have also spoken of how conservation itself can be interpreted as a dominant way of visualising the proper use of nature, through metropolitan secular projects for national improvement backed up by concrete and often oppressive state interventions. In addition, there may be multiple indigenist-regional narratives around resource use and resource rights that map on to particular notions of regional spaces nested within a larger nation, imaginings of which are not unified either. In this paper, the particular contours of a regional Odiya eco-nationalism connect with ideas of territorial sanctity and the abhorrence of 'infiltration' by immigrants, as systematically incorporated into successive laws and state policies. In a larger sense then, the paper seeks to draw critical attention to the injustice of misrecognition, through misrepresentation and framing that environmental discourses might engender, although its precise effects may vary from one context to another.

FOOTNOTES

1. All villages names have been changed to preserve anonymity.
2. A *gram* or village panchayat is the lowest unit of the three tier system of elected local government in India.
3. Map courtesy Sudhanshu Behera. Prepared with Arc GIS Software, 2011.
4. Refer to Gupta and Sharma (2008) for an interesting perspective on how state restrictions impact upon cross border movements of coastal fisherfolk in South Asia
5. Interview with Retired Deputy Director Fisheries, Cuttack, April 2012.
6. http://www.odishafisheries.com/website/production/fish/sector.htm (Accessed 6 Jan 2016).
7. http://www.odishafisheries.com/website/production/fish/sector.htm (Accessed 6 Jan 2016).
8. Interview with retired Deputy Director, Fisheries, Cuttack, April 2012.
9. Interviews with fishers at Matsyapalli village, April 2012.
10. http://www.ndtv.com/india-news/odisha-unveils-roadmap-to-boost-fish-production-1207287 (Accessed 6 Jan 2016).
11. The Citizenship Amendment Act 1985. Text paraphrased fromhttp://lawmin.nic.in/legislative/textofcentralacts/1985.pdf (Accessed 6 Jan 2016).
12. The Citizenship Amendment Act 2003. http://indiankanoon.org/doc/949775/ (Accessed 6 Jan 2016).
13. Focus group discussion with women in Matsyapalli village, November 2011.
14. http://www.iczmpodisha.org/aim_and_objective.htm (Accessed 7 Jan 2016).
15. Implemented through the Fisheries and Animal Resources Development Department, this component aims to provide alternative livelihoods options to 80 fishing villages comprising 600 self-help groups (SHGs) and more than 9000 families in both coastal stretches. The sarpanch of Matsyapalli gram panchayat told us that there ought to be about 300 SHGs in Kendrapara district, but approximately 60 SHGs had been formed (at the time of the interview in 2012). He also reported that money came to each SHG through phased instalments, which was 'demoralising' for members.
16. Interview with environmentalist, Operation Kachhapa, March 2012.
17. http://eprints.cmfri.org.in/3331/1/Special_Publication_No_32.pdf (Accessed 6 Jan 2016).
18. http://investodisha.org/download/Odisha_Fisheries_Policy_2015.pdf (Accessed 6 Jan 2016).
19. Gill nets are vertical net panels (about a km or two in length and around 15–20 ft deep), propped up by weights against a wall. Trawl nets on the other hand are meant to scrape the bottom of the sea, though mid-water trawling is also practised in certain parts.
20. Interview with retired Deputy Director, Fisheries, Cuttack, April 2012.
21. Interview with Behera, self-styled leader of the *Kaibartas* and prominent environmental campaigner, Paradip, April 2012.
22. Conversations with fishing net knitters, Paradip, April 2012.
23. Interview with Secretary, OTFWU, Cuttack, April 2012.
24. Interview with ADF Marine Fisheries, Cuttack, April 2012.

REFERENCES

1. Baviskar, Amita. 2003. States, communities and conservation: the practice of ecodevelopment in the great Himalayan national park. In *Battles over nature: science and the politics of conservation*, ed. Vasant Saberwal and Mahesh Rangarajan, 267–300. Delhi: Permanent Black.
2. Baviskar, Amita, Kavita Philip, and Sinha Subir. 2006. Rethinking Indian environmentalism: industrial pollution in Delhi and fisheries in Kerala. In *Forging environmentalism: justice, livelihood and contested environments*, ed. Joanne Bauer, 189–257. New York: M.E. Sharpe.
3. Chatterji, Joya. 2007. *The spoils of partition: Bengal and India, 1947–67*. New Delhi: Cambridge University Press.
4. Chhotray, Vasudha, and Roger Few. 2012. Post-disaster recovery and ongoing vulnerability: Ten years after the super-cyclone of 1999 in Orissa, India. *Global Environmental Change* 22: 695–702.
5. Chhotray, Vasudha. 2016. Nullification of citizenship: Negotiating authority without identity documents in coastal Odisha. Contemporary South Asia. Revise and resubmit.
6. Dove, Michael R. 2006. Indigenous people and environmental politics. *Annual Review of Anthropology* 35: 191–208.
7. Fraser, Nancy. 1997. From redistribution to recognition? dilemmas of justice in a 'Postsocialist' Age. In *Justice interruptus: critical reflections on the 'Postsocialist'*

condition, 11–40. New York: Routledge.

8. Fraser, Nancy. 2000. Rethinking recognition. New left review.http://newleftreview.org/II/3/nancy-fraser-rethinking-recognition. (Accessed Feb 2015).
9. Forsyth, Tim. 2006. Are environmental social movements socially exclusive? an historical study from Thailand. *World Development* 35(12): 2110–2130.
10. Gupta, Charu and Mahesh Sharma. 2008. Contested coastlines: Fisherfolk, nations and borders in South Asia. New Delhi: Routledge.
11. Hajer, Maarten A. 1995. The politics of environmental discourse: Ecological modernisation and the policy process. Oxford: Clarendon Press.
12. Holston, James. 2008. *Insurgent citizenship: disjunctions of democracy and modernity in Brazil*. Princeton: Princeton University Press.
13. Jayal, Niraja Gopal. 2010. Balancing political and ecological values. *Environmental Politics*10(1): 65–88.
14. Jayal, Niraja Gopal. 2013. *Citizenship and its discontents: an Indian history*. Cambridge: Cambridge University Press.
15. Kapur, Ratna. 2012. *Makeshift migrants and Law: gender, belonging and postcolonial anxieties*. London, New York and Delhi: Routledge.
16. Kudaisya, Gopal. 1997. Divided landscapes, fragmented identities: east Bengal refugees and their rehabilitation in India, 1947–79. *Singapore Journal of Tropical Geography* 17(1): 24–39.
17. Low, Nicholas, and Brendan Gleeson. 1998. *Justice, society and nature: an exploration of political ecology*. London and New York: Routledge.
18. Li, Tanya Murray. 2000. Articulating indigenous identity in Indonesia: Resource politics and the tribal slot. Comparative Studies in. Society and History 42 (1): 149–179.
19. Li, Tanya Murray. 2004. Environment, indigeneity and transnationalism. In *Liberation ecologies: environment, development, social movements*, ed. Richard Peet and Michael Watts, 339–370. London: Routledge.
20. Mawdsley, Emma. 2004. India's middle classes and the environment. *Development and Change*35(1): 79–103.
21. Neumann, Roderick P. 2004. Nature-state-territory: toward a critical theorisation of conservation enclosures. In *Liberation ecologies: environment, development, social movements*, ed. Peet Richard and Watts Michael, 195–218. London: Routledge.
22. Peluso, Nancy Lee, and Peter Vandergeest. 2001. Genealogies of the political forest and customary rights in Indonesia, Malaysia and Thailand. *The Journal of Asian Studies* 60(3): 761–812.
23. Rangarajan, Mahesh. 2001. *India's wildlife history: an introduction*. Delhi: Permanent Black and Ranthambhore Foundation.
24. Roseberry, William. 1994. Hegemony and the language of contention. In *Everyday forms of state formation: revolution and the negotiation of rule in Mexico*, ed. Gilbert Michael Joseph and Daniel Nugent, 354–366. Durham and London: Duke University Press.
25. Saberwal, Vasant, and Mahesh Rangarajan. 2003. Introduction. In *Battles over nature: science and the politics of conservation*, ed. Vasant Saberwal and Mahesh Rangarajan, 1–28. Delhi: Permanent Black.
26. Sahu, Geetanjoy. 2008. Mining in the niyamgiri hills and tribal rights. *Economic and Political Weekly* 43(15): 19–21.
27. Salagrama, Venkatesh. 2006. *Trends in poverty and livelihoods in coastal fishing communities of Odisha state*, 490. India: FAO Fisheries Technical Paper.
28. Sampath, V. 2005. An overview of the international and national fisheries policy and related issues: A desk research.http://odishafisheries.com/File/Review%20of%20International%20and%20National%20Fishery%20Policies-dec04.pdf. Accessed 20 Aug 2012.
29. Schlosberg, David. 2004. Reconceiving environmental justice: global movements and political theories. *Environmental Politics* 13(3): 517–40.
30. Scott, James. 1998. Seeing like a state: How certain schemes to improve the human condition have failed. New Haven and London. Yale University Press.
31. Sharma, Mahesh. 2012. *Green and saffron: Hindu nationalism and Indian environmental politics*. Ranikhet: Permanent Black.
32. Sinha, Subir, Shubhra Gururani, and Brian Greenberg 1997. New traditionalist discourse of Indian environmentalism. *Journal of Peasant Studies* 24(3): 169–204.
33. Sinha, Subir. 2012. Transnationality and the Indian Fishworkers' movement, 1960s–2000.*Journal of Agrarian Change* 12(2,3): 364–389.
34. Singh, H.S. 2002. *Marine protected areas in India: status of coastal wetlands and their conservation*. Gujarat: GEER Foundation.
35. Sivaramakrishnan, K., and Gunanel Cederlof. 2012. Ecological nationalisms: claiming nature for making history. In *Ecological nationalisms: nature, livelihoods and identities in south Asia*, ed. Gunnel Cederlof and K. Sivaramakrishnan, 1–40. Ranikhet: Permanent Black.
36. Sridhar, Arati, Basudev Tripathy and Kartik Shanker. 2005. A review of legislation and conservation measures for sea turtles in Odisha, India. Indian Ocean Turtle Newsletter. 1(1): 1-17 .
37. Tietze, U. 1985. *Artisanal marine fisherfolk of Orissa*. Cuttack: Vidyapuri.
38. Walker, Gordon. 2012. *Environmental justice: concepts, evidence and politics*. London: Routledge.
39. Wright, Belinda, and Biswajit Mohanty. 2006. Operation kachhapa: an NGO initiative for sea turtle conservation in odisha. In *Marine turtles of the Indian subcontinent*, ed. Kartik Shanker and B.C. Choudhary, 290–302. Private Limited: Hyderabad Universities Press.
40. Young, Iris M. 1990. *Justice and the politics of difference*. Princeton: Princeton University Press.

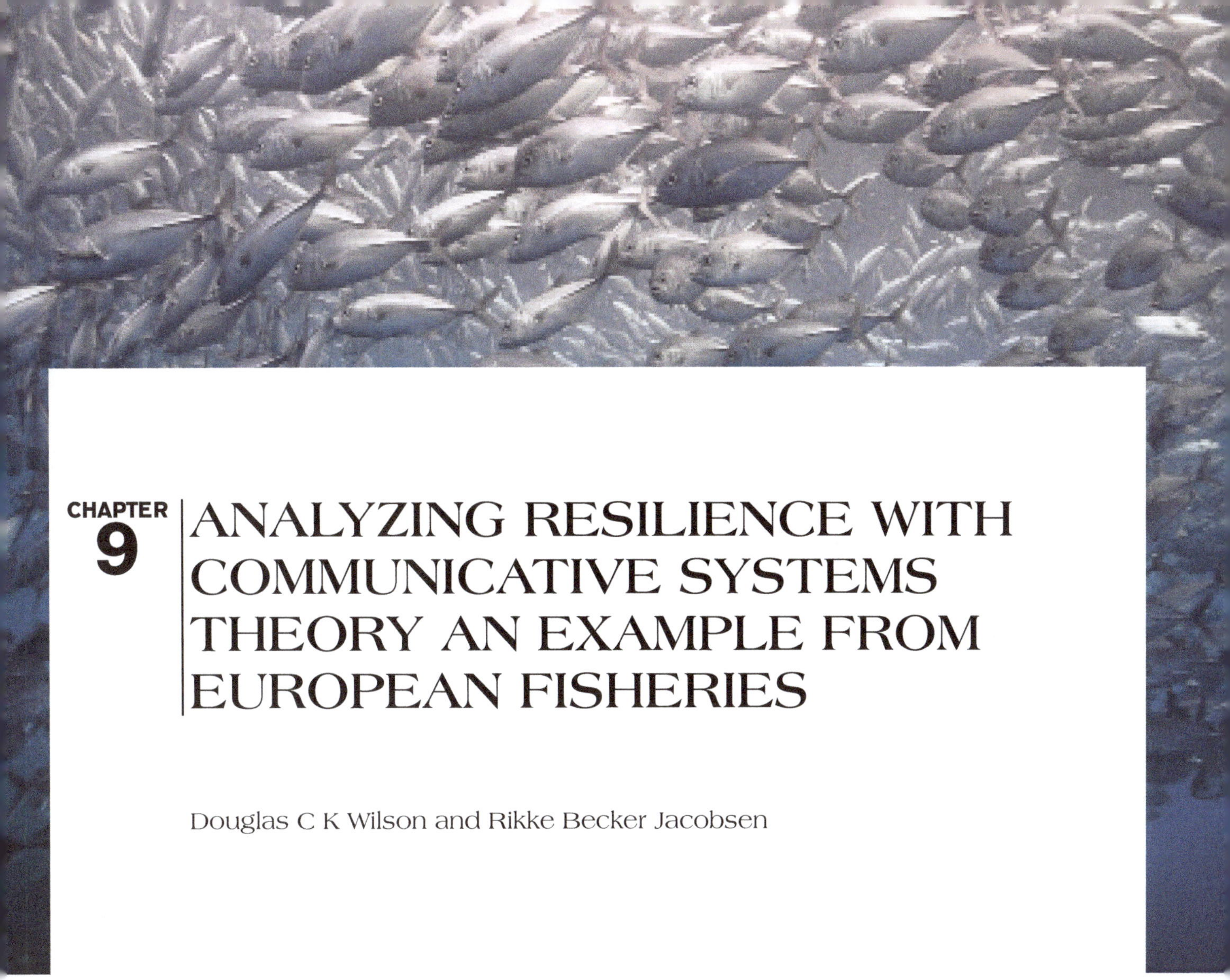

CHAPTER 9 ANALYZING RESILIENCE WITH COMMUNICATIVE SYSTEMS THEORY AN EXAMPLE FROM EUROPEAN FISHERIES

Douglas C K Wilson and Rikke Becker Jacobsen

INTRODUCTION

In recent years there have been a number of exciting theoretical developments around the idea of the resilience of social-ecological systems (SES). Many are inspired by theories of complex adaptive systems in both natural science and economics (for instance Holland1992; Miller and Page2007. Using agent-based approaches they illuminate SES by analysing the relationships between institutions, the strategies that emerge from the incentives and norms of agents, and ecological systems (Folke2006; Ostrom2005,2007). However, while resilience theorists recognizes the importance of the characteristics of communication, especially in the form of collective learning (Armitage2005; Folke2006; Wilson2002) systematically theorized communicative approaches in resilience thinking are lacking; a major reason being the difficulty of developing such approaches within an agent-based framework. This paper helps address this lack by examining science-policy "boundary objects" using Habermas' (1987) communicative systems theory (CST). We suggest that "communicative resilience" is an essential part of "social–ecological resilience". We illustrate the approach through a discussion of one complex type of SES: multi-species, mixed fisheries (MSMF). The paper begins with a description of the MSMF problem, then outlines the theoretical approach, and finally describes research based on communicative resilience that examines two types of boundary objects that have been important in MSMF in Europe. The conclusion suggests communicative resilience can be improved by not trying to reduce discussions of complex systems to simple categories designed for institutional convenience.

MULTISPECIES MIXED FISHERIES AS COMPLEX SOCIO-ECOLOGICAL SYSTEMS

A multispecies fishery refers to fisheries where more than one species is vulnerable to being caught. In many cases a multispecies fishery is also a mixed fishery, meaning that more than one combination of fishing vessels and gears are being used. Mixed fisheries generally involve four major kinds of actors. The fishing industry is very diverse but generally holds maintaining long-term sustainability as a strategic goal while facing short-term pressure to cover operating costs. Conservation ENGOs wish to protect marine environments, biodiversity and sustainability. To some degree managers' incentives match both of the other main groups when it comes to conservation objectives; they are interested in both profits

---

Citation: Douglas C K Wilson and Rikke Becker Jacobsen, Analyzing resilience with communicative systems theory an example from European fisheries, DOI: 10.1186/2212-9790-12-1.

and sustainability and, sometimes biodiversity. Scientists are called upon to provide advice on ecological conditions on which policies will be built.

MSMF are managed by a varied array of measures. The two most basic approaches are 'input' control over the amount and type of fishing activities and 'output' control over the amount of fish caught. In Europe the two main systems for implementing these are referred to respectively as 'effort management' and 'quota management'. Effort management makes it hard to transparently divide fishing opportunities among fishers or countries. Fishers also quickly find ways to catch more fish with a given effort, a problem called 'effort creep'. However, effort management has weaker incentives for discarding unwanted fish because a vessel may land whatever fish it catches in its allotted time. Effort is also not necessarily applied to a specific fish stock and this simplifies scientific monitoring and estimations of fish abundance.

Quotas are an amount of fish, so they are easily divisible and the relationship between quotas and catches is in principle direct. Quotas, however, are based on a total allowable catch (TAC) that must be defined based on a fairly precise estimate of how many fish are available to be sustainably caught from a particular stock. These estimates can be especially difficult in MSMF because of biological interactions among species and the data complexities that arise from having many different fleets and gears catching fish. All quota systems face problems with discarding fish, for example to bring higher quality fish to market. In MSMF an important driver of discarding is catching fish that a fisher does not have a quota for or that is otherwise protected. Discarding also creates data problems because discarded fish are not recorded. Quota management has historically been based on a common pool quota given to a fishing fleet, but more recently granting an individual quota (IQ) to fishers has become popular. These systems eliminate economically inefficient and dangerous practices such as racing to fish a common quota. When made freely transferable (ITQ) the market will allocate fish with the highest economic efficiency among management alternatives (Hannesson,2004).

MSMF management has evolved in complex ways. In Japan the strategy selected was to control effort, through reductions in licenses, shortening seasons and imposing closures. The emphasis was on the simplicity and effectiveness of these measures compared to a quota system within a large coastal fishery (Nagaski and Chikuni,1989). Effort creep nullified the effectiveness of the input controls (Yagi2001) so a quota system was introduced for several major species in 1997. Managers were still concerned about the scientific accuracy needed for a quota system, and an effort approach was retained for some stocks. The New England groundfish fishery in the USA experienced serious decline in important stocks as early as the 1960's. The first generation of measures combined input and output control, introducing catch quotas, minimum fish size, minimum mesh size and seasonal closures of spawning areas and later individual trip limits. Direct limitations on both catch and fishing effort came in 1994–1995 with a range of new measures including individual stock TACs for five stocks and an aggregated TAC for seven others. Recently sector management has been introduced. This involves self-organising groups of fishers who enter into a contractual relationship with management in which they develop their own plans to fish sustainably for a total allowable catch based roughly on their individual histories. These contracts then excuse them from some other restrictions, including the days-at-sea effort control system.

Countries using ITQ systems have addressed the discarding problem in several ways, all of them complex and multilayered. In Canada vessels are required to have a quota for any species they might catch and discarding is banned with some few exceptions. The larger vessels carry observers on board but discarding remains a problem on smaller vessels (Rescan and Wilson2009) because fishers find it easier to dump than to go through the complex system of trading for the quota they would need to legalise their catch (Caddy & Agnew2004). In Canada and New Zealand temporary closures of areas with high discards are used to back up the discard ban. This measure has led to more selective fishing (Branch and Hilborn2008), as fishers seek to avoid the automatic closures resulting from an exhausted cod quota (Caddy & Agnew2004). New, inexpensive surveillance technologies are increasingly being used to monitor fishing activities at sea.

Sanchirico et al. (2005) review how ITQ systems in MSMF in five countries have addressed the general problem of balancing catches with allocated quotas. The key is to give fishers the flexibility to work within the system. They found that it was helpful to have multiple instruments in place to give fishers different ways to match quotas and catches. All the systems required ongoing changes and elaboration in response to ecological and economic conditions (Sanchirico et al.2005). In addition to effort and quota control, management also makes use of myriad technical measures and protected areas. ITQ systems help address the allocation problem, but systems based on strong property rights are also problematic from the viewpoint of cumulative effects because they are a poor fit with measures that decrease the value of quotas. Murray et al. (2010) argue that the relentless accumulation of management measures brings about a loss of flexibility and the erosion of community.

THEORETICAL APPROACH

Resilience thinking is rooted ideas about complex adaptive systems (Holland1992) first developed in ecology. Berkes et al. (2003) suggest that resilience has been defined two basic ways in ecology. The first defined resilience as the ability to return to this steady state. More recently the focus has been on how a once stable ecological regime, i.e. a particular structure of ecological processes, can change into a new stable regime. Resilience is defined as the magnitude of the disturbance needed for such a change. They, along with others (Folke2006; Gunderson and Holling2002; Low et al.2003) emphasize that resilience thinking is a step away from an equilibrium concept of systems towards one that assumes change and seeks to explain stability. Redundancy, homogeneity, diversity and adaptability are commonly associated concepts as in, for example, Folke's (2006) argument that poor management results in reduced resilience when landscapes become homogenized and more vulnerable to disturbances.

Social resilience is a much less developed concept with very general definitions, such as Adger's (2000) "the ability of groups or communities to cope with external stresses and disturbances as a result of social, political and environmental change" (347) being common. Too often theorizing social resilience has been muddied by analogies from natural ecology (Wilson2010). Capacity for adaptive social learning is perhaps clearest dimension of social resilience found in the literature and great emphasis is placed on it by Armitage (2005), Berkes et al. (2003), Gunderson and Holling(2002), Hahn et al. (2006), and Wilson(2002) among many others.

By seeing society and its material basis as a set of interconnected systems ecological and social resilience can be thought of as aspects of SES resilience. Both kinds of resilience depend on one another, but one cannot assume that one guarantees the other (Folke2006). Folke(2006:259) cites Carpenter et al.2001 in respect to three core ideas within SES resilience: "(1) the amount of disturbance a system can absorb and still remain within the same state or domain of attraction, (2) the degree to which the system is capable of self organization (versus lack of organization, or organization forced by external factors), and (3) the degree to which the system can build and increase the capacity for learning and adaptation. "Central to the argument in this paper is the idea that adaptive processes result from a system's self organization (Folke2006, Norberg and Cumming,2008) and, hence, "the concept of resilience in relation to social–ecological systems incorporates the idea of adaptation, learning and self-organization in addition to the general ability to persist disturbance" (Folke2006:259).

Using a version of CST inspired by Habermas(1987;1991), and further developed elsewhere (Wilson and Jentoft,1998; Wilson2003,2009a), we offer the concept of "communicative resilience" as the part of the resilience of SES that makes adaptive social learning possible. We define communicative resilience as the degree to which people within an SES are able to express themselves by drawing on the full range of background understandings that allow communication to be meaningful. We argue that many institutions, i.e. the shared cognitive, normative and regulative understandings that maintain stable patterns of social interaction (Wilson and Jentoft,1998), systematically restrict such expression. These restrictions can and do inhibit social learning by blocking the criticism and/or contextualization of understandings of the social or ecological environment.

If any institution is going to maintain a stable pattern of behaviour it must simultaneously coordinate two often contradictory kinds of interaction: strategic action in pursuit of actors' objectives and communicative action for developing mutual understandings of situations. "Mutual understanding" is not agreement; it means that participants' perceptions of a situation are close enough for responsive interaction to occur. Agent-based theories of institutions, while recognising the importance of communication, conceptualise it as one more strategic resource available to the agent (Ostrom2005). For example, Ostrom et al. (1994:33) consider 'the way actors acquire process, retain, and use knowledge, contingencies and information' to be one of four key characteristics of an actor. CST, in contrast, focuses on the conditions of communication at the institutional level.

CST considers communication through the contrasting logics of the 2nd person (you and I communicate with each other) and 3rd person (I/you/he/she observe a system of communications) perspectives. The "mutual" in "mutual understanding" depends on the 2ndperson communication and begins with the standards of "rational communication". Habermas' famous ideal speech situation, that there is "no manipulation involved in the communication; and that everything communicated is open to question about its validity" (White1988, 56), expresses such standards. The power of these counterfactual ideals derives from the fact that during 2nd person interaction we use them to evaluate communicative actions. Mutual understandings, even in situations of differential power, require an element of persuasion. Your boss can *tell* you what to say and do but he or she must *persuade* you to think something is actually true. These standards guard access to what Habermas calls the "lifeworld", the background of shared meanings – language, culture, socialization – that make communication possible.

Rational communication also requires that claims be oriented toward types of mutual understandings. For example, a claim of fact is oriented toward consensus about its truth; a claim of value is oriented towards an agreement about the rightness of its general applicability, and a claim of interest is oriented towards possible compromise. These orientations are background assumptions in communication that help us make sense of what is being said. An illustration is that if someone says "let's just split the difference" in respect to interests this makes perfect sense, but saying the same thing in respect to facts is at best an admission that the facts are unknown.

The 3rd person perspective is utterly different. From this perspective we begin with the question of strategy. When we observe the communicative system from the 3rd person we often see persuasion as a veneer over manipulation or draw attention to how the voices of the powerless have been silenced. As social scientists seeking to understand a discourse we look at how people link facts, values and interests, not at how they keep them rationally separate. It is this huge difference between the logic of the 2nd and 3rd person perspectives that is the heart of CST.

To relate these two perspectives CST takes an entirely different angle from agent-based approaches and describes social systems in communicative terms. The 3rd person view of mutual understanding begins by observing the social-psychology of building and reproducing institutions at the small group level. We account to each other for what we do and say and these accounts iteratively reproduce patterned social interaction. Scott and Lyman(1968:46) tell us that accounts 'shore up the timbers of fractured sociation' and observe that ways of accounting for behaviour become routine within subcultures and groups. The ethnomethodologists showed through repeated observation and experiments that norms and rules do not so much provide a cookbook for action as the language of accounting for behaviour after the fact (Heritage1984). Weick(2001) describes how organizations faced with new situations act first and make sense out of their actions later. Competing accounts of reality are the material of strategic action in the communicatively conceived social system.

CST develops the systemic aspects of the 3rd person perspective through the idea of "generalized communications" which reconceptualises as mechanisms of communications those dynamics most closely associated with social power. Prestige and influence are generalized communications that contribute to mutual understanding through pre-understandings about whose voices will carry the most weight.

They operated in highly lifeworld-dependent institutions such as science or communities. Markets, private bureaucracies, and governments, however, make heavy use of money and authority,[a] generalized communications that Habermas(1987) calls "steering media". Money and authority create mutual understandings while shortcutting communications dependent on the lifeworld. Steering media do this in a mechanical and automatic fashion facilitated by a nearly complete marginalization of the lifeworld. Authority requires access to the lifeworld only to create an understanding of to the content of the order, the related sanction and the recognition of a condition that triggers the order's application. One can use money to buy a soda in a place where one knows nothing of the language or culture, the needed mutual understanding is automatically provided. Steering media are critical to institutions that operate on higher scale levels and/or require more predictable outcomes. For steering media to operate some form of coercion or pressure must be in place to replace the persuasion that the vastly reduced access to the lifeworld makes no longer possible. This coercion must refer to some objective reality. For example you cannot get the soda without paying, and the existence of the soda must also be verified. More relevant to the discussion at hand, authority requires a verifiable, legal "rule condition" (Ostrom2005), to support its ability to replace the richer communications lifeworld access makes possible.

In institutions where steering media are the primary mechanisms of communication, accounting becomes formal accountability focussed on these objective realities. The contrast of the two perspectives is important here. What we call "objectivity" from the 3rd person is the same condition of sanctioned reality that we refer to as "transparency" from the 2nd person. The scientific method that we rely on to produce objective facts is nothing but a set of procedures for making knowledge as transparent, i.e. as verifiable by the community at large, as possible (Wilson2009a). Accountability is not possible without this dual condition. To hold people accountable we must *see* what they have done. This duality means that accountability acts as a sort of lubricant between the often contradictory goals of strategic actions in pursuit of interests and communicative actions in pursuit of mutual understandings. If certain fish stocks are to be avoided in an MSMF what is required is both a commitment by the fishers to selective fishing and evidence that selective fishing actually taking place, without both of these in place the attempt to create a new pattern of behaviour will grind to a halt.

Within environmental management these objective realities are increasingly 'boundary objects'. Boundary objects are such things as models, indicators, environmental action plans and joint research products that focus discussions across the boundary of science and policy (Cash et al.2002). In complex SES, developing mechanisms for accountability almost always involves science because of the complexity of human activities and of monitoring their impacts on ecosystems. When a boundary object is working it pulls together concepts that are acceptable to the various parties involved in its development. In complex SES, the elements of institutional grammar that Ostrom calls "conditions," are often boundary objects. Conditions are the 'variables that define when and where an institutional statement applies' (2005:149). From the CST perspective these are the objective referents that authority depends on to shortcut the lifeworld.

The need for objective rule conditions means that institutional imperatives heavily sway the content of boundary objects often resulting in less access to the lifeworld for both scientists and stakeholders seeking to raise questions. This is because in complex SESs, environmental managers often seek boundary objects that give simple messages with little nuanced content even when the science is complex and/or uncertain. An example is the setting of TACs. ICES scientists very frequently deal with uncertainties by inserting rich textual explanations of data and analysis below the tables where TACs are reported, but the European Commission has no good way to use this textual information and so must simply adopt the reported number (Wilson2009a). Rule conditions have to be created even under conditions of complexity and uncertainty. The threat is that, once in place, a rule condition can be reified into a new "objectivity" that stands beyond effective contextualizing, questioning or explaining and hence is not objective because it is not transparent. Rule conditions anchor a steering medium that works by marginalizing the lifeworld, i.e., the same communicative resources that are needed to generate the critiques, nuances and caveats that allow a rich reflection of the realities in the social and natural environment. Institutions that are crucial to the resilience of SES are grounded by science/policy boundary objects that are used to literally truncate meaning itself – including meaning needed for accurate science. No institution is more dependent on full lifeworld access than science. Theories rooted in a 3rd person perspective have noted the creation of scientific "black boxes" but have depicted them as the result of strategic interactions (e.g. Latour1987), something that can only illuminate aspects of the particular forms the boundary objects take. From the 2nd person perspective within CST it is the shortcutting of the lifeworld that seals the box around science/policy boundary objects.

CST's 2nd person perspective opens up the possibility of the amelioration of these problems through institutional design. This is what communicative resilience adds to the overall resilience of SESs. The mixture of generalized and rational communications found in institutions is variable and can be changed by design. More rational communications allows the communication of richer and more nuanced pictures of both ecological and social situations. It is the foundation of social learning, the self-organization and adaptation of social institutions.

Once understood, CST is a coherent description of social reality, it is rooted in everyday experience and some of its tenants have survived systematic tests (Wilson1998). However, if communicative resilience based in CST is going to be useful for diagnosing problems and informing institutional design then it must offer a methodological approach rooted in the 2ndperson, 3rd person distinction that is central to the theory.

Standard social science methods can be used to alternate between the perspectives to build up a picture of the central diagnostic question: the impact the 3rd person communicative system is having on the 2nd person lifeworld. Qualitative, in-depth interviews and focus groups, when correctly carried out, straddle the 2nd and 3rd person. The interviewer, who is seeking to understand the subjective experience of the respondent, listens and questions while evaluating statements through the screen of rational communications. But the discussion itself is oriented to the 3rd person because what is being asked is a description of the SES of interest. This output is subjected to a discourse analysis; a 3rd person exercise that maps the way facts, interests, and values are linked in discursive themes while trying to minimize the question of the validity of the claims being made. Then institutional factors that are blocking claims people wish to make can be identified. The first case study below of an MSMF boundary object stops at this point.

However, as the second case below demonstrates, quantitative methods used to understand the subjective viewpoints of respondents through various sorting exercises (Brown,1980; Coxon,1999) can contribute as a further iteration that clarifies the relationship between the two perspectives by rendering respondents' 2nd person perspectives into comparable forms. The elements being sorted are usually statements of facts or value but they can also be physical or natural objects. They must emerge from the discourse analysis and not from some *a priori*theory. Kapasa et al. (2005), for example, using Q sorts and consensus analysis, show that Zambian fishers who use a type of gear that, without scientific confirmation, is considered "destructive" tend to have both the highest levels of local ecological knowledge and are the most likely to hold opinions that are contrary to a strong consensus agreeing with the ecological opinions and policies of the local Department of Fisheries. Their use of illegal gear, however, blocks their full participation in the local fisheries co-management programme. In another example Wilson(2010) combines agreement with statement sets and social network analysis. The degree to which pairs of respondents agree with each other about science while disagreeing about policy is used as a measure of rational communication. The analysis found that the only network pairs having higher agreement on science than policy were fellow members of a fisheries co-management committee.

The boundary objects case studies that follow seek to apply CST through this same set of methods.

BOUNDARY OBJECTS IN EUROPEAN MULTI-SPECIES MIXED FISHERIES

This section draws on field data to explore two kinds of boundary objects in European MSMF. The two are nearly opposites in terms of the role played by the boundary objects in the creation of rule conditions. Recovery plans for depleted stocks are sets of various legal acts and their discussion as observed here rarely touched on specific rules. The definitions of fisheries and fleets, however, are directly used as the basis of both scientific models and management rules. The field data comes from the observation of nine working groups of European fisheries scientists and three meetings of Regional Advisory Council (RAC) working groups. RACs are stakeholder fora on which both the fishing industry and conservation ENGOs sit. The three RACs included here are the North Sea, Baltic and Southwest Waters RAC. The RACs' remit is to provide advice on fisheries management to the European Commission's Directorate General for Marine Affairs and Fisheries (DGMARE). In addition to observing meetings, interviews were also carried out with fishers in 12 fishing ports, six fisheries scientists, five government fisheries managers and three focus groups. All of these were in-depth, open ended interviews. The port interviews also included a pile sorting exercise to see how fishers classify vessels. Information from a short email questionnaire done with twelve RAC members is also used.

Recovery and long-term management plans

This set of boundary objects involves scientists, managers, conservation ENGOs and the fishing industry. The first recovery plan began in February 2001. Initially they were sets of emergency measures, but general guidelines came in 2004. While individual measures are legally enacted, recovery plans are not a single legal act. A general stakeholder consensus supports the plans. However, this shared ideal of seeking the recovery of a stock leaves many issues of disagreement in the details. In spite of these differences in details and emphasis, even this broad, symbolic consensus has proven important because it has allowed stakeholders with conflicting interests to develop a cooperative agenda, especially in the RACs (Wilson2009b).

The following is from a public talk by the Director of Conversation Policy at DGMARE early in the process of recovery plan development. The focus is on the scientific input into the process of developing recovery plans:

Q1. What do we need for such recovery plans? As they will normally contain targets for recovery or long-term management, usually expressed in terms of biomass, we need clear precautionary and limit reference levels for the stocks concerned. They will also contain harvest rules; that is rules for the fixing of TACs over several years ... [later in the document]. It means a concentration of resources, at least in the short term, on the analysis of the most depleted fish stocks (Farnell2003).

For DGMARE recovery plans are strategic mechanisms that are used to concentrate fisheries management capacity on a subset of the most serious problems with depleted stocks. There is an emphasis on biomass targets and the precautionary approach (PA).

Two central questions about the recovery plans are their relationship to long-term management plans (LTMPs) and effort management. Q2 presupposes a rather smooth movement from recovery plans to long-term management plans (LTMP) based on harvest control rules (HCRs). These are rules in a form 'for Y condition X change in harvesting pressure is required' and are one of the aspects of LTMPs that all stakeholders see as useful for differing reasons. The following is unequivocal about LTMPs while being more circumspect about effort management:

Q2. Interviewer: So, you see the recovery plans as sort of a starting point for a general reform, a general move towards effort management and long-term management? Commission Administrator: Certainly to long-term management; to effort management in justified cases. I think it's clear there won't be a general signing up to effort management as the normal, or as a normal instrument for fisheries management, but I think there is recognition that we will need to manage effort in order to make particular efforts to stop the serious degradation of the stock collapses.

The movement from recovery plans to more general long-term planning is a priority for the fishing industry. The following is from a member of an industry RAC member:

Q3. [LTMPs] remove political interference in the process of setting allowable catches by removing the ability of Member States to question the output given that the plan would have been agreed through Council. The output is clinical and precautionary thus affording greater security to the stocks.

Recovery plans are to be driven by technical considerations through HCRs. For the industry an important aspect of this is reducing uncertainty in business planning and stability of supply.

Some members of the industry see positive outcomes from the recovery plans because they provide a way of contributing their knowledge about technical conservation methods, selective fishing, and closed areas and gave individual fishers a way to benefit from responsible fishing. However, both the industry and the ENGOS were engaged late in the process. A RAC staffer described them as "sprung upon" the industry based on an "ICES philosophy". The International Council for the Exploration of the Sea (ICES) is an independent multilateral organisation. It produces the official scientific advice for fish stocks in the northeast Atlantic and adjacent seas and DGMARE is its biggest client. Recovery plans were initially developed by DGMARE and ICES and then presented for comment. In the development of science – policy boundary objects scientists are not seen as simply neutral facilitators no matter how hard they may try to keep their role as doing science but not management. The failure to involve stakeholders early limited the range of their responses. For one thing, the plans were already based on individual fish stocks as their basic unit, while the industry wanted this unit to be fisheries and the ENGOS wanted a more ecosystem-based perspective. A RAC staff member:

Q4. We had a really important meeting where we laid down our ideas for long-term management plans, and the most important thing that came out of that was that it was the fisheries that the plans ought to relate to, not the stocks. Now, that's been diluted since, whenever we discuss management plans with the Commission they still try to refer to stocks.

Of course, many fishers simply do not agree with the recovery plans. The industry sees both scientists and managers as interpreting the precautionary approach too narrowly in the development of the plans. Many in the industry felt that if they had been included earlier in the process then the objectives would have been more realistic. They echoed the classic MSMF complaint about protecting one threatened stock by limiting access to all the others:

Q5. Industry RAC Representative: Opportunities to take other species in a mixed fishery are lost when a so called recovery stock is taken. The abundance of cod makes it almost impossible for cod to be avoided and thereby additional days to catch other species are lost.

Conservationists also strongly support the move towards a longer term approach. One conservationist RAC Member argued that the LTMPs being developed are improving under adaptive management and developing them for all stocks should be a priority. Conservationists are also link the recovery plans to a more ecosystem-based approach to management.

Q6. Conservationist RAC Member: It appears that a requirement to protect/restore ecosystem functioning is missing from the current application, and this must be incorporated into processes in the future, if recovery is to be successful.

Perhaps the greatest challenge the Commission faces as it tries to develop recovery plans is that, particularly in their long-term aspects, these plans are not a clean fit with official EU fisheries decision-making. The actual legal decisions are made year-by-year by the Council of Ministers, who have been reluctant to bind themselves to the details of long-term plans.[b] This includes the kinds of if 'X condition then Y action' HCRs that would avoid short-term political decision-making.

The discourse analysis on recovery plans reveals a number of ways that access to the lifeworld was enhanced or reduced by institutional factors. The legal complexity and ambiguity allowed the stakeholders to discuss the plans while avoiding conflicts over specific rule conditions. This meant that the emerging RACs were able to use recovery plans as a critical part of inventing themselves as stakeholder fora that attracted ongoing participation from the fishing industry and ENGOS and meaningful attention from the Commission. Other institutional factors limited lifeworld access. The exclusion of the industry and ENGOs from the early development of the recovery plans limited the issues they were able to raise effectively when they did get involved. The premising of the recovery plans on individual fish stocks led to the definition of the MSMF problem as simply an expanded single species management issue. This made it difficult for the ENGOs to draw effective attention to taking a more ecosystem based approach to the problem and for the both the ENGOS and the industry to effectively point out the limitations of basing management on single stocks.

Fisheries and fleets

A fishery is an amalgam of fish stocks, fishing gears, fishing fleets and métiers; where 'fleet' usually refers to geography and vessel size, 'métier' is a term of art for sections of fleets with similar gears and fishing patterns. DGMARE's Data Collection Framework (DCF) defines métiers according to gear classes (e.g. trawls, dredges), gear groups (e.g. bottom trawls, pelagic trawls), gear types (e.g. bottom otter trawl, bottom pair trawl), main target species, and net mesh size or other selective devices (Ulrich et al.2012). Attempts at effort management beginning in 2003 for cod used days at sea limits that ignored the DCF and defined métiers in terms of gear type and mesh size. Reforms in 2009, responding to protests from fishers, led to categories at the level of the Member States over broader gear/mesh size categories, thus allowing for more flexibility, at least for national fisheries managers (Ulrich et al.2012). Vessel size and vessel power are also common bases for regulation, especially in effort management. These categories are used as the basis of both science and regulation, defining vessels' access rights to the fishery.

Discourse analysis

DGMARE must allocate fish among Member States. In MSMFs this means considering all the different fleets that fish for part of a stock either directly as a target species or indirectly as bycatch. To try to arrive at an objective way to address this problem DGMARE turned to the fisheries scientists:

Q7. Commission Fisheries Scientist: We want to reduce fishing mortality for cod. … The approach is to break down of the data from each of the fleets, to separate out catches of each of the species. Look at the activity of the fleet and then use a sort of optimisation technique where you put a score on the outcome that you want to have.

DGMARE is looking for a solution that 'optimises' the distribution of fishing quota and effort allocations among the various fleets. The optimisation would provide objective rule conditions for fish allocation decisions among the Member States. For example, scientists were asked to develop such an optimisation model for the 20-odd fleets and 60-odd métiers that fish for cod in the North Sea. This optimisation (Q7) requires precise data on what the fleets are catching (Ulrich et al.2012). At the time of these discussions ICES' position was that the NS cod stock could not take any fishing pressure and, citing data problems for this complex array of fleets, they were unable to provide DGMARE with advice about how to fish for other stocks without damaging the cod further (ICES2003). This reluctance on ICES' part was not well received. A high-level fisheries manager for a Member State, in fact, suggested that if ICES would not develop this advice they needed to find scientists who would.

The North Sea RAC also defines 'fisheries' in order to develop LTMPs. The fisheries they define bear no resemblance to those the defined by the DCF or the allocation models.

Q8. Interviewer: Is defining a fishery a problem? RAC Staff Member: We didn't find it difficult; actually we defined the five key fisheries for the North Sea without any difficulty at all. ... Interviewer: How does this relate to the fleet definitions, that the Commission uses. RAC Staff Member: Not at all, I mean, we simply, I think sat down at an executive committee meeting or working group meeting and said: You know, what the key fisheries that we need to have management plans for, so we, we re-invented the wheel in that respect, it didn't relate to gear type at all. Interviewer: Didn't at all? RAC Staff Member: No. Interviewer: So, did the RAC see there is a problem that all data gathering is based on a different system? RAC Staff Member: I think it didn't specifically, didn't have its attention specifically drawn to that particular problem.

Stakeholder recognize that definitions are necessary for both promulgating and analysing the impacts of management measures, especially as they would like fisheries, albeit much more broadly defined, to be the units of management. However, the definitions of fleets and métiers used in management are driven by bureaucratic needs for allocation and data collection rather than by the realities of fishing itself. The imposition of regulations on a class of vessels makes it much more difficult for a vessel to shift to another class.

This kind of flexibility was often an issue that came up when we were discussing the categorisation of boats for management in the fishing ports. Most fishers preferred categories that were large enough to allow seasonal flexibility. A member of a local fishers' organisation in Denmark told us:

Q9. It is self-evident that if you have to operate with those "days at sea", you have to say that the rules cannot be the same for a trawler and a net boat. Of course they cannot be the same. But once you've been put into boxes, it is just so difficult to come and say 'I would actually like to try and do something different from what I used to do'. You cannot do that. It's over. If you have not been in Kattegat fishing, then you will never come into Kattegat fishing.

Another challenge in elaborating this kind of boundary object is that fishers' behaviour, and hence the definitions of fleets and métiers do not remain static. This becomes more of a problem as the definitions become more detailed and specific. Nor are stocks and fisheries independent categories. A French fisher:

Q10. Here, it is a bit difficult. The métier we do is a function of the fish that pass by this area. So one year you are going to fish sea bream and there are years when you will catch hake and other years when you will catch whiting. It is never the same. One cannot define from one year to another what métier we are going to do, how many fish we are going to catch.

In Scotland in the 1960's and 1970's there was an explosion in multipurpose boats, and the fishers would shift according to season. But these days more and more boats stay with one kind of fishery instead of shifting according to season. The following quote from a Scottish fisher demonstrates how this is a progressive process as different kinds of management measures interact with one another:

Q11. In 1999 the boat that I had sunk and I was, it was a sort of crossroad what we were going to do. I had to build a new boat, and I built the new boat looking at what I was entitled to catch, what was the best, most efficient way to catch my quota, and that's what I came up with. So we are getting more and more, the groups are getting more and more polarised as time goes by, getting more and more difficult to jump. If I were to say later that I want to catch langoustine, that's a sort of no go. If I say I'd better go catch langoustines, I would then have to look at ways for me to exchange my demersal quota for a langoustine quota, and then I would have the expenses of making my boat capable of pursuing that split. So when it gets at it, I may not bother.

Some argue that this growing focus of the industry on particular species has reduced the ecological resilience of the fisheries system by making it difficult for fishers to fish for one species when another becomes less valuable or more difficult to catch:

Q12. Industry RAC Member: For bureaucratic, management and control reasons there has, over the years, been a move towards single species fisheries. However, this reduces the resilience of the fishing industry as the flexibility to change target species, gear *et cetera* in response to stock depletion or changing markets is removed. It is important not to define the fleets and fisheries into artificial categories that are either unnecessarily restrictive or devoid of meaning.

While fishers will argue that fishing flexibility provides ecological benefit, conservationists argue that it makes monitoring and controlling of fishing behaviour that much harder. They are more sympathetic to the industry view when it is part of a broader shift towards an ecosystem approach:

Q13. Conservation NGO RAC Member: I am an advocate of the ecosystem approach – which I believe needs to be integrated into the recovery plans if they are to deliver. The plan should allow, if that›s what the industry requires to be viable, the option to switch between fisheries, provided that the overall take of all species and individual species is within the limits determined by scientific assessments.

Not all RAC stakeholders agree that defining fisheries and

fleets is even a requirement. Indeed, this same advocate of the ecosystem approach believes that if we did put effective limits on the overall take the general knowledge already held about fishing activities would be sufficient.

Defining fisheries and fleets is contentious because these definitions are critical components of boundary objects used as rule conditions. The process limited lifeworld access for all stakeholders. Defining fisheries for highly complex optimization models demanded that ICES scientists provide advice that they were reluctant to and sometimes refused to offer. This involved institutional pressure great enough so one manager expressed a wish for other, more cooperative, advice givers.[c] This same approach to fisheries definitions undercuts the ENGOs ability to raise ecosystem issues effectively. Fishers, both their representatives on the RACs and skippers in ports, have a much more complex idea of fisheries than are reflected in either modelling or regulations but have no way effectively to include them in discussions where more official definitions are already established in monitoring, allocation modelling and regulations. These issues were further clarified using the pile sorts.

Pile sorts

A pile sorting exercise was carried out in all 12 fishing ports (Jacobsen et al.2009). Fishers were given the names of local vessels on cards, asked to place them in piles of similar boats "in any way that makes sense", and then asked to explain how they made their category choices. These pile sorts were then used as the basis of a cluster analysis in order to report general results. Every port used fishing technique, which loosely corresponds to the DCF's "gear group" and boat size as important criteria. All of the main regulatory and data gathering categories appear in some form in the fishers' categories except for mesh size.

The great difference is that fishers recognize many more dimensions than the regulators do. The pile sorts documented thirteen other categories relevant to the fishers with considerable variation found among the ports. Many categories revolve around fishing area, broad categories of target species and time and distance from port. Other categories that appeared included professionalism, processing possibilities on board, a specialized fishery, quota ownership, taxation, cooperation, working alone on board, oil consumption, environmental impact, fishing intensity, production, and treatment by management. Conflicts between gear groups, local port rules, and fuel costs were important determinates of categories in addition to national or EU fishery regulation. In general, fishers' prefer categories that are large enough to allow seasonal flexibility while at the being precise enough to appreciate their ways of working. For example, coastal fisheries exhibit a preference for flexible gear categories over species categories (Jacobsen et al.2009).

CONCLUSION

Both global and European experiences with MSMF suggests that the management of such fisheries is a dynamic process in which management institutions will need to shift constantly in the face of changing ecological and bio-economic realities. How, within these kinds of SES, can resilience be understood and improved? DGMARE's current suggestions for reforming the CFP (CEC2009) suggests that the answer is 'getting the incentives right'. When institutions recognise the norms and strategies of actors and use them as a way to build toward the common good then these institutions are more resilient. This is the critical insight about resilience that agent-based theories offer.

A communication-based approach provides other sorts of diagnostics for improving SES resilience. One of the sets of boundary objects we examined increased communicative resilience because it provided a focus of discussion, but did so by avoiding the discussion of specific rule conditions. The other set dealt directly with rule conditions and decreased communicative resilience by creating different sets of categories that limited the raising of validity claims because only a sub-set of categories could be effectively referenced.

For recovery plans, all four of the main players are in basic agreement. Recovery plans are needed and they should lead to LTMP based on harvest control rules for the species involved. The key role the plans played was to create a platform for different stakeholders to work out a common strategic goal in the midst of their many differences. Recovery plans were particularly important in the early development of the RACs as multi-stakeholder management institutions (Wilson2009b). The abstraction of the discussion of recovery plans as broad strategies rather than specific rules requiring defined rule conditions was critical for this. They gave the RACs a way to develop a shared lifeworld without having to commit to specifics, this reduced each groups need to interact strategically.

The definitions of fisheries, fleets and métiers are boundary objects that are much more directly involved in more precise needs for accountability. The discussions were always already rooted in the definitions of fisheries and métiers found either in the DCF, in individual effort control regulations or allocation models. This made the discussion of complexity effectively impossible, even discussions among scientists and managers came to an impasse as DGMARE's need for optimizing mixed fishery models for allocation led to placing demands on scientists they were not ready to meet. Raising the broader issues of concern to the industry and ENGOs was even less possible. The growing narrowness of definitions of fleets and métiers in management was seen as a very serious problem in all of the fishing ports. They have truncated discussion of actual fishing operations while creating incentives and licensing regulations that directly inhibit fishers' ability to change how, where and for what they fish. This is the clearest link in the case studies between communicative resilience – the ability to raise issues requiring rich communications – and the overall resilience of the SES, which depends on adaptive flexibility in respect to the environment. It also illustrates the complementarity of the agent-based and communicative approaches: the communicative failure meant an inability to discuss what it meant to "get the incentives right".

One interesting aspect of comparing the regulatory categories with the fishers' categories revealed in the pile sorts is the concept of decomposable systems. Simon(1962) argued that decomposability is a more common characteristic of complex systems than one would presuppose because of limits on the how many subsystems can simultaneously interact. Furthermore, and this is the point emphasised by Ostrom(2007), nearly decomposable systems are also much easier for us to understand. Ostrom(2007) therefore argues that we need to use tools to decompose SES in order to arrive at effective and well contextualised diagnoses of complex problems. A corollary of this is a temptation when creating boundary objects to create decompositions that are driven more by bureaucratic logic and the needs of easy monitoring than by the categories that best reflect conditions on the ground. The end result is a reification of imposed categories that not only oversimplify the local situation but actually change the local fleets to reflect the imposed and now rigid categories.

Hence a key issue revealed by this linkage of CST with boundary objects is how to make accountability effective in the midst of complexity. The basic warning is not to seek accountability through the simplification of reality in ways that create black boxes beyond effective questions and caveats. The classic way to approach the question of downward accountability is the development of nested institutions where decisions are made at the lowest feasible level as defined by ecological processes (Ostrom1990). This is a useful standard approach but it does not answer the question of how to develop boundary objects for effective accountability for which sufficient lifeworld resources are available to handle complexity. CST suggests that these boundary objects must be simple in a specific way: they must allow effective transparency and direct accountability while avoiding limits to lifeworld access that undercut the boundary objects.

One example of such a system is "results-based" fisheries management (Ulrich et al.2012), which takes various forms but which emphasizes small scale self-management initiatives characterized by simple but rigorous accountability based on demonstration by fishers that specific, measurable limits on ecological impacts have been honoured while not imposing any further rule conditions on fishing operations. The boundary object of interest is a contract that specifies how the fisheries rights holders will demonstrate compliance with limits. The sector management being adopted in New England approaches this form, although requirements for contracts still involve attempts at operational control (NOAA2009). Similar ideas are being implemented in Australia, Canada and some coastal fisheries in Europe, often in conjunction with transferable access rights.

Communicative and strategic actions within institutions are linked, but each has its own logic. Where this linkage is clearest is in the need to develop science policy boundary objects that are transparent and allow effective rule conditions. SES resilience is as much a function of what we have called communicative resilience as it is of strategic action in response to the right incentives. The analysis presented here suggests that considering both types of action, and the systems for objectivity/transparency and accountability that lubricate them, may allow us to get a fuller and more systematic picture of resilience within complex SES.

ENDNOTES

[a]The term "power" is the translation found in Habermas(1987) for what we think is more clearly expressed as "authority" as Habermas links the concept to offices.

[b]Since this research was carried out the European Parliament has been given a "co-decision" making role in fisheries.

[c]This is not a realistic desire. ICES' independence is protected by the fact that it also provides advice to Norway, Russia and Iceland with whom the EU negotiates over shared stocks. These countries would strongly resist the EU replacing ICES with another source of official advice.

ACKNOWLEDGEMENTS

The research presented here resulted from both field work and discussions within three EU Framework Programme research projects: The 6th Frameworks supported Understanding the Mechanisms of Stock Recovery (UNCOVER) Project no. 022717; A Framework for Fleet and Area Based Fisheries Management (AFRAME) Project no. 044168; and the 7th Framework supported Judgement and Knowledge in Fisheries Involving Stakeholders (JAKFISH) contract no. 212969 This work does not reflect the Commission's views and in no way anticipates its future policy in this area.

REFERENCES

1. Adger WN: Social and ecological resilience: are they related? *Progress in Human Geography* 2000,24(3):347–364. 10.1191/030913200701540465
2. Armitage D: Adaptive capacity and community-based natural resource management. *Environmental Management* 2005,35(6):703–715. 10.1007/s00267-004-0076-z
3. Berkes F, Colding J, Folke C: Introduction. In *Navigating Social-Ecological Systems: Building Resilience for Complexity and Change*. Edited by: Berkes F, Colding J, Folke C. Cambridge: Cambridge University Press; 2003.
4. Branch TA, Hilborn R: Matching catches to quotas in a multispecies trawl fishery: targeting and avoidance behavior under individual transferable quotas. *Canadian Journal of Fisheries and Aquatic Science* 2008, 65: 1435–1446. 10.1139/F08-065
5. Brown S: *Political Subjectivity: Applications of Q Methodology in Political Science*. New Haven: Yale University Press; 1980.
6. Caddy JF, Agnew DJ: An overview of recent global experience with recovery plans for depleted marine resources and suggested guidelines for recovery planning. *Reviews in Fish Biology and Fisheries* 2004, 14: 43–112. 10.1007/s11160-004-3770-2
7. Carpenter SR, Walker BH, Anderies JM, Abel N: From metaphor to measurement: resilience of what to what? *Ecosystems* 2001, 4: 765–781. 10.1007/s10021-001-0045-9
8. Cash D, Clark W, Alcock F, Dickson N, Eckley N, Jaager J: Saliency, credibility, legitimacy and boundaries: linking research, assessment and decision making. In *School of Government*. Edited by: Kennedy JF. Harvard University Faculty Research Working Paper; 2002. Series RWP02–046 Series RWP02-046
9. CEC: *GREEN PAPER: Reform of the Common Fisheries Policy*. Brussels: DG MARE; 2009.
10. Coxon APM: *Sorting Data: Collection and Analysis*. Thousand Oaks CA: Sage Publications; 1999.
11. DC Wilson: *The Paradoxes of Transparency: Science and the Ecosystem Approach to Fisheries Management in Europe*. Amsterdam: University of Amsterdam Press;

2009a.

12. Farnell J: *Reform of the CFP and its implications for scientific advice and research. Presentation at European Fisheries and Aquaculture Research Organisation Meeting*. Rhodes; 2003. 4 June 2003 4 June 2003
13. Folke C: Resilience: The emergence of a perspective for social–ecological systems analyses. *Global Environmental Change* 2006, 16: 253–267. 10.1016/j.gloenvcha.2006.04.002
14. Gunderson LH, Holling CS: *Panarchy: understanding transformations in human and natural systems*. Washington: Island Press; 2002.
15. Habermas J: *The Theory of Communicative Action: Volume II Lifeworld and System: A Critique of Functionalist Reasoning*. Boston: Beacon; 1987.
16. Habermas J: *The Philosophical Discourse of Modernity: Twelve Lectures*. Cambridge MA: MIT Press; 1991.
17. Hahn T, Olsson P, Folke C, Johansson K: Building knowledge generation and organizational innovations: the role of a bridging organization for adaptive comanagement of a wetland landscape around Kristianstad, Sweden. *Human Ecology* 2006, 34: 573–592. 10.1007/s10745-006-9035-z
18. Hannesson R: *The Privatization of the Oceans*. Cambridge: the MIT Press; 2004.
19. Heritage J: *Garfinkel and Ethnomethodology*. Cambridge: Polity Press; 1984.
20. Holland JH: *Adaptation in Natural and Artificial Systems*. Cambridge: MIT Press; 1992.
21. ICES: *ACFM Annual Report*. Denmark: Copenhagen; 2003.
22. Jacobsen RB, Damalas D, Delaney A, Wilson DCK: *Stakeholder perceptions of groups of boats. Deliverable 6.1 & 6.2 of the Improved Fishery Management Through Fleet- and Area-Based Assessments (AFRAME) project*. 2009. EC Framework Programme 6: Contract no: 044168 EC Framework Programme 6: Contract no: 044168
23. Kapasa C, Malasha I, Wilson DC: *Experience Based Knowledge and Fisheries Management in the Mweru-Luapula System. Annex 11 of the Final Report of the Knowledge in Fisheries Management (KNOWFISH) Project*. 2005. EU Framework Programme 5. Contract number: ICA4-CT-2001–10033 EU Framework Programme 5. Contract number: ICA4-CT-2001-10033
24. Latour B: *Science in Action*. Cambridge: Harvard University Press; 1987.
25. Low B, Ostrom E, Simon C, Wilson J: Redundancy and diversity: do they influence optimal management? In *Navigating Social-Ecological Systems: Building Resilience for Complexity and Change*. Edited by: Berkes F, Colding J, Folke C. Cambridge: Cambridge University Press; 2003:83–114.
26. Miller JH, Page SE: *Complex Adaptive Systems: An Introduction to Computational Models of Social Life*. Princeton NJ: Princeton University Press; 2007.
27. Murray G, Johnson T, McCay BJ, Danko M, St K, Martin ST: Creeping enclosure, cumulative effects and the marine commons. *International Journal of the Commons*2010,4(1):367–389.
28. Nagaski F, Chikuni S: *Management of multispecies resources and multi-gear fisheries*. 1989. FAO Fisheries Technical Paper/305. http://www.fao.org FAO Fisheries Technical Paper/305. http://www.fao.org
29. National Oceanographic and Atmospheric Administration (NOAA): *NOAA Fisheries Service Fact Sheet*. 2009. Commercial Fisheries News. · Special Supplement · August Commercial Fisheries News. · Special Supplement · August
30. Norberg J, Cumming GS (Eds): *Complexity Theory for a Sustainable Future*. New York: Columbia University Press; 2008.
31. Ostrom E: *Governing the Commons*. New York: Cambridge University Press; 1990.
32. Ostrom E: *Understanding Institutional Diversity Princeton*. Princeton: Princeton University Press; 2005.
33. Ostrom E: A diagnostic approach for going beyond panaceas. *Proceedings of the National Academy of Sciences* 2007,104(39):15181–15187. 10.1073/pnas.0702288104
34. Ostrom E, Gardner R, Walker J: *Rules, Games, and Common Pool Resources*. Ann Arbor: The University of Michigan Press; 1994.
35. Sanchirico JN, Holland D, Quigley K, Fina M: *Catch-Quota Balancing in Multispecies Individual Fishing Quotas. Resources for the Future*. 2005, 05–54. Discussion Paper Discussion Paper
36. Scott MB, Lyman SM: Accounts. *American Sociological Review* 1968, 33: 46–62. 10.2307/2092239
37. Simon HA: The Architecture of Complexity. *Proceedings of the American Philosophical Society* 1962,106(6):467–482. Dec. 12, 1962 Dec. 12, 1962
38. Ulrich , Clara , Douglas CK, Wilson J, Rasmus Nielsen , Francois Bastardie , Stuart AR, Bo SA, Ole RE: Challenges and opportunities for fleet- and métier-based approaches for fisheries management under the European Common Fishery Policy. Ocean and Coastal Management. *Maritime Studies* 2012. In press – accepted manuscript In press – accepted manuscript 10.1016/j.ocecoaman
39. Weick KE: *Making Sense of the Organization*. Malden MA: Blackwell Publishing; 2001.
40. White S: *The Recent Work of Jurgen Habermas: Reason, Justice and Modernity.*. Cambridge: The Cambridge University Press; 1988.
41. Wilson DC: Markets, networks, and risk: an analysis of labor remuneration in the lake Victoria fishing industry. *Sociological Forum* 1998,13(3):425–456. 10.1023/A:1022127424920
42. Wilson J: "Scientific uncertainty, complex systems, and the design of common-pool institutions". In *The Drama of the Commons*. Edited by: Ostrom E, Dietz T, Dolsak N, Stern PC, Stonich S, Webe EU. Washington: National Academies Press; 2002:327–359.
43. Wilson DC: *Stakeholder Participation in Recovery Plans*. 2009b. Deliverable No. 30 of the Understanding

the Mechanisms of Stock Recovery (UNCOVER) Project. EU 6th Framework Project No 022717 (SSP 8) Deliverable No. 30 of the Understanding the Mechanisms of Stock Recovery (UNCOVER) Project. EU 6th Framework Project No 022717 (SSP 8)

44. Wilson DCK: Considering critical dualism: a response to Fikret Berkes. *MAST*2010,9(1):51–54.

45. Wilson DCK: *Network Structure and Agreement about Science: Applications to Results Based Management. ICES Annual Science Conference. Session P Reversing the burden of proof: results based management of fisheries*. Nantes; 2010.

46. Wilson DC, Jentoft S: "Structure, agency and embeddedness: sociological approaches to fisheries management institutions". In *Alternative Management Systems*. Edited by: Symes D. Oxford: Blackwell Science; 1998:63–72.

47. Wilson DC, Nielsen JR, Degnbol P: Conflict and scale: a defense of community approaches in fisheries management" Chapter 11. In *The Fisheries Co-management Experience: Accomplishments, Challenges and Prospects*. Edited by: Wilson DC, Nielsen JR, Degnbol P. Dordrecht: Kluwer Academic Publishers; 2003.

CHAPTER 10 SUSTAINABLE SMALL-SCALE FISHERY

Crisol Méndez-Medina, Birgit Schmook and Susannah R. McCandless

Keywords: Small-scale fisheries; Fishing cooperatives; Fisheries governance; Common-pool resources; Fisheries conservation

INTRODUCTION

In Latin America, most artisanal fisheries are degrading rapidly, provoking widespread concerns about overexploitation of fisheries resources (Defeo and Castilla 2005). Many of these fisheries began as open access regimes under national jurisdiction. Initially, products from artisanal fisheries supplied domestic markets. Once national markets were integrated into the global economy, fishery products became export commodities, which often resulted in overexploitation of the resource (Defeo and Castilla 2005; Orensanz et al. 2013). Concerns over how to manage fisheries resources sustainably have resulted in intensive research interest in successful artisanal fisheries (Defeo and Castilla 2005; Gallardo et al. 2011; Gutiérrez et al. 2011; Orensanz et al. 2013).

One salient conclusion has been that rights-based management systems tend to function better for sustainable artisanal fisheries. Case study examples from Latin America illustrate the benefits of such systems, such as the Seris in Sonora, Mexico (Basurto et al. 2012; Orensanz et al. 2013), and the Juan Fernández Archipelago lobster fishery (*Jasus frontalis*) in Chile (Orensanz et al. 2013). Other rights-based systems studied include the Territorial Use Rights for Fishing in North Central Chile loco abalone fisheries (*Concholepas concholepas*) (Defeo and Castilla 2005; Gallardo et al. 2011; Orensanz et al. 2013), and the concessions of Central Baja California, Mexico (Orensanz et al. 2013; McCay et al. 2014). The Punta Allen spiny lobster fishery (*Panulirus argus*) in the Mexican Caribbean has also been extensively profiled (Seijo and Fuentes 1989; Schlager and Ostrom 1992; Seijo 1993; Defeo and Castilla 2005; Seijo 2008; Sosa-Cordero et al. 2008; Brenner 2010). On the other hand, rights-based management alone is not the key to successful and sustainable fisheries. Research has shown that if divergent interests among stakeholders are not reconciled, serious conflicts may arise, threatening both the resource base and local livelihoods. Examples of less successful rights-based artisanal fisheries include that for the sea urchin (*Loxechinus albus*) in Chile (Defeo and Castilla 2005), and the Galapagos sea cucumber (*Isostichopus fuscus*) of Ecuador (Orensanz et al. 2013). The Juan Fernández lobster fishery in Chile shares some characteristics with the Punta Allen fishery. Both share informal territorial access rights. Local fishers designate

Citation: Crisol Méndez-Medina, Birgit Schmook and Susannah R. McCandless, sustainable small-scale fishery, DOI 10.1186/s40152-015-0026-9.

individual fishing spots, and in both cases non-invasive fishing gear is employed. Both are single-species lobster fisheries, with the benefit of high commercial value. Both developed their fisheries in geographic isolation, a characteristic that forced them to adapt and respond to environmental perturbations as a group. Unlike in Punta Allen, however, in the Juan Fernández case, the government has not recognized the system of traditional rights. This renders their system vulnerable to the incorporation of disruptive, externally imposed rules (Ernst et al. 2013).

Gutiérrez et al. (2011) analyzed 130 cases of fisheries with community-based co-management, and found them to be a key management strategy across countries with different degrees of economic development (Punta Allen was included in the study). The authors found that the attributes contributing to co-management success are the presence of strong leadership, followed by the implementation of community quotas, social cohesion, and protected areas limiting access to non-group members. All these elements are present in Punta Allen. However, the presence of these attributes does not fully explain the success of the case study, either.

The example of the Maine American lobster fishery (*Homarus americanus*) (Ernst et al. 2013) contributes a few additional elements to the analysis of successful artisanal fisheries. The Juan Fernández, Punta Allen and Maine fisheries all feature informal territorial rights. Maine lobster fishers access a common territory, in contrast to the two other cases, where fishers hold rights to specific fishing locations individually (Ernst et al. 2013; St. Martin 2001). In Maine, a well-rooted conservation ethic among local fishers in the industry resulted in laws that heavily support and facilitate resource conservation. For example, these laws established minimum and maximum lobster size measurements for extraction, a prohibition on the extraction of "berried" lobsters (egg-carrying females), and the use of non-invasive fishing gear. The most important aspect of the above-mentioned laws is their compatibility with existing territorial systems (Acheson and Brewer 2003), a characteristic that the Maine fishers share with those of Punta Allen. This successful integration of local rules with law and policy—in conjunction with factors such as relatively easy access to high-market-value species and isolated geographic conditions—have resulted in a very particular form of social organization. That organizational form represents the essential ingredient in the maintenance of a system of rights that sustains the life of the fishery developed by the Vigía Chico Cooperative of Punta Allen in the Mexican Caribbean. Tracing the evolution of these distinctive conditions in the Punta Allen case thus has the potential to inform both artisanal fisheries management policy, and community practice in far-flung locales.

Setting the scene

In Mexico, fisheries are listed as secondary activities in State development strategies because of their minor contribution to the Gross Domestic Product (GDP). Nonetheless, small-scale fisheries make significant economic contributions. According to official data, in Mexico 70 % of fisheries resources are fully exploited, 10 % are in potential development and 20 % are in or close to collapse (Arreguín-Sánchez 2006). Artisanal fisheries in the country represent a huge management challenge. Catch levels are not fully recorded, and poaching and violation of closed seasons occur along the coasts of Mexico. The role of artisanal fisheries in the worldwide fishery crisis is often ignored or regarded as marginal compared to the effects of industrial fishing (Defeo and Castilla 2005). Nonetheless, small-scale overfishing can lead to serious local environmental and economic consequences.

In the Southern Yucatan Peninsular state of Quintana Roo, fishing has major social, political and commercial significance. Along the state's 900-km coastline, social life is organized by a system of artisanal fisheries. Experts consider these fisheries to be highly organized, second only to those of Baja California, nationally (Hidalgo and Méndez 2007). The system is dominated by cooperatives formed by local fishers, with little involvement of the outside commercial sector (Sosa-Cordero and Ramírez-González 2001). The fishery resources that historically have had the greatest economic significance for the state are the spiny lobster (*Panulirus argus*), shrimp (*Farfantepenaeus brasiliensis*), and the queen conch (*Strombus gigas*). Spiny lobster production has significant economic importance for Quintana Roo. From 2000 to 2008, 165.1 tons were produced (counting both tails and live lobsters). The catch represented six percent of the state's fishery production; fisheries, in turn, represent 40 % of the state's total gross product (Sosa-Cordero 2011).

There are three fishing areas in the state: North, Central and South. Each area has different levels of development, habitat characteristics and types of fishing gear employed in lobster catching. In the Central area, the focus of our study, the predominant types of gear used are*sombras,* or shades[a], and *jamo* [b], a type of net (Seijo 2008). This area includes the bays of La Asención and Espíritu Santo, both located with the boundaries of the Sian Ka'an Biosphere Reserve.

Spiny lobster abundance was a decisive factor and one of the principal drivers of the formation of the first cooperatives in Quintana Roo in the 1950s (César and Arnáiz 1986). After the 1970s, the activity lost importance in Federal development strategies (Sosa-Cordero and Ramírez-Gonzalez 2011). When conservation emerged as an important issue in Federal environmental development concerns in the 1990s, lobster fishing was identified as a threat to the Mesoamerican Barrier Reef. Fisheries became a key target of State strategies for coastal management (Velez et al. 2014). These changing socio-environmental circumstances make the lobster fishery of Punta Allen a valuable case study. Despite these apparently adverse conditions and the increasingly bleak scenario of many fisheries globally, the cooperative has emerged as a success story.

The Vigía Chico lobster cooperative of Punta Allen is located in what is today the Sian Ka'an Biosphere Reserve. As an instance of a sustainable artisanal fishery coexisting with marine conservation, the case has been frequently analyzed. The establishment of property rights over marine space and the efficacy of cooperative members' harvesting techniques have drawn the attention of many researchers (Seijo and Fuentes 1989; Seijo 1993, 2008; Defeo and Castilla2005; Sosa-Cordero et al. 2008; Brenner 2010). To better understand the cooperative's success, we employed a historical perspective and an ethnographic lens. This combination of approaches facilitates an analysis of the community's interaction with changing state policies and institutions.

The lobster fishery of Punta Allen has been studied previously using an institutional analysis framework (Seijo 1993; Cochran 1998). The framework analyzes how collectively managed fisheries function, how group members are accepted or rejected, and whether possession of these rights to common property permit or prohibit the holder's sale of the resource. Institutional analysis is a well-recognized analytical approach. It defines institutions as the shared conceptions that humans use in recurring situations, which are organized by rules, norms and strategies (Ostrom 2007). By the term rules we refer to:

"The result of implicit or explicit efforts by a set of individuals to achieve order and predictability within defined situations by: (1) creating positions (e.g., member, convener, agent, etc.); (2) defining how participants enter or leave positions; (3) agreeing on which actions participants in these positions are required, permitted, or forbidden to take; and (4) stating which outcome participants are required, permitted, or forbidden to affect." (Ostrom 1986:5).

We can see examples of these operational rules when fishers specify what types of fishing gear are permitted within a fishing ground (Schlager and Ostrom 1992:251). The rules that create and enforce property rights regimes play a primary role in institutional analysis. Rules can also be understood as the variety of rights and their allocations to particular stakeholders or categories of stakeholders (Poteete 2010). We understand "rights" as particular actions that are authorized (V Ostrom and John 1976). Rights are a product of rules, and thus not equivalent to rules (Schlager and Ostrom 1992). The institutional analysis framework has been invaluable in understanding the organization of fisheries and other governance regimes around the world (Schlager and Ostrom 1992; Jentoft 2004; Basurto and Coleman 2010; Gallardo et al. 2011; Basurto et al. 2012).

By employing a historic and ethnographic approach to the Punta Allen case, we seek to deepen the understanding of how this common property right of use has come into existence. The factors affecting contextual emergence of common property rights of use remain a vital question (Kadekodi 2004), and the principal concern of our paper.

METHODS AND STUDY AREA

We conducted our case study on the Javier Rojo Gómez fishing colony—the official name of Punta Allen—within the perimeter of the Sian Ka'an Biosphere Reserve, Quintana Roo, Mexico (Fig. 1). Demarcation of the Sian Ka'an Biosphere Reserve began in 1982. It was gazetted in 1986, in a process led by the Research Center of Quintana Roo (CIQROO) with support from the National Council of Science and Technology (CONACYT). At the time of its demarcation, the only human settlements recognized within its boundaries were Punta Allen and Punta Herrero. Today María Elena, another fishing camp belonging to the Cozumel fishing cooperative, is also recognized.

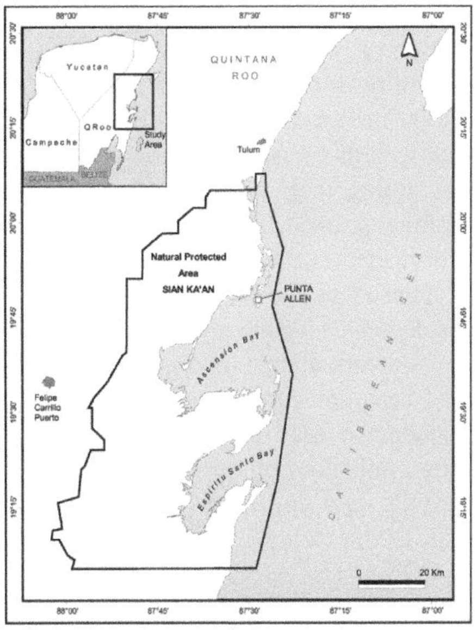

Figure 1: Study area, showing location of the Sian Ka'an Biosphere Reserve.

The Sian Ka'an Biosphere reserve has an area of 652,193 hectares, of which approximately 120,000 are marine. The marine portion includes an 110 km-long continuous barrier reef, which makes it a focal point for the preservation of coastal ecosystems. This position also makes the territory of Sian Ka'an a priority focus for the tourism industry. The reserve encompasses a transition zone between terrestrial and marine ecosystems, containing an extraordinary diversity of environments in good condition. However, its poor soils are not suitable for agricultural activities. Due to a warm and humid climate, the region features mesic forests, flooded forests and mangroves (INE 1996). Hurricanes, northerlies (*nortes*), and droughts periodically affect the area.

For many years, the extraction of gum (*chicle*) was the main productive activity in the region, together with timber extraction and copra plantations for coconut oil production. Between 1960 and 1980, timber extraction faltered before an increasing scarcity of commercial timber species, due to overexploitation, and the chicle bubble popped with the emergence of synthetic gum. Copra production declined precipitously due to "lethal yellowing", a viral disease of the coconut palm. Given this grim economic scenario, local populations began to exploit marine resources through fishing. To ensure their resource use rights, in 1968 they established the "Vigía Chico Cooperative Society for Fishery Production" (Sociedad Cooperativa de Producción Pesquera Vigía Chico), followed in 1970 by the "Javier Rojo Gomez Fisheries Camp" (Punta Allen) in Ascension Bay (Brenner 2010). Today Punta Allen is highly dependent on the extraction of fishery resources (primarily spiny lobster) and tourist activities. In Ascension and Espiritu Santo Bays (located at the center and south of the Sian Ka'an Reserve, respectively), the use of*casitas,* or capture shades, and *jamo* as fishing gear currently dominate (Seijo 1993; Sosa-Cordero et al. 2008).

Ethnographic fieldwork was conducted from February to May 2013 and in July of the same year. The first author conducted in-depth interviews with 30 members of the cooperative, using a snowball sampling technique (Biernacki and Waldorf 1981). Guest et al. (2006) recommend conducting

30 to 50 interviews to allow the ethnographic researcher to achieve saturation of information on a particular topic, while capturing variability, for the proposes of thematic analysis. The first author interviewed current leaders of the lobster cooperative, as well as founding members, and accompanied some of them during their working hours (both off-season and in the fishing season). During ethnographic fieldwork, she spoke with leaders of the cooperative, and with some of the families opposed to the current Board of Directors, to triangulate the degree of legitimacy of the shared access rules. Ethnographic fieldwork additionally included interviews with residents who were not current cooperative members. The first author conducted four semi-structured interviews with officials of the National Commission of Natural Protected Areas (CONANP), and one informal anonymous interview with an official of the National Committee for Fisheries and Aquaculture (CONAPESCA). She also interviewed members of non-governmental organizations working in the study area. Finally, she interviewed three fishermen who no longer live in the community, but who had been founding members of the cooperative.

We conducted a review of the cooperative's historical documents, including records of the assemblies and meetings of the Board of Directors during the cooperative's first 20 years of existence. We triangulated this information with a review of historical literature covering 100 years of public policy in Mexico. This enabled us to situate our understanding of the process of cooperative decision-making over time. We also traced how informal institutions and relationships between Punta Allen's Vigía Chico Cooperative and a variety of state-level stakeholders developed.

HISTORICAL REVIEW OF COASTAL SETTLEMENTS

The lighthouse keepers: the basis of social organization

The first settlers of the eastern coast of the Yucatan Peninsula (presently the State of Quintana Roo) were descended from the Chontal, a Mayan ethnic group. The Chontal brought their prowess in navigation from the State of Tabasco, enabling them to master the coast and coastal waters (César and Arnaíz 1990). During pre-Columbian times, salt production and fishing were the main productive activities in Yucatan. Dating well back into the pre-Colombian era, cultural exchange and trade existed throughout the Caribbean (Thompson 1979). During the era of conquest English and French pirates from the Caribbean islands repeatedly attacked the coast of Quintana Roo, making the rapid settlement of the area by other Europeans impossible. Nevertheless, the Maya were able to establish trade agreements and alliances with the pirates. These relations were retained during subsequent periods of repopulation.

Throughout the nineteenth century, the Yucatán Peninsula witnessed numerous battles between incoming settlers and the Mayan residents. The most significant conflict in the area's history was the Caste War (1847 to 1901), which lasted more than half a century. At the end of the Caste War the population, who had fled the region for decades, began to return. Repopulation began in the south of the state and was made official with Mexican state promotion of the establishment of Rancho San Miguel (now Cozumel Island). Subsequently the municipality of Isla Mujeres was established on the eponymous island—named for its cultural importance as a site of Mayan women's rites of passage—and adjacent mainland. Regional resettlement occurred very slowly, as the territory was isolated from the rest of the country, and navigating the Caribbean waters remained difficult because of long hurricane periods (César and Arnaíz 1990).

At the beginning of the 20th century, the Mexican government established a Navy presence along the coast. The Marine Corps recruited local fishermen from nearby Isla Mujeres (César and Arnaíz 1992), who took advantage of local food resources like sea turtles, and practiced small-scale fishing while simultaneously protecting Mexico's southern border. The establishment of lighthouses was another piece of the government's strategy to secure and populate Mexico's southern border, thereby integrating this marginal territory into the nation-state. Those in charge of the lighthouses became the region's first non-Mayan settlers, and extracted copra. These settlers earned their living as farmers, fishers, and hunters, and created a form of social organization in the territory based on extractive and farming practices (César and Arnáiz 1986;1990).

The presence of lighthouse keepers along the coast was essential for the consolidation of the social organization of most of the fisheries in Quintana Roo, and remains so today. Punta Allen was part of that early-20th century network of lighthouses. Among the records of the first lighthouse keeper we found the first formal proposal of how life and work in the settler community should be organized. The Punta Allen fishers' cooperative initially operated out of the lighthouse keeper's ranch, and one of the current cooperative leaders is that founder son.

Fisheries in Quintana Roo: the early years

During the early decades of the 20th century, the population living along the coast of Quintana Roo remained fairly isolated from the rest of the country. Socio-economic exchanges happened through contacts with Cuban and Belizean fishermen, who arrived by sea. Local populations traded lobster with the Cubans and Belizeans for scarce staples like sugar, alcohol, salt and oil, transforming a product used for local consumption into a commercial commodity with high economic value. During these early years of the lobster fishery anyone could harvest lobster without restrictions; most extraction happened in areas close to the coast and reefs. Given resource abundance, lobster fishing was easy and required little effort. As commercial exploitation increased, interviewees say that the Cubans explained to one of the fishers how to use capture shades (*sombras*) to make lobster fishing more economically attractive and labor-efficient.

At the national level, fisheries were consolidated during the 1930s and 1940s. Nationwide, the 1940s and 1950s were of great importance for the development of the fishing industry. For Mexico, the Second World War boosted the fishing industry, creating an increased demand for seafood products for the U.S. domestic market (Cruz-Ayala and Igartúa-Calderón 2006). The introduction of diesel engines into local fishing practices represented an adaptation of military technologies to civilian uses after WWII. During the 1940s and 1950s, the government also promoted nationwide migration from urban centers to coastal areas. As part of a territorial reorganization plan, government policy concomitantly supported major investments in port infrastructure (Soberanes Fernández 1994; Cifuentes-Lemus and Cupul-Magaña 2002). In Quintana Roo, however, fisheries did not gain economic importance until after Hurricane Janet, in 1955 (César and Arnaíz 1990). After Janet hit the coast, most of the region's coconut palms were destroyed. Copra extraction ceased to be economically viable, and the territory's economic identity had to be redefined. The first fishing cooperatives of Quintana Roo were formalized during this time. Although initial efforts to encourage the creation of productive groups were made during the government of Lázaro Cárdenas (1934–1940)[c], it was not until the mid-1950s to the late 1960s that extractive practices were organized by Cooperative Societies for Fisheries Production (César and Arnáiz 1986).

DRIVERS OF SUCCESS

Social organization

The Punta Allen community was established under particular socio-geographical conditions. State development plans did not include significant infrastructure development in the area, so the first inhabitants had to travel long distances on rough roads to reach major regional urban centers like Felipe Carrillo Puerto and Chetumal. After copra plantations began to decline in the 1950's, fisheries came to the fore as an important economic activity in Quintana Roo. During the early years of settler community formation, this isolation from the rest of the territory was decisive for the practices of production and consumption enacted by fishing families and the few other individuals who lived in the community.

The first settlers belonged to three extractive traditions: copra, the quintessential coastal agriculture; fishing, practiced by fishers in Cozumel and Holbox, some of whom who traveled to Punta Allen to continue plying their trade; and finally, other agricultural and forestry traditions. One of the most important activities in these first years of the settlement (1950's-1970's) was crocodile hunting (both of the American crocodile, *Crocodylus acutus* and Morelet's crocodile,*Crocodylus moreletii*). A few lobster fishers participated in this activity and sold crocodile meat and skin in Chetumal. Also, some new settlers worked exclusively in the manufacture of fishing shades and cutting chit palm (*Thrinax radiata*), used to construct the first shades and for roofing houses.

In this isolated settlement, social life was organized collectively; the few women living there washed and cooked as a group. The group's representatives in charge of marketing the product outside the community were responsible for bringing back basic supplies. Some fishers' wives accompanied their husbands on fishing trips, and prepared food along the way.

During the early years of the cooperative in Punta Allen, lobster was distributed to the city of Chetumal by sea, or was transported overland via a truck whose winding route ran from the Vigía Chico Carrillo harbor and could take 2 or 3 days because of the hostile terrain. Once the distributors managed to get out to the main road, the product was brought to Merida, Chetumal and Felipe Carrillo Puerto. Along with lobster, they transported turtle meat, one of the most popular seafoods at that time. Some trips went as far as the Mexico City.

Control over access to common resources was organized from the early days of the cooperative into three production groups, centered on three individuals with high leadership capacity. Though the early leaders had different views on how the fishery should be organized, the aforementioned conditions of geographic isolation, scarcity, and rough roads forced them to combine their efforts to market the product and purchase the products necessary for Punta Allen families' subsistence. These leaders started families in the town, and today remain the power holders in the community. The first leader, a founder of the community, came from a family of Spanish origin. He had little formal education, but was highly skilled in the management of economic resources. The second leader was the descendant of fishers who had participated in the formation of the Cozumel cooperative, which he left due to a conflict. He was the driving force behind the formation of the cooperative as the local mode of labor organization. The third leader was known in the community for his conciliatory capacity. He came from the area's farming tradition and was the individual who conducted negotiations and mediated between the community, NGOs and the government during the early years of the community and cooperative.

At this time, a local fisher deployed the *sombra* technique introduced by the Cubans. The new technology involved investment in materials, manpower and time during the lobster harvest, and thus disputes over resource ownership emerged. The first fisherman to place a capture shade into the sea was followed by other fishers, who noted where he was placing them with the intent of later stealing his catch. Another fisher followed suit, placing his own sombras, but they suffered the same fate, and his catch was stolen. These events were the genesis of the marine tenure system characteristic of Punta Allen. The catch-stealing problem was resolved by a fisherman who had previously worked as a farmer in the municipality of Felipe Carrillo Puerto. He decided to establish a fishing zone as a "*parcela* [d]", bounding an area of marine space. Other fishers imitated him and established that each fisher would have an assigned area in the sea for lobster fishing, in which he could place shelters to catch lobsters, just as a farmer would install infrastructure on his land. People in Punta Allen do not say that they fish lobster; instead, they harvest lobster. This apparently minor linguistic shift carries important symbolic weight in their fishing practice.

Once the cooperative was formed, extractive practices continued to be organized in working groups. In the process of delimiting lobster fields, fishers who belonged to one of the three major groups managed to defend their chosen marine territories and consolidate their ownership.

"When we began to divide the sea… it was terrible, there were fights in the cantina. So that it wouldn't go on for too long, I made groups… you know, don't grab this plot just for yourself, divide it among four; so, when someone fights with you, you are going to fight as a foursome…whereas if you were alone, they would plaster you in the cantina. They made groups… [so now] you are talking about how you are going to fight six people, and it [acts as a deterrent]… I do not know if you understand how the strategy works…" (Rodrigo, founder, member of the Board of Directors, personal interview, April 20, 2013). As the years passed and the cooperative matured, the families of these three leaders became the community leaders. They were responsible for negotiations with governmental institutions, participation in academic projects, and decision-making on how access to resources in the zone was allocated. Since the cooperative's inception, Mexico's Federal Cooperative Law (1938) served as a basis for regulating practices within the community. However, this was not an easy task. The low level of formal educational level of the majority of the fishers in the cooperative and the isolated geographic conditions spawned constant trips to the State capital (Chetumal) to ask government authorities to serve as arbitrators in many disputes. These iterative learning interactions with the State began to confer legitimate power on these leaders, who held administrative positions on the Board of Directors.

"Before, no, we did not carry out our responsibilities because we didn't know about the performance of our activities, our responsibility... we always went and were told 'that's what conciliation and arbitration are for, conflicts.... The surveillance guy, that's his job to see that all the committees work well, and [bring to the Board's attention] those that don't, those are your duties'..." (Rodrigo, founder, member of the Board of Directors, personal interview, April 20, 2013).

This reinforcement of the legitimation of local authority figures via state backing helped ensure that interpretations of the Cooperatives Act would become instrumental to the regulatory practices of the community. This process further promoted the cooperative as the dominant form of community social organization. Notably, tourism, which has become an important economic activity in the community, is regulated by these same cooperative rules. The original leaders of the lobster cooperative continue to hold power in almost all spheres of public life, and the descendants of these leaders sustain this system of concentration of power.

Governmental organizations collaborate with the cooperative because of its implementation of sustainable extractive practices and its participation in the policing of marine conservation at a low cost to the State. Local fishers have firmly supported the surveillance system because the Ascension Bay fishing concession, which was granted by the State, provides them with security over the resource and protects their internal system of regulation.

In 1992, the cooperative's exclusive right to the fishery resource was eliminated by changes in State policy during the Salinas de Gortari government (1988–1994). At that time, the Fisheries Act was modified to include active private sector participation in fishing extractive activities. Species reserved for exploitation by cooperatives disappeared from the Act and a new regime was designed to grant concessions, permissions and authorizations, whereby these rights of use could be transferred to private investors. Additionally, early bids began for the sale of the state-owned enterprises Ocean Garden and Mexican Fishery Products (Soberanes Fernández 1994; Cifuentes-Lemus and Cupul-Magaña 2002). Even though cooperatives lost exclusive fishing rights under these changes in national fisheries legislation, the cooperative continues to be the only way the Punta Allen community can conceptualize the social organization of labor. The cooperative was able to maintain and consolidate power, because since its early days the founding leaders had respected state regulations, and used them to legitimize their decisions. Through their repeated consultations with regional authorities, and long-time adherence to State regulations, these leaders had strengthened their connections with state agencies and actors in ways that continued to benefit them, and offer legitimacy, despite the changes in legislation.

Responses to socio-environmental perturbations

When Hurricane Gilbert hit the coast of Quintana Roo in 1988, just 2 years after the formalization of the Biosphere Reserve, the course of local fisheries changed unexpectedly. The model of fishing had shifted at the federal level, with the 1983 implementation of the Exploration and Assessment of Fisheries Resources of the Exclusive Economic Zone and Territorial Waters program, a multi-agency and industry partnership[e]. Its intent was to estimate the resources that Mexico could extract from its exclusive economic zones and territorial waters. With this measure, the State changed its discourse from one of resource protection and regulation to one of deregulation. For each newly inventoried marine resource, trade values were assigned, and changes were introduced to fisheries legislation to encourage the participation of the private sector, as well as foreign companies (Soberanes Fernández 1994). The State also conducted an inventory of state-owned assets and commenced their sale. Multiple interviewees reported that before 1988, state-owned Mexican Fishery Products and Ocean Garden had been economically important, and responsible for most of the processing and distribution to packing companies in Quintana Roo (Soberanes Fernández 1994). State-sponsored deregulation forced local fisheries to identify new ways to survive in the national and international market.

The Punta Allen cooperative faced one of its worst financial crises in the 1990s. The cooperative had enjoyed a boom in productivity a few years before Hurricane Gilbert struck in 1988 with devastating impact. Several fishers recalled needless and excessive spending of the cooperative's financial resources in the mid-1980s. Given resource abundance and a seemingly endless flow of money, people in the community were not concerned about investing their earnings or consolidating family savings. The cooperative's leaders did not worry about money management, and expenses were often much higher than actual income. The cooperative directors decided to build a packing plant in the town of Tulum, 55 km distant, to receive, package and market all products from the fishing cooperatives in the State. The ice factory and baler were built first, followed by the marketing offices.

After the ice factory and baler were completed, bank managers provided the credit for the marketing offices in a single payment. Construction began shortly before Hurricane Gilbert hit the northern coasts of Quintana Roo. Lobster extraction plummeted. Debts became unsustainable, and cooperative members learned about the financial situation of the cooperative the hard way. Before the 1988 financial disaster, the accountant and directors had kept the cooperative's finances secret. Given this situation, no one wanted to take charge of the bankrupt cooperative, and the responsibility fell to the cooperative's president. The cooperative's debt far exceeded its productive capacity, and many members left the community. Others, in despair over their low incomes, sold lobsters outside of the cooperative. This clandestine selling became such a problem that it seriously threatened the cooperative's continued existence (for a related example, see

McCandless and Emery 2008).

The Board of Directors began to crack down on violators, initiating a wave of expulsions of all members who violated the internal rules. The membership rolls shrank from approximately 120 to 70. In 1993, 5 years after Hurricane Gilbert, the cooperative managed to pay off their debt to Ocean Garden, as well as to the Bank of Mexico. The cooperative also decided not to accept new members, including former cooperative members who had left the cooperative during the crisis. The children of former members also lost the right to aspire to membership. After the debt crisis, only the children of current, active members could become members, thereby effectively limiting resource access to a subset of the community. Rights to lobster fields could only be sold to members of the cooperative, and the cooperative regulated all transactions. The selling of a "lobster field" did not refer to the marine parcel itself, but only to the improvements: the "shades" the previous rights-holder had constructed. The cooperative re-emerged as a result of the restructuring of the internal rules of operation. The reorganization had implications for the way fishery resource property rights were protected, which subsequently permitted the cooperative to comply with the requirements to obtain a fishing concession.

Changes in focus and fishing gear

In the 1950s, 1960s and 1970s, area residents harvested, turtle, shark and caiman, as well as scaly fish. Although these were not primary economic activities, they represented a significant portion of the revenues generated during the off-season for lobster. However, shortly after the Sian Ka'an reserve was gazetted, government efforts shifted, following international trends set at the 1992 Earth Summit in Rio. The new policies focused on the diversification of productive activities to lessen the impact of extractive practices on area ecosystems. The government's intention was to put an end to fishing as a primary commercial activity, and replace extractive practices with environmentally friendly activities. As part of this policy, during the 1990s, the National Commission of Natural Protected Areas (CONANP) engaged in an ongoing campaign to convince fishers to stop fishing for turtle and caiman and abandon their fishing nets.

The perceived panacea for unsustainable resource use was tourism. The government promoted tourist cooperatives running sightseeing tours, and some of the fishers created the first tourist cooperatives in Punta Allen and Vigía Grande. The Gaytanes and Las Boyas tourist cooperatives followed. More recently, area residents formed the Nativos de la Bahía and the women's Orquídeas de Sian Ka'an cooperatives (though at the time of writing the latter is not yet operating tours). After a long process of negotiations and exploration into tourist activities, the fishers decided to withdraw the use of fishing nets. The rationale for this change was the nets' negative impacts on the population of high commercial-value sport-fishing species such as ladyfish (*Elops saurus*), snook (*Centropomus spp.*), permit (*Trachinotus falcatus*), and dolphinfish (*Coriphaena hippurus*).

In the 1990s, tourism was already complementing family incomes, but it was not until 2000 that tourism became the most important economic activity in the community of Punta Allen. Some fishers combined fishing and tourism, while other cooperative members simply abandoned fishing. Over time, a series of tourism cooperatives emerged to regulate the area's resources, strengthening the local community's decision-making capacity. Influenced by external buyers, the remaining lobster fishers decided to change their fishing gear, leaving the hook behind and shifting to the *jamo*. The change in fishing gear was a response to the observed increase in commercial value of the undamaged product harvested with the *jamo*.

Academics have also been important actors in this renewed fishery. At the beginning of the 21st century, the cooperative requested support rom some of the universities and research centers working in the region. Using new technologies, a map of all the lobster fields was created. Previously, the local fishers had delimited the boundaries of their lobster fields with buoys or stakes. Their profound knowledge of the marine space allowed them to find the exact locations of their traps; however, they had never been able to visualize the shape of each field or its actual dimensions. Academics taught them how to use global-positioning systems (GPS), allowing the fishers to improve their surveillance systems and resource usage. Researchers systematized the knowledge about lobster population behavior and performed bathymetry[f] that was subsequently incorporated into a database.

"...Through SISIERRA [a national research program][g], a map was made using GPS, and it was from that point that we learned the shape of our lobster fields... if you were my neighbor along my boundary, I knew you were my neighbor and where you were, but I didn't know about the others; when we did the mapping in 2001, each of us saw. This is something that we had always wanted, that information could be provided to us (by the academics)... so when this young man showed us the GPS, boom! Our eyes opened...the objective was to map, but now we use the GPS to locate the lobster shades... ."(Ramiro, founder, member of the Board of Directors, personal interview, March 6, 2013).

Through these collaborations, the fishers continued to expand and formalize their knowledge of their primary resource, and the environment in which lobster fishing takes place. This knowledge gave them greater negotiating power and control relative to other stakeholders involved in territorial management. All of these processes resulted in a very particular way of understanding territory. The fishery's continued advancement within a conservation area, and the establishment of the Biosphere Reserve, add additional layers of regulation on permissible resource use within those boundaries. The capacity of the Reserve to formally grant and preclude access came to establish a kind of invisible lock that protected the local cooperative and their property rights over the fishery resources, whilst excluding others.

The fishing concession as avenue to cementing success

As mentioned earlier, the Fisheries Act of 1999 recognized two types of usage rights for fishery resources: permits and concessions (DOF 1999). That the Punta Allen cooperative obtained the latter is key to understanding their success and re-emergence. The radical difference between permits and concessions lies in the temporality of rights and forms of access. Under the Fisheries Act, concessions are granted based on an evaluation of the results of technical and economic studies, as well as the amount and predicted recovery time of the investment. It can be granted for up to 50 years. Concessions give rights over benthic resources (such as lobster) in defined geographic areas (Ramírez Félix and Manzo Monroy 2004). Permits are granted for smaller investments and do not require technical and economic studies. For permits, only the nature of the activities is evaluated, and the ownership of the required fishing gear and vessels must be demonstrated (1999 Fisheries Act regulations). While concessions, as a type of right-to-use, are regulated under Mexican law, their application is not an everyday practice in Mexican fisheries, because this kind of access right is not easy to obtain. Though policy at the federal level showed a trend towards deregulation of resources and the inclusion of private capital in extractive practices, in 1993 the fishery cooperative in the Punta Allen community obtained an exclusive 20-year concession for lobster fishing in the Bay of Ascension. This achievement is a significant example of the Punta Allen community's success—most fishing cooperatives in Mexico are not sufficiently well-organized or economically strong enough to obtain such a concession.

The Vigía Chico cooperative received this concession because of their internal organization and marketing capacity, as well as the support of various non-governmental and academic organizations. The cooperative's receipt of this concession is highly significant. The cooperative can then confer rights, also known as concessions, on its authorized users. They provide local fishers with tenured security over their resources and the opportunity to develop future strategies to continue fishing in the area. Those rights, though, are not guaranteed, and must be actively husbanded.

"...There is a lot of private initiative behind these concessions, so even if you've had it, it could happen. Although I have lived here for years, I will go into tourism, but if I lost my concession...it could happen even if I have the concession that I could lose it, if I didn't care for it; it has a lifespan, it expires...it's not as simple as I want it and it is given to me. You have to fulfill certain requirements...but there are interested parties...the director [of the Reserve] already mentioned to us that there are people after them [for the concession]... then anything we fail to do..." (Ramiro, founder, member of the Board of Directors, personal interview, April 13, 2013).

The cooperative's concession allows local fishers to use this renewable rights-granting mechanism to reinforce their systems of resource policing, and exert ongoing pressure on State authorities to support them in this process. The concession also gives local stakeholders the mechanism to obtain greater control over the organization of extractive practices, via participation in the defense of national sovereignty occurring within their local territory.

"...We cannot detain a person because we are civilians, so when we go about our operation, we bring the military, as clearly we are civilians... [The military] are entitled to detain you; we cannot put anyone in jail and much less detain them, like, say, a pirate.... We operate by boat, dragging and towing the boat, and bring them up here...with the Marines.... Since we go out on operations, we have to assemble three elements. We are going to go prepared, because we can't go find them [pirates] and then later bring the Marines; [it has to be] all at once, three marines or four to set out..." (Rodrigo, founder, member of the Board of Directors, personal interview, April 20, 2013).

The maritime space of the Bay of Ascension, granted in concession as a common property to a productive social organization, effectively becomes the locus of interaction between various stakeholders. Tourism cooperatives, fishing cooperatives, non-governmental organizations, state agencies and entities and academia have negotiated a shared understanding of the resource—and the space it occupies—resulting in sustainable management practices.

Throughout the process of settlement, the Punta Allen community developed an awareness of the importance of adhering to state policies in order to pursue their livelihood activities. The cooperative creatively introduced and adapted these policy prescriptions into their local system of rules. Specially, three significant strategies were developed. First, the cooperative was not only designed and perceived as an organization focused on the productive aspects of fisheries, but also as the primary social institution in the community. Second, cooperative members recognized the Sian Ka'an Biosphere Reserve as an effective organizing structure to support their exclusive management of the marine space. Finally, fishing concessions were effectively deployed as a strategy to control extractive resources access.

CONCLUSIONS

To fully understand the success of the Punta Allen lobster fishery we carefully reviewed the settlement history of the community, which yielded a particular social structure. Today, we can trace the positive impacts of forms of spatial and institutional organization linked to particular lobster fishing strategies. These forms of organization have shaped the life of Punta Allen, and now extend beyond the fishery to influence other economic sectors. In Punta Allen, the cooperative, in tandem with the community, has developed a robust, well-tended and strictly enforced system of rights. Through them, the cooperative and community ensure sustainable resource use: they have fostered tourism as an important new dominant economic activity, while still profiting from lobster extraction. We identified several factors enabling this particular set of socio-environmental relationships.

One of these factors was the isolated conditions incomers had to confront during the early days of settlement. Difficulties in marketing the fishing catch, which was necessary to purchase goods needed for survival, forced the villagers to address community issues collectively. The strong community bonds motivated the three leaders, who originally departed from a top-down decision-making approach, to find ways to integrate the existing knowledge and worldviews of community members into collective forms of action. These initial leaders' capacity and willingness to be advised on the legal mechanisms to exercise the rights and obligations as a cooperative also proved critical. Understanding the inner workings of these legal mechanisms allowed the community to effectively integrate federal laws and regulations into local policy and practice. This bridging between governmental laws and local rules gave symbolic and legal authority to the community's modes of social organization, thereby strengthening locally established systems of sanctions and community rights, as in the Maine lobster fisheries case analyzed by Acheson and Brewer (2003). Cooperative working relationships with regional representatives of various state institutions and NGOs also advanced the community and cooperative's legitimacy in the eyes of government officials.

The ongoing presence and interest of academics in the community has helped to legitimize the state's strategy of enforcing conservation measures while allowing extractive activities. It has simultaneously positioned academics as mediators between the state and local actors. For example, at the time of writing, local fishers and academics are working together to generate a resource inventory, to enable reliable monitoring of natural resources and generate valid indicators for sustainable resource use. This resource inventory will provide the basis for greater scientific management capacity, further legitimating the cooperative's activities and establishing a better position for negotiations with other involved stakeholders. Their position may come to resemble that of the concessionaires of Central Baja California, Mexico (see Orensanz et al.2013; McCay et al. 2014). At the same time, the group's locally and academically co-produced knowledge of the characteristics of the resource and the institutions involved in its regulation makes users aware of which practices might affect both the resource and the group's existence over the long term. This helps the community keep its informed interests at the forefront of negotiations with government agencies and buyers.

Market influence fostering changes in fishing technique and equipment have also repeatedly influenced the fishery, increasing the value of the spiny lobster as a product and improving fishers' income. More recently, changes in gear have reduced impacts on the area's marine ecosystem. Furthermore, the biogeographic conditions of the bay imposed challenges on inhabitants, forcing them to increase their degree of cooperation and improving their capacity to respond to perturbations, such as hurricanes. This gave the community the ability to understand the significance of cooperation for achieving goals. The ecological and economic perturbation following the aftermath of Hurricane Gilbert offered a window into the potential socioeconomic consequences of a livelihood scenario without fishing. That crisis also forced cooperative members to rethink the system of rules that allow or prohibit group membership and to strengthen its system of internal rules.

The success of this particular community is influenced by a complex of socio-environmental factors not shared by all artisanal fishing communities globally. Even given its particularity, an analysis of the historical foundations of this particular case allows us to identify indicators worth taking into account in fisheries management more broadly. Punta Allen's Vigía Chico Cooperative increasingly serves as a model for other fishing cooperatives, as well as for NGOs and government bodies. These visitors are gradually developing efforts to reproduce elsewhere some of the mechanisms that have maintained the strength of the Punta Allen cooperative over more than three decades. We hope that our detailed analysis contributes to these efforts.

Today, fishing in Mexico presents many challenges. Fishing permits, when granted, apply to the entire Mexican coast. Government monitoring efforts are inadequate. The result is an institutional apparatus that has been unable to meet the challenges of fisheries management. However, despite the policy of deregulation, the undoing of legislation privileging the cooperative model, the requirements of formal conservation, and the pressure exerted by private initiatives to access high-commercial-value resources, the Punta Allen cooperative persists. Its continued success offers a clear example of how locally generated, tended and enforced rules can be legitimated and strengthened by strategic engagement with State regulations and regulators. Together with the support of ongoing innovation and collaboration with outside experts, these practices can support successful management.

The Punta Allen cooperative has managed to engage with available governance mechanisms and technologies and collectively fit them to their local context. Their adaptability and tenacity has permitted them to successfully negotiate the shifting terrain of a succession of state actors, policies, and priorities, as well as market and environmental shifts. Equally important has been how the different stakeholders interact in a territory of economic, ecological and strategic importance at the regional and national level.

This case, and other studies of successful artisanal fisheries reviewed here, show the multiple benefits that accrue when different stakeholders work tenaciously to reconcile their interests. Through direct negotiations, legitimacy derived from adherence to regulatory standards and scientific management, processes of exclusion, and skillful engagement of different parties in the performance of everyday practices, Punta Allen's success attests that well-organized local groups can secure viable fisheries and coastal livelihoods.

ENDNOTES

[a]The shelters or shades are made with concrete and represent an artificial refuge for lobsters. This fishing gear allows the selective capture of live lobster, decreasing damage to juveniles and egg-bearing females, which are returned to the sea during fishing season (Seijo 1993).

[b]The *jamo* is a net that is used to collect lobsters without injuring them. It resembles a butterfly net, but is woven from very strong fishing line.

[c]The time during which this president held office was called *Cardenismo*, and was characterized by its policy of support for the working class, land distribution, the promotion of the cooperative movement and the nationalization of property (see León and Marván 1999; Semo 1993).

[d]"*Parcela,*" or parcel, is a term that farmers use to designate their agricultural plots.

[e]In 1983, during the presidential tenure of Miguel de la Madrid, the "Program of Exploration and Assessment of Fisheries Resources of the Exclusive Economic Zone and the Territorial Waters" was initiated with the participation of the Secretary of the Navy, the National Council of Science and Technology, the National Cooperative Conference of the Mexican Republic, the National Chamber of the Fishing Industry, the state-owned company Mexican Fishery Products, the National Autonomous University of Mexico and the National Polytechnic and other universities in the province. Its purpose was to estimate fishery resources for allocation (Soberanes Fernández 1994).

[f]Bathymetry is the study of the relief of the seabed and its depth contours.

[g]SISIERRA (Sistema de Investigación Justo Sierra) was the name of a research program founded by CONACYT, the

National Council of Science and Technology (2001).

ACKNOWLEDGEMENTS

We wish to thank all the members of the Cooperative and the community who took the time to share their valuable work and stories with the first author. We want to acknowledge Biol. Angel Omar Ortiz, Director of Uaymil Wildlife Protected Area and Sian Ka'an Biosphere Reserve, for granting permission to conduct this research, and to Biol. Yadira Gómez for all the facilities provided to complete fieldwork. Our thanks to the Mexican National Council for Science and Technology (CONACYT), for partially financing the primary author's fieldwork through a grant provided by its Basic Science Project, award 129323: Responses and experiences of social innovation in the structural crisis of wage employment. The manuscript benefited from feedback from two anonymous reviewers, and we are especially grateful to editor Maria Gassalla for her constructive comments, which helped to substantially improve the paper. Finally, thanks to Holger Weissenberger for generating the map of the study area.

AUTHORS' CONTRIBUTIONS

CMM and BS designed the research; CMM performed the research; CMM and BS drafted the manuscript; SRM improved and added to the text. All authors read and approved the final manuscript.

REFERENCES

1. Acheson, JM, and JF Brewer. 2003. Changes in the territorial system of the Maine lobster industry. The commons in the new millennium: challenges and adaptation. In The commons in the new millennium: challenges and adaptation, ed. N Dolšak and E Ostrom, 37–59. Cambridge, MA: MIT Press.
2. Arreguín-Sánchez, F. 2006. Pesquerías de México. In Pesca, acuacultura e investigación en México, ed. P Guzmán Amaya and DF Fuentes Castellanos, 13–36. Mexico City: Centro de Desarrollo Rural Sustentable y la Soberanía Alimentaria, Cámara de diputados, LIX Legislatura/Congreso de la Unión, Mexico, Comisión de pesca.
3. Basurto, X, and E Coleman. 2010. Institutional and ecological interplay for successful self-governance of community-based fisheries. Ecological Economics 69: 1094–1103.
4. Basurto, X, A Cinti, L Bourillón, M Rojo, J Torre, and AH Weaver. 2012. The emergence of access controls in small-scale fishing commons: a comparative analysis of individual licenses and common property-rights in two Mexican communities. Human Ecology 40: 597–609.
5. Biernacki, P, and D Waldorf. 1981. Snowball sampling: problems and techniques of chain referral sampling. Sociological Methods & Research 10: 141–163.
6. Brenner, L. 2010. Gobernanza ambiental, actores sociales y conflictos en las Áreas Naturales Protegidas mexicanas. Revista Mexicana de Sociología 72(2): 283–310.
7. César, AA, and SM Arnáiz. 1986. Estudios socioeconómicos preliminares de Quintana Roo. Sector pesquero. Puerto Morelos, Quintana Roo, Mexico: CIQRO.
8. César, A and S Arnaíz. 1990. Quintana Roo. Biblioteca de las entidades federativas. Mexico City: Centro de Investigaciones Interdisciplinarias en Humanidades.
9. César, A, and S Arnaíz. 1992. El Caribe mexicano. Una introducción a su historia. México City: Centro de Investigaciones de Quintana Roo.
10. Cifuentes-Lemus, JL, and FG Cupul-Magaña. 2002. Un vistazo a la historia de la pesca en México. Administración, legislación y esfuerzos para su investigación. Ciencia ergo sum 9(1): 112–118.
11. Cochran, K. 1998. Fishery Co-management: the case of the Punta Allen spiny lobster fishery, 2–5. Salt Lake City, Utah: American Agricultural Economics Association Annual Meeting.
12. Cruz-Ayala, M, and LE Igartúa-Calderón. 2006. La transformación de la legislación pesquera en México: un acercamiento en el contexto político-económico (1925–1992). In Pesca, acuacultura e investigación en México, ed. P Guzmán Amaya and DF Fuentes Castellanos, 233–247. Mexico City: Centro de Desarrollo Rural Sustentable y la Soberanía Alimentaria, Cámara de diputados, LIX Legislatura/Congreso de la Unión, Mexico, Comisión de pesca.
13. Defeo, O, and JC Castilla. 2005. More than one bag for the world fishery crisis and keys for co-management successes in selected artisanal Latin American shellfisheries. Fish Biology and Fisheries 15: 265–283.
14. Diario Oficial de la Federación (DOF). 1999. Reglamento de la Ley de Pesca. September 29, 1999. Mexico City
15. Ernst, B, J Chamorro, P Manríquez, JL Orensanz, AM Parma, J Porobic, and C Román. 2013. Sustainability of the Juan Fernández lobster fishery (Chile) and the perils of generic science-based prescriptions. Global Environmental Change 23(6): 1381–1392.
16. Gallardo, G, W Stotz, J Aburto, C Mondaca, and K Vera. 2011. Emerging commons within artisanal fisheries. The Chilean territorial use rights in fisheries (TURFs) within a broader coastal landscape. International Journal of the Commons 5(2): 459–484.
17. Guest, G, A Bunce, and L Johnson. 2006. How many interviews are enough?: an experiment with data saturation and variability. Field Methods 18: 59–82.
18. Gutiérrez, NL, R Hilborn, and O Defeo. 2011. Leadership, social capital and incentives promote successful fisheries. Nature 470(7334): 386–389.
19. Hidalgo, H, and A Méndez. 2007. Diagnóstico organizacional y de necesidades en las comunidades pesqueras en el Sistema Arrecifal Mesoamericano. Livingston, Guatemala: Mar Fund.
20. Instituto Nacional de Ecología (INE). 1996. Serie Programas de Manejo: Programa de Manejo de la Reserva de la Biosfera Sian Ka'an. Mexico City: SEMARNAP.
21. Jentoft, S. 2004. Institutions in fisheries: what they are, what they do, and how they change. Marine Policy 28: 137–149.

22. Kadekodi, GK. 2004. Common property resource management. Reflections on theory and the Indian experience. New Delhi: Oxford University Press.

23. León, S, and I Marván. 1999. La clase obrera en la historia de México: en el cardenismo (1934–1940). Mexico City: Siglo XXI-Instituto de Investigaciones Sociales, UNAM.

24. McCandless, SR and MR Emery. 2008. Partial Power, Partial Knowledge: Accounting for the Dis-Integration of a Costa Rican Cooperative. Society and Natural Resources 21: 310–323.

25. McCay, BJ, F Micheli, G Ponce-Díaz, G Murray, G Shester, S Ramirez-Sanchez, and W Weisman. 2014. Cooperatives, concessions, and co-management on the Pacific coast of Mexico. Marine Policy 44: 49–59.

26. Orensanz, JM, A Cinti, AM Parma, L Burotto, S Espinosa-Guerrero, E Sosa-Cordero, C Sepúlveda, and V Toral-Granda. 2013. PART I: Latin American rights-based fisheries targeting sedentary resources. In Rights-based management in Latin American, ed. JM Orensanz and JC Seijo, 1–69. Rome, FAO: Fisheries and Aquaculture Technical Paper, No. 582.

27. Ostrom, E. 2007. Institutional rational choice: An assessment of the institutional analysis and development framework. In Theories of the policy process, ed. PA Sabatier, 21-64. Boulder, CO: Westview.

28. Ostrom, E. 1986. An agenda for the study of institutions. Public Choice 48: 3–25.

29. Ostrom, V, and R John. 1976. Commons foundations for policy analysis. Journal of Economic Issues 10(4): 839–857.

30. Poteete, AR. 2010. In Analyzing the politics of natural resources: from theories of property rights to institutional analysis and beyond, ed. I Vaccaro, EA Smith, and A Shankar, 59–79. Cambridge, MA: Cambridge University Press.

31. Ramírez Félix, E, and E Manzo Monroy. 2004. Comparación entre el uso de dos derechos de acceso pesquero, concesiones y permisos, en la pesquería de erizo rojo de mar, Strongylocentrotus franciscanus. Ciencias Marinas, Universidad Autónoma de Baja California 30(4): 547–560.

32. Schlager, E, and E Ostrom. 1992. Property-rights regimes and natural resources: a conceptual analysis. Land Economics 68(3): 249–262.

33. Seijo, JC. 1993. Individual transferable grounds in a community managed artisanal fishery. Marine Resource Economics 8: 78–81.

34. Seijo, JC. 2008. The Punta Allen lobster fishery: current status and recent trends. Rome: FAO: Case Studies in Fisheries Self-Governance, No. 504.

35. Seijo, JC, and D Fuentes. 1989. The spiny lobster *(Panulirus argus)* fishery of Punta Allen, Mexico. In Fisheries credit programmes and revolving loan funds: case studies, vol. 312, ed. U Tietze and P Merrikin, 89–100. Rome: FAO: FAO Fisheries and Aquaculture Technical Paper.

36. Semo, I. 1993. El cardenismo revisado: la tercera vía y otras utopías inciertas. Revista Mexicana de Sociología 55(2): 197–223.

37. Soberanes Fernández, JL. 1994. Historia Contemporánea de la Legislación Pesquera en México. In El regimen jurídico de la pesca en México, ed. M González Oropeza and MA Garita Alonso, 1–26. Mexico City: UNAM.

38. Sosa-Cordero, E. 2011. La langosta, pesquería emblemática de Quintana Roo. In Riqueza biológica de Quintana Roo: un análisis para su conservación, ed. C Pozo, N Armijo Canto, and S Calmé, 221–227. Mexico City: CONABIO, ECOSUR, Gob. Quintana Roo, PPD.

39. Sosa-Cordero, E, and A Ramírez-González. 2001. Dinámica de las pesquerías artesanales. Ecofronteras 14: 13–15.

40. Sosa-Cordero, E. and A. Ramírez-González. 2011. Pesca Marina. In Riqueza biológica de Quintana Roo: un análisis para su conservación, ed. C Pozo, N. Armijo Canto and S. Calmé, 183-189. Mexico City: CONABIO, ECOSUR, Gob. Quintana Roo, PPD.

41. Sosa-Cordero, E, MLA Liceaga-Correa, and JC Seijo. 2008. The Punta Allen lobster fishery: current status and recent trends. In Case studies in fisheries self-governance, ed. R Townsend, R Shotton, and HC Uchida, 149–162. Rome: FAO.

42. St. Martin, K. 2001. "Making Space for Community Resource Management in Fisheries," The Annals of the Association of American Geographers 91(1): 122-142.

43. Thompson, EJ. 1979. Historia y religion de los mayas. Mexico City: Siglo XXI.

44. Velez, M, S Adlerstein, and J Wondolleck. 2014. Fishers perceptions, facilitating factors and challenges of community-based no-take zones in the Sian Ka'an biosphere reserve, Quintana Roo, México. Marine Policy 45: 171–181.

CHAPTER 11 MANAGEMENT OF COLD WATER CORALS

Jannike Falk-Andersson, Naomi S Foley, Claire W Armstrong, Sybille van den Hove, Thomas M van Rensburg and Rob Tinch

Keywords: Valuation, Deliberative, Cold water corals, Ecosystem services, Precautionary principle

INTRODUCTION

The many dismal experiences from fisheries management (Myers and Worm 2003, Jackson et al. 2001) have led to the realization that in order to secure sustainable utilization of marine resources, a broader ecosystem management approach is needed. This involves widening the focus from single stocks to ecosystems, including human activities and values (Garcia et al.2003, Aanesen et al. 2012). Stakeholders and citizens use ecosystem resources and benefit from their services in different ways. Interests may conflict with respect to what is seen as appropriate use of the resources. Furthermore, resource conflicts may arise in cases where the resource is a common public good, externalities may occur, and legal rights and obligations are poorly defined (Armstrong and van den Hove 2008, Glenn et al. 2010). The management changes required will therefore be most effectively achieved if a wide range of stakeholders become involved in issues pertaining to the sustainable use of marine resources.

This paper addresses the particular case of cold water corals (CWCs), which are often badly damaged by certain fishing practices, particularly bottom trawling. There is a conflict between current, known fishing values and future, uncertain values associated with conservation, biodiversity protection and other uncertain or unknown opportunities, including fishing. There is also significant uncertainty regarding the roles and functions of CWCs in marine ecosystems and the on-going and future costs of damage. This creates challenges for policy formation because improving the knowledge base requires resources, takes time, and may be unfeasible with the tools currently available to marine scientists. While decisions therefore have to be made without full scientific understanding, policy can nevertheless be informed by better knowledge of stakeholder and citizen perceptions, beliefs, and attitudes towards the issues at stake. The potential role of stakeholder input in decision making processes will be discussed here.

The almost total lack of public familiarity with CWC and its relation to ecosystem functions and services makes this resource difficult to value using standard stated preference valuation techniques, such as contingent valuation or discrete choice experiments. Recent attempts have been made (LaRiviere et al. 2014), but the interpretation of values

Citation: Jannike Falk-Andersson, Naomi S Foley, Claire W Armstrong, Sybille van den Hove, Thomas M van Rensburg, Rob Tinch, Management of cold water corals, DOI: 10.1186/s40152-015-0023-z.

elicited under such conditions of unfamiliarity, uncertainty and ill-defined preferences is problematic (see Wattage et al. (2011) for more detailed discussion). In order to support decisions, reducing peoples' values to an uncertain monetary entity may not be sufficient. To delve more deeply into the ways in which people understand and reason about policy decisions that have an impact on CWC, we therefore employed a deliberative approach (Wilson and Howarth 2002, Lo and Spash 2013). Although well-developed and integrated in decision support (Söderholm 2001, Morgan 1996), studies that apply deliberative techniques to the precautionary principle by involving citizens and stakeholders are still rare (see Mason et al. (2010) for one example). We combined group discussions and individual questionnaire responses in three focus groups carried out in Norway, to explore attitudes towards CWC and related services and values. Questions focused on individual and group thinking regarding the trade-offs between conflicting values; in particular, opinions regarding trade-offs between future potential uses versus current fishery values were studied.

Since this study looked into Norwegian CWC, we consider the Norwegian populace at large as the relevant population. Both stakeholder groups and individual citizens are of interest (Kahane et al. 2013). Fishers have a particular stake since regulations may affect their activity directly, and since they believe CWC are important to fish (Armstrong and van den Hove 2008). However, both fishers and non-fishers may have an interest as citizens in maintaining fish populations and securing other CWC values, including non-use values. CWC – the "rainforests of the sea" - may also be of more global interest, but for practical reasons we limited attention to the Norwegian population.

To tease out nuances regarding attitudes to CWC in the context of the scientific uncertainty about their extent, condition and functions, perceptions regarding protection of CWC were studied in two different contexts:

- Permanent protection of known CWC grounds because they provide benefits or represent values now and in the future (albeit with uncertainty about these benefits), and
- Temporary protection of candidate areas for the purpose of obtaining information on the presence or absence of cold water corals.

Both issues were discussed in the focus groups as well as in questionnaires answered by each participant during the meetings, in order to gauge differences between individual perceptions and group deliberations.

The objective of the study was to investigate the use of deliberative approaches to provide information relevant to decision making and management, especially in the context of the implementation of the precautionary principle. Insights gained are useful in assessing how CWC values can best be captured and communicated.

BACKGROUND

Cold water coral and the precautionary principle

Bottom trawling has been the main threat to cold water corals. Together with other drivers such as oil/gas extraction, climate change and ocean acidification, corals and other benthic habitats are experiencing increasing anthropogenic pressures (Foley et al. 2010). CWC is a slow-growing, reef-building organism found in all the world's oceans (MacIsaac et al. 2001). Though known to both fishermen and scientists for centuries, the degree of understanding of these organisms and their ecosystem functions is limited (Foley et al. 2010). The damage caused to CWC by fishing pressure is reflected in the substantial estimated declines in CWC coverage, including in Norwegian waters, where damage has been estimated at 30–50 % (Fosså et al. 2002). Additionally, their slow growth implies that they do not completely recover from damage, which, at least in the context of reefs, is largely irreversible on human-relevant timescales (Foley et al.2010). The marine protected areas established in order to protect cold water corals cover 2445 km^2 or 0.3% of the Norwegian Exclusive Economic Zone. Furthermore, Norwegian legislation demands that special care should be taken when fishing, in order to avoid damage to corals (Anonymous 2004).

Uncertainty regarding ecosystem functions and values in combination with irreversibility of damage is generally viewed as a trigger for implementation of the precautionary principle in management (UNESCO 2005, The European Environmental Agency, The 2001), though precautionary action is not always taken in practice. There are many different definitions of the precautionary principle, resulting in misunderstandings that have been identified as one of the barriers to its use (The European Environmental Agency, The 2001). Key elements of the principle relevant to this study can, however, be broadly agreed to include: the need for decision makers to anticipate potential harm before it occurs or becomes likely; the justification to act to prevent or minimize the harm; and accounting for the advantages and disadvantages of action and inaction (Tinch et al. 2011). The latter implies an exploration of values - these may be considered in terms of monetary costs and benefits, but this is not a necessary requirement.

Cold water coral and valuation

Although the uncertainty and irreversibility associated with CWC damage should be sufficient to invoke the precautionary principle, the realities of policy formation in most of the world means that the cost of action is an important criterion. Information about the costs and benefits (or more generally advantages and disadvantages) of action and inaction, about the consequent trade-offs, and about the ways in which people consider and value the issues at stake, form important inputs into the policy process (Morris et al. 2000). Identification and sometimes valuation of ecosystem services can be used to make such trade-offs explicit. For many of these services their value is not reflected in any market, either directly or indirectly, making market-based and revealed preference methods inapplicable. A number of stated preference methods have been developed to elicit individuals' valuation of these services (see for example Liu et al. (2010), Defra (2007)). LaRiviere et al. (2014) apply a discrete choice model to study how the effect of signals regarding respondents' knowledge affects willingness to pay in the case of protection of cold water coral in Norway. The results showed that the Norwegian public were in fact willing to pay for cold water coral protection in Norway, and that the respondents' beliefs about their own knowledge affected this willingness to pay. The study does however not give insight into the more complex perceptions regarding management and option values. Glenn et al. (2010) used a discrete choice model to estimate the Irish population's willingness to pay for protection of CWCs. Although the study did not find a statistically significant willingness to pay, it did reveal a strong preference for a ban on bottom trawling where CWC are thought to exist. Furthermore, a large percentage of those surveyed claimed to value conservation of corals. Respondents stated that they would like to see CWC protected for future generations, for their role as a perceived essential fish habitat, for their existence value and also for the option to use or see them in the future (Glenn et al. 2010).

One possible explanation for this latter seemingly contradictory result is the public's lack of familiarity with CWC, limited visibility and very limited potential for first-hand exposure. Other explanations are rejection of various features of the study, including the payment mechanism. In effect, the restrictive nature of the stated preference valuation instrument may prevent it from capturing in monetary terms the values and attitudes that are held by the respondents and that they express in response to non-monetary elicitation techniques. A discursive approach, on the other hand provides a setting in which individuals become exposed to the ideas of others, which facilitates discussions and thoughtful consideration of public (not just private) values (Morgan1996). It is in this light that we apply a deliberative approach to valuation of CWC; deliberation allows more time for learning and thereby for preference or attitude formation relating to an unfamiliar issue.

Ecosystem services and values

The first synthesis report of The Economics of Ecosystems and Biodiversity (TEEB 2010) introduced the idea of a tiered approach to values in environmental assessment and management. This considers the integration of value evidence in decision making as existing on different levels, from basic and general societal recognition of values integrated in social norms, through formal demonstrations and assessments of values, to full use of value evidence as key components of management tools. These tiers of recognising, demonstrating and capturing value require various levels of identification and methods for measurement of values arising from ecosystems. The current dominant framework is based on the identification of the ecosystem services provided to humans. Describing ecosystems and their functions in terms of the services people benefit from is intended to help articulate the importance of ecosystems and raise awareness of the many ways in which they support and provide value for human life and well-being. This translation is intended to offer a framework for representing public values for conservation in a more utilitarian form, rather than a more holistic environmental view (Barkmann et al. 2008). Many frameworks for assessing ecosystem services have been developed, for example by TEEB (2010), in national ecosystem assessments (seehttp://catalog.ipbes.net/) and in particular the Common International Classification of Ecosystem Services (CICES 2013), though there is as yet no agreement on a single approach. The Millennium Ecosystem Assessment (MEA 2005) framework divides ecosystem services into four categories (that remain, broadly speaking, the basis for more recent frameworks): provisioning, regulating, cultural and supporting services. Figure 1 illustrates the MEA framework adapted for CWC, starting with the more indirect supporting services that can potentially feed into the direct provisioning, regulating and cultural services.

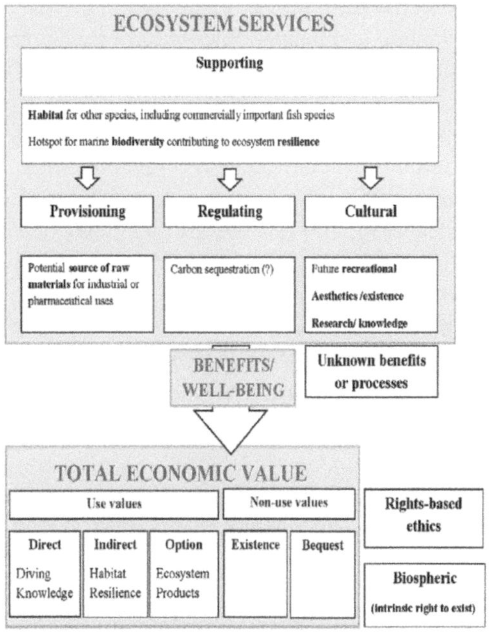

Figure 1: Constituents of MEA and TEV (grey boxes) for CWC and factors that are not part of these concepts (white boxes).

Provisioning services are products used by humans that are obtained directly from ecosystems. As of today CWC provide no direct provisioning services, apart from a few cases of mining and collection of corals to make jewellery. However, CWC are potentially a source of original raw materials for industrial and pharmaceutical uses (Foley et al. 2010). The unusual characteristics of deep-sea organisms, their unique adaptations that enable them to survive in dark, extreme temperatures and highly pressurized environments, offer unique opportunities, making them the subject of considerable excitement in the scientific community, with many potentially interesting commercial possibilities (Armstrong et al. 2010, Arico and Salpin 2005).

Regulating services are benefits obtained from the regulation of ecosystem processes such as climate and water regulation, erosion control and water purification. CWC could offer regulating services such as the sequestration of carbon, providing a climate change mitigation benefit (Foley et al. 2010).

Cultural services are non-material benefits obtained through, for example, recreation, aesthetic enjoyment, inspiration and learning. To date there is no direct recreational use value for CWC, apart from the unique case of Trondheim fjord, Norway, where they are found at depths accessible for recreational diving. Future recreational use may be possible, for example employing submersibles. Indirect aesthetic use is possible via films and books (Foley et al.2010). Part of the willingness to protect CWC reflects the charismatic nature of these resources (Armstrong and van den Hove 2008) and that people want to preserve them simply for their existence or for future generations (Glenn et al. 2010). CWC provide educational and scientific services for many people involved in learning about the marine environment. CWC could also provide archives recording intermediate to sub-surface water temperatures and salinity and serve as a good climate change proxy (Lutringer et al. 2005, Puglise et al. 2005).

Supporting services are those necessary for the production of all other ecosystem services such as nutrient cycling, primary production and habitat. Supporting services related to CWC specifically include habitat, biodiversity and resilience (Figure 1). These deep-sea structures have been shown to harbour high biodiversity (Freiwald et al. 2004) and to support important habitat functions, notably with regards to fish nurseries, refugia to hide from predators and feeding areas (Freiwald et al. 2004). Both are likely to contribute to greater resilience of ecosystems and thereby the capacity to cope with disturbance (Foley et al. 2010). CWC seem to attract fish concentrations (Husebø et al. 2002), in turn also attracting fishermen (Armstrong and van den Hove 2008).

The total economic value (TEV) framework (Pearce and Moran 1994) has been used to value the benefits or well-being generated by provisioning, regulating and cultural services (Figure 1), with supporting services being valued through their contribution to the other services (Defra2007). TEV divides economic values into use values, non-use values and option values (Pearce and Moran 1994). Use values arise from direct or indirect use of a resource, while non-use values are derived from the value of knowing that something exists and/or that other people, including future generations, may benefit from a resource. Option values are a third category capturing uncertainty regarding future use or provision of the resource: what is it worth giving up today in order to keep those future uncertain options available? A related concept is quasi option value which refers to the value of future information which can be protected by delaying an irreversible development (Perman et al. 2011). In this case, it would relate to willingness to pay to avoid irreversible consequences of development today, given the expectation of future increase in knowledge related to the resource (Perman et al. 2011). An important goal of this study was to find out if focus group participants support the use of temporary CWC protection measures for the purpose of obtaining information on the presence of cold water corals in the future.

Given that research on CWC is relatively recent and the degree of knowledge is limited, many of their services are unknown or potential services, i.e. option values. One example is bioprospecting: there is a value *now* to delaying or avoiding decisions that could reduce deep-sea genetic diversity, due to the unknown *future* potential for these genetic resources to contribute to human wellbeing (Armstrong et al. 2012).

Furthermore, since damage is essentially irreversible within a time frame relevant to humans, due to the low growth rate of CWC, the concept of quasi option value is important. While the benefits of preservation are uncertain today, they could become more certain through time as information increases (Foley et al. 2010). Valuation of option and quasi-option values is closely linked to the precautionary principle, as the valuation applies to processes of complex social-ecological systems, characterised by non-linearity and thresholds and also by uncertainty, ignorance and ambiguity for which it is not possible fully to quantify scientific uncertainty (UNESCO 2005).

The TEV framework, as usually applied, is grounded in values derived from individual preferences, and is therefore focused on utilitarian values associated with an ecosystem. This does not mean that only personal use is covered – the non-use category covers altruistic values – but the key reference point remains the preferences of individuals, expressed in monetary terms which is then aggregated. Abson and Termansen (2011), however, stress that it is only appropriate to apply the TEV framework when it captures the values that humans associate with the ecosystem services of concern. TEV has been argued to be incompatible with rights-based ethics as a basis for decision making, because concerns for doing the right thing, meeting social norms, and recognition of an intrinsic right to exist independent of human benefit, have been found to be important features of human attitudes towards environmental issues (Spash 2006). On the other hand, these concerns influence human preferences and their expression, and non-use values – part of TEV – may be concerned with unselfish motives. Although rights based ethics may seem to require absolute positions (the rejection of any trade-off), in fact, because of the existence of an irreducible plurality of values, absolute 'rights' of different people often conflict and in such cases trade-off or compromise is literally inevitable (van den Hove 2006). Discursive, deliberative based approaches are one way of dealing with a plurality of values and world views that depart from individual-based TEV. Appropriate choices depend on contexts, and 'problems' when applying stated preference methods may be interpreted as a sign that people feel uncomfortable with the use of individual preference based methods in a given situation. This is often associated with rejection of the implied property right or other forms of protest bidding. Figure 1 illustrates the constituents of MEA and TEV in terms of CWC, as well as the factors that are not included in these concepts. In this paper, we have chosen to discuss the results in terms of the TEV concept and the intrinsic right for something to exist, since we were especially interested in exploring the support for protecting option values.

Deliberative approaches

Many environmental issues are characterised by complexity, uncertainty, large temporal and spatial scales, and irreversibility (van den Hove 2000). These features may favour deliberative methods, as these methods allow for greater clarification and interaction. Deliberative methods that focus more on negotiation and consensus building may also be viewed as better suited to valuing public goods, including many environmental services (Howarth and Wilson 2006, Sagoff 1998, Kumar and Kumar 2008). In the academic discussion on environmental planning there has also been a tendency towards communicative rationality where agreement is reached based on shared validity criteria (Gezelius and Refsgaard 2007) and plurality of values makes it more appropriate to speak of compromise rather than consensus solutions (van den Hove2006).

There are a number of different deliberative approaches to valuation, from loosely defined 'group-based approaches' to more codified methods of deliberative polling, citizens' juries, focus groups and deliberative monetary valuation (DMV). Table 1 illustrates some of the diversity of deliberative methods applied in valuation. Discursive approaches can be used to derive monetary values, as in DMV, though this has been criticised for example by Niemeyer and Spash (2001) who compare DMV with cost-benefit analysis, and find the same theoretical limitations in both, arguing especially against the "allure and simplicity of a single number as the basis for determining policy" (op. cit. p.583). Non-monetary methods can also be used to allow expression of preferences in units other than money, and may focus on direct expression of preferences for outcomes (Tinch et al. 2011).

Table 1: Summary of deliberative methods to valuation (Perman et al. 2011 , Turner et al. 2010 , Niemeyer and Spash 2001)

Method	Description
Deliberative Polling	A random opinion poll prior to an invitation to one or several meetings. Respondents are given the opportunity to listen to and question expert witnesses, as well as discuss the issue within the group before being given the original questionnaire once again.
	A costly approach, where the number of attendees must be substantial if the results are to meet the standards of regular opinion polls.
Citizens' Juries	Consists of a smaller number of representatives. Less resource demanding than polling, and deemed to be equally legitimate.
Focus Groups	Selection of the public representing a specific type, and seeking a 'focus' view for that type. Replicates of a number of different types then represent a range of typical views.
Deliberative Monetary Valuation (DMV)	Applies more stringent stated preference methods to a discursive approach. Small groups that in total represent society are given concrete valuation problems for discursive evaluation.

We used a *focus group* approach to deliberation: our objective was not to determine a single numerical 'value', but rather to explore wider attitudes and preferences with regards to CWC and their management. Furthermore, with such a complex and unfamiliar set of issues, use of groups in which individuals shared key characteristics, such as the same interest or employment, eased the discussion, and allowed some consensus or compromise reaching, and clear identification of points of disagreement.

METHOD

Focus groups were held in Norway in October 2011. Three groups were chosen to examine similarities and differences in opinion of people with varying interests, backgrounds and connections to the ocean: a commercial user group (6 fishers), a recreational user group (7 sailors), and a group with no specific connection to the ocean (7 singers from an inland choir). The fishers lived in a small coastal community, the sailors were coastal city dwellers, while the choir came from a small inland community. The three groups may be seen as a continuum from stakeholder to citizen (Soma and Vatn 2010). All participants within each group were adults and already knew each other through the activities they had in common. This helped to ensure free and open discussions. The groups were reached via contact with their respective organisations. The choir and the sailors club were offered a token contribution to the club. Members in these two clubs participated voluntarily and received no personal payment. Each focus group involved a facilitator guiding the discussion and an assistant taking notes on the discussion as basis for subsequent analysis. The small group sizes allowed sufficient documentation of the discussions. The focus groups took place in meeting rooms at hotels and participants were offered refreshments.

A protocol was prepared as an aide to prompt and engage participants in discussion. The protocol was open to provide flexibility but a number of objectives with related questions were set to steer the discussion. The general format was as follows: the researchers explained the background for the study and that the aim of the focus groups was to explore with the participants their perceptions and opinions regarding the topic. Then the participants were asked to introduce themselves. Throughout the focus group a discussion of a topic was followed by the participants filling in the corresponding section in a questionnaire. Thus, prior to filling in the questionnaire, the participants shared their perceptions and opinions.

The first part of the focus group was kept broad, to ease participants into the discussion and to get information on their uses of the sea, opinions of the state of the sea in general, and knowledge of the products and functions the sea provides. After this, since CWC is largely an unknown good, some general information on CWC was given along with a digital presentation, on their biology, what is known about CWC (they provide habitat for fish and CO_2 storage, and their high biodiversity is a potential source of genetic resources for bio-prospecting), the threats to them (trawling, oil and gas industry, mining, ocean acidification), and alternatives for their management (closures, gear restrictions and eco-labelling). Then respondents were asked whether they had heard of CWC prior to participation in the focus group, whether they thought people in general would have heard about CWC, their knowledge of deep-water fisheries, as well as attitudes to management of CWC (should they be protected and why/why not, how should they be protected). One of the goals of this study was to obtain information about how respondents think about full protection of known coral grounds versus temporary protection of sites thought to have CWC. In order to make sure that respondents fully understood this, they were presented with two different options for closures: 1) closure of areas only after corals have been documented in an area, and 2) closure of a wider range of areas while waiting for areas to be mapped for corals. It was made very clear that the former would imply that corals could be damaged forever (irreversible action) while waiting for mapping, while the latter would be a precautionary action to ensure damage could not occur while obtaining information on the presence of cold water corals, but would have greater short-term implications for the fishing industry.

Key words from the discussion were noted on a black board and the administrator took notes as the basis for qualitative analysis. These notes were later compared to the answers in the questionnaires. The use of scientific terminology was avoided and the presentation was given in Norwegian. Care was made to not use value laden terms such as "services" and "values" in order to avoid causing a bias in the discussion, particularly with respect to reasons for protection. During the discussions the moderator specifically encouraged less active participants to contribute, and underlined that there was no right or wrong answer. If needed, the moderator probed into issues that had not surfaced during the discussions. In all the focus groups the participants were vocal, showed interest and contributed enthusiastically to the debate.

After discussions, the corresponding sections in the questionnaire were filled in. The questionnaire listed values corresponding to TEV that CWC represents, as well as their intrinsic right to exist, but without using scientific jargon. Including an individual questionnaire along with focus groups is a practice that is not uncommon (Morgan 1996). It allows a check for bias that can arise due to differences among participants in terms of how dominant they are in the discussions. Further rationales for combining individual and group interviews point to the greater depth of the former and improved breadth of the latter (Crabtree et al. 1993), and to the ability to explore specific opinions further (Duncan and Morgan 1994). A number of questions were posed in the questionnaire with alternative answers which were elicited through a discrete visual analogue scale, in addition to options of not knowing or not caring. The questionnaire asked whether the respondents agreed that CWC should be conserved because of their potential for bio-prospecting (direct use-value), as habitat for fish (indirect use-value), carbon storage (indirect use-value), future generations' use of CWC (bequest value), because they have the right to exist independent of humans (intrinsic right to exist), and/or because of the large uncertainties regarding the values CWC represent (option value). The respondents were also asked whether they would prefer closing areas for one year, 2–5 years, five years or until mapping of CWC is completed. These questions gave us preferences regarding which of the values related to CWC the respondents emphasised, as well as attitudes to precautionary action. The respondents rated the questionnaire statements from disagree (1–2), neutral (3) to agree (4–5) or «don›t know/ don›t care». The data were pooled and the percentage of respondents that «agreed» was calculated for each statement. The survey responses are displayed in simple graphs, since the low number of participants does not justify further statistical analysis. In this study we only compare the group and survey responses with respect to reasons why cold water corals should be protected and how this should be carried out, as these are the most relevant issues related to management.

RESULTS

The following presents attitudes relating to CWC based on deliberation and questionnaire findings. The results from the surveys are compiled from the three focus groups, totalling 18 responses, as two surveys were incomplete.

The first issue was to address the protection of known coral areas. The reader is reminded that the objective was to elicit the values focus group participants associated with CWC and other factors considered important in determining whether they should be protected. The emphasis here is on the protection of cold water corals because they may provide services and represent values now or in the future, for coral areas that are known, even if there is still uncertainty about the specific level of benefits. During the discussions, fishers recognised the indirect use-value that CWC habitat provides for fish stocks, including commercial species such as redfish and halibut. For this reason, they would be glad of coral protection, but were concerned about the size of buffer zones around the coral areas which would prohibit them fishing close to the corals. Furthermore, the fishers stated that they had become so efficient that the fish needed some protection, and that they already tried to avoid CWC areas, in part due to loss or damage to fishing gear when fishing on coral. Both the fishers and some of the choir were opposed to the protection of CWC merely for their beauty (existence value). However, some members of the choir felt this might change if it was possible to see the corals. The remaining members of the choir felt, for them, it was enough to know of the existence of CWC (non-use value/ existence value) to warrant protection, but they thought that the general public would need to be able to see the corals to justify protection. The sailors had a similar attitude. The discussions also revealed that the fishers were opposed to closures based on suspicions of their existence (option value) and emphasized that they must have a function in the food chain in order to be protected (indirect use-value).

There were important differences between the questionnaires and the group discussions. While use-values were emphasised in the discussions, participants favoured non-use and intrinsic values of CWC in the questionnaires. The questionnaires showed that 89% would like to protect CWC for future generations (bequest values) and 83% for their intrinsic right to exist. Similarly, a high percentage, 82%, wanted to protect corals for their option values. Figure 2 shows the aggregated results from the questionnaires for the three focus groups on the reason why CWC should be protected.

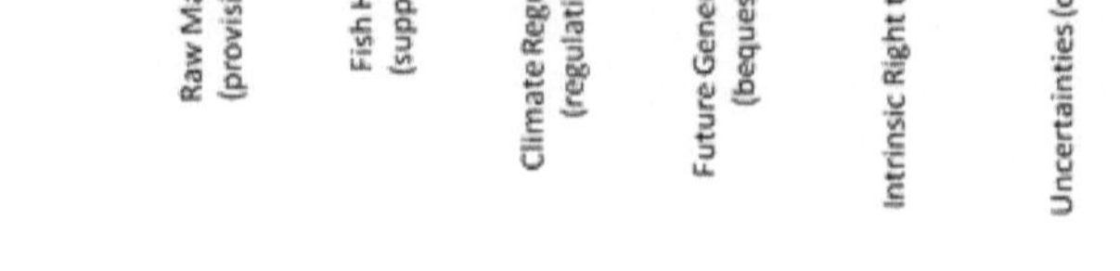

Figure 2: Reasons that CWC should be protected, questionnaire results.

The second major goal was to evaluate attitudes towards temporary closures specifically for the purposes of obtaining information about the presence of CWC, thereby protecting quasi option values. The fisher and sailor focus groups showed a general opposition to the idea of closures in order to map coral areas and to closures based on the potential for coral presence. All of the groups expressed concern about the economic consequences to industry, particularly fisheries, from temporary closures. While some suggested that coral areas should be managed based on the best available knowledge of their services today, the importance of a proper justification for closures based on knowledge of the ecosystem services CWC provides was stressed. Some suggested that a one year closure could be acceptable if compensation was provided for loss of income. Others felt more surveying was required before closures – again, in effect, rejecting the research premise that the closures would be a necessary precautionary action in order to allow a programme of surveying. While being aware of the economic consequences of closures to fishers, the inland choir did not show as much opposition to the idea of closing all areas until the mapping was done. The fishers and the sailors found it unreasonable to close an area to fishing for a year to ascertain whether corals were present when fishing had occurred for 30 years, on the grounds that the damage would already have been done in these areas.

The question posed in the questionnaire was '*while mapping is taking place would you be willing to close areas for (a) 1 year; (b) 2–5 years; (c) 5 years; (d) until mapping is complete*'. The results from the questionnaires supported the outcomes from the discussions. The majority of the *agree* statements were for a closure of one year, and only 47% of those surveyed agreed to this. As the time increased the willingness to support closures to increase knowledge decreased as shown in Figure 3. Approximately 75% of those surveyed were opposed to the idea of closing areas for 2–5 years, with only 6% support for a 5 year closure of potential coral areas. The majority, 61%, also disagreed with a closure until mapping was complete. The inland choir made up all of the 17% of those willing to close areas until the mapping was complete.

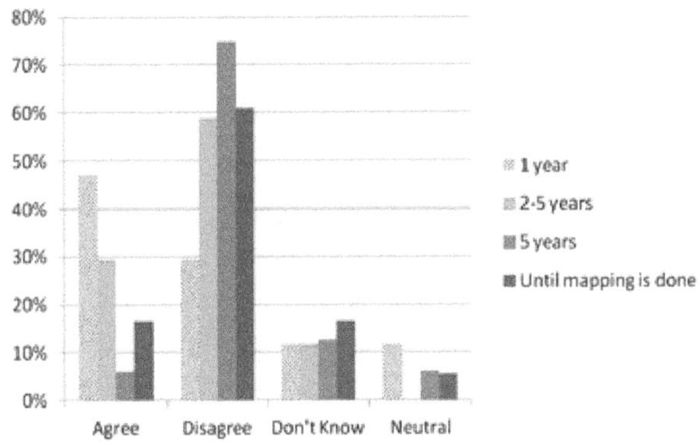

Figure 3: Responses to time closures while searching for CWC. Indication of quasi option values.

DISCUSSION

Equality in exclusion

Although fishers were in favour of measures to protect known CWC areas, they stressed that if they were to be excluded, the same would have to apply to all types of fishing, including recreational fishing, as well as the oil industry. The attitude that protection of CWC should also apply to all types of fishing and industries was also reflected in responses by Norwegian fisheries organisations to the draft regulation of CWC in Norwegian waters (Armstrong and van den Hove 2008).

Irreversibility, precaution and perceptions of harm

While participants generally recognised the potential services and values associated with CWC, most were not willing to close areas with a view to obtaining more information on their presence and functions due to the consequences for fisheries. Thus, when faced with trade-offs, protecting the known existing use values of the fisheries seemed to weigh more substantively than protection of uncertain potential values. This suggests that they did not necessarily support a precautionary approach to managing CWC. The discussions, however, also reflected that the issue of irreversibility may not be straightforward. For example, some mentioned that local loss of corals could be acceptable if their existence was secured globally. Finally, while the fishers stated in discussions that they had become so efficient that protection was needed and emphasised the supporting services of known coral areas, they rejected the hypothesis that there could be any option values associated with CWCs in fished areas, on the grounds that the damage had already been done. Thus, we cannot conclude that our respondents necessarily reject a precautionary approach outright, since the additional damages of continuing trawling while mapping are perceived by some to be limited. Essentially, this is rejecting a premise in the research method, and serves to illustrate that this problem of premise-rejection is not limited to stated preference instruments – though with the difference that it is more easily detected in a discursive setting. Rejection of the research premise may in particular be likely when user groups have extensive experience with certain aspects of a resource. Resolving such differences of opinion would require adjusting the premises in the research questionnaire, which could be achieved through repeated focus groups, or through methods other than focus groups, for example by applying weights in a multi-criteria mapping (http://www.multicriteria-mapping.org/) or social multi-criteria evaluation (Munda 2004). However adjusting the premises in the survey for each focus group was not possible in this present study since an aim of this study was to contrast the findings of each of the three focus groups.

One particularly important element to account for when considering precautionary action, is that there exists a fundamental asymmetry at play in situations of potentially high and irreversible impacts. In such situations it may actually be more appropriate to focus on avoiding false negatives (i.e. avoiding a failure to anticipate damage) rather than avoiding false positives (i.e. avoiding action to avoid damage that turns out to be unnecessary) (European Environmental Agency, The 2001). If suspicions of irreversible and high damage trigger a precautionary policy measure (e.g. temporary closure) and more intensive research, then at some point the research may show that this was a false positive (e.g. the corals and/or their services were not at risk) and the precautionary measure can be cancelled. The losses would be economic and social benefits from not having fished those grounds during the time it took to show that there was no cause for concern. If no precautionary action is taken, but more research shows, only much later, that there was indeed real cause for concern (e.g. fishing leads to destruction of CWCs), then irreversible damages will have occurred. Participants in this study seem to have responded either as if such asymmetry between economic damage to fishermen and irreversible ecological damages to an ecosystem did not exist, or that such damage was not significant, for example because damage is local and global presence of coral continues.

This is an interesting result, which to some degree differs from attitudes to, for instance, GM foods in Europe (Nelson 2001), hazardous waste in the US (Hadden 1991) or marine mining in Australia (Mason et al. 2010) where risk or uncertainty leads to precautionary attitudes. However, it must be noted that in addition to different methods being applied in the above studies, in our analysis the respondents are confronted with a trade-off that is not just related to an 'anonymous' beneficiary of an economic activity, but to fisheries. Fishers in Norway are often associated with marginalised coastal communities that have strong social and political support. However, when asked in the questionnaire whether a fund to protect CWC should be used to compensate lost income to fishers, the majority disagreed or were neutral. There was somewhat more support for this kind of compensation if the fishers could document sustainable practices.

Group discussion vs. questionnaire

The potential role of CWC as providers of services that have use values was emphasised in the information given and the discussions that followed. Despite this, however, the questionnaire reflected that bequest values and the intrinsic right to exist (non-use values) were seen as the most important reasons to protect CWC. The latter was also found in the CWC survey by Glenn et al. (2010), and other studies on factors determining people's preferences and environmental concern (Freeman 2003, Barkmann et al. 2008). This may suggest that focus groups are less likely to reflect on some non-use and ethical concerns than stated preference methods. The range of topics that participants feel comfortable discussing may also matter. Some topics may have been more difficult to discuss among some categories of participants than others. It could be that focus group participants are reluctant to deal with sensitive topics in a discussion setting compared with a survey (Morgan, 1996). However, while respondents may not feel comfortable arguing for protection based on non-use values, it is also possible that they did not think of this argument and it was not probed further.

The different emphasis in the questionnaire responses compared to the group discussions may be due to participants being presented with clear categories which could be non-exclusively evaluated in the questionnaire. However, it is possible that the wording of the questions, including terms such as "future generations" and "intrinsic rights" could trigger respondents to lean in a particular direction. This is partly because these wordings inevitably bring new information: it is impossible to ask fully neutral questions about "intrinsic rights" because the very act of posing the question makes respondents think about these rights, whereas this may not arise so readily, or with such clarity, in a general discussion.

As found in other studies (see e.g. Álvarez-Farizo and Hanley (2006), Dietz et al. (2009)) group discussion settings may encourage social behaviours and thought processes in a way that the questionnaires do not. Questionnaires – especially written rather than face to face – encourage inner reflection, while the discussion encourages consideration of how others will respond to ones› arguments, and concern with 'image'. Participants may experience pressure to conform to norms of group behaviour (Maier 1967), which may, among other things, lead to strategic behaviour. There could be paradoxical results: for example, concern with what others view as realistic and solid bases for arguments may lead people to focus their statements more on tangible, short-term values rather than on values reflecting their «true» preferences. Dietz et al. (2009) found that people were less likely to consider future generations and environmental benefits of mitigating CO_2 emissions in a group setting compared to individually, and more likely to consider scientific evidence behind global warming when determining willingness to pay in a group setting. Individual questionnaires resulted in a focus on personal considerations in terms of environmental values and beliefs, while in groups people acted like policy analysts evaluating side costs and benefits, feasibility and efficiency of implementation (Dietz et al. 2009).

'Minimal group identification' (i.e. division by ostensibly trivial common features) is sufficient to encourage group members not only to favour other members of the group, but to discriminate against other groups (Tajfel 1970). A focus group setting, involving a group of individuals sharing key common characteristics, is perfectly designed for encouraging such behaviours, and this could influence expressed values. Indeed there is some evidence for this from the comments given by commercial fishers, most of them not bottom trawlers, who questioned why they should be excluded from certain areas, but not the oil industry nor the tourist fishers. Further research in this area would be warranted.

Limitations of the study

This study suffers from typical limitations associated with focus group studies. At a first glance the sample size in this present study appears small. Focus groups normally involve structured discussions among 6 to 10 homogenous strangers in a formal setting, and most studies consist of 4 to 6 focus groups (Morgan, 1996). However, group segmentation by gender, socioeconomic status or geographical location was not central to the research in this present study. Segmentation and replication would have greatly multiplied the number of groups required and in our view are not necessary given the research goals. Instead far more important for the aims of this study was an assessment of the views and attitudes of the different stakeholder/citizen groups. Also the nature and complexity of the topics under discussion required a high level of participant involvement for which small groups are deemed more effective than larger groups (Morgan, 1996). For this reason three different stakeholder groups were thought to be sufficient to meet the research goals of the study.

Furthermore, in order to not bias or influence the discussions, the facilitator avoided specifying the services and values connected to CWC, and only informed about current knowledge regarding their biology. This could have biased the results comparing discussions with the questionnaire results, as the latter specified both use- and non-use values, as well as intrinsic right to exist, while the presentation only mentioned ecological facts that can be translated into use-values. Despite these potential biases, we believe some important insight has been gained from this study concerning different stakeholder and citizen perspectives on the precautionary management of CWC.

CONCLUSION

This study uses focus groups coupled with questionnaires to contrast the known current benefits of fishing versus unknown future benefits associated with CWC. The work also investigates ideas embodied in the precautionary principle by comparing support by respondents for closure of areas confirmed to have CWC versus closures with a view to gaining information about the presence of CWC.

In general the results show that respondents believe CWC are valuable and should be protected. The study also reveals that different values are expressed spontaneously in focus groups compared with the answers given in the questionnaires. In the questionnaires, emphasis was given to non-use values such as existence and bequest value, despite the fact that use-values, particularly as habitat for fish, dominated the focus group discussions. Thus, to secure legitimacy in forming environmental policies involving such complex environmental public goods, the two approaches of gaining insights should be combined. Furthermore, deliberation to capture attitudes regarding such goods could benefit from probing specifically into the issue of non-use values.

The groups expressed reservations in both the discussions and the questionnaires towards temporary closures as a means of protecting quasi option values due to the economic consequences to fisheries. Thus the respondents reject the precautionary principle in so far as they do not condone using temporary closures as a means of gaining information about possible future benefits linked to CWC. Two reasons were given: first, some participants believed quasi-option values in the future to be small because they held the view that most CWC have already been destroyed. Secondly, many participants have positive associations with fishing communities.

Whatever the reasoning, a message here is that using temporary closures as a means of implementing the precautionary principle may receive little support. This leaves policy makers with a dilemma since the precautionary principle functions as a mechanism which allows the gathering of information needed to justify the policy in the first place, in particular in situations of irreversibility and/or high stakes. However, this study also showed that informing the public about the issues at stake, including a broad discussion regarding the plurality of values the ecosystem represents, is important when evaluating the implementation of precautionary measures.

ACKNOWLEDGEMENTS

The authors are grateful for comments from reviewers of the paper. Claire Armstrong and Jannike Falk-Andersson further acknowledge financing from the Research Council of Norway over the project “Habitat-Fisheries interactions: Valuation and bio-economic modeling of cold water corals”, grant # 216485.

AUTHORS' CONTRIBUTIONS

TR, JFA, CA and NF designed the focus group study. JFA and CA carried out the focus groups. JFA, NF, CA, SH, TR and RT contributed to analysis and writing the manuscript.

REFERENCES

1. Aanesen, Margrethe, Claire W Armstrong, and Luc van Hoof. 2012. The changing environment of fisheries policy in Europe. Marine Policy 36(5): 1172–1177. doi:http://dx.doi.org/10.1016/j.marpol.2012.02.002.
2. Abson, DJ, and M Termansen. 2011. Valuing Ecosystem Services in Terms of Ecological Risks and Returns. Conservation Biology 25(2): 250–258. doi:10.1111/j.1523-1739.2010.01623.x.
3. Álvarez-Farizo, Begoña, and Nick Hanley. 2006. Improving the Process of Valuing Non-Market Benefits: Combining Citizens› Juries with Choice Modelling. Land Economics 82(3): 465–478. doi:10.2307/27647723.
4. Anonymous. 2004. Chapter XIII. Protection of coral reefs (in Norwegian). Forskrift om utøvelse av fisket i sjøen. In FOR-2004-12-22-1878, ed. Industry and Fisheries Department of Trade. Norway: Industry and Fisheries Department of Trade.
5. Arico, S, and C Salpin. 2005. Bioprospecting of genetic resources in the deep seabed: scientific, legal and policy aspects. United Nations University - Institute of Advanced Studies (UNU-IAS) Report.
6. Armstrong, CW, and S van den Hove. 2008. The formation of policy for protection of cold-water coral off the coast of Norway. Marine Policy 32: 66–73.
7. Armstrong, C, N. Foley, R. Tinch, and S. Van den Hove. 2010. Ecosystem Goods and Services of the Deep. HERMIONE FP7.
8. Armstrong, Claire W, Naomi Foley, Rob Tinch, and Sybille van den Hove. 2012. Services from the Deep: Steps towards valuationof deep sea goods and services. Ecosystems Services 2: 2–13.
9. Barkmann, J, K Glenk, A Keil, C Leemhuis, N Dietrich, G Gerold, and R Marggraf. 2008. Confronting unfamiliarity with ecosystem functions: The case for an ecosystem service approach to environmental valuation with stated preference methods. Ecological Economics 65(1): 48–62. doi:10.1016/j.ecolecon.2007.12.002.
10. CICES. 2013. Common International Classification of Ecosystem Services. Accessed 8 Aughttp://cices.eu.
11. Crabtree, BF, MK Yanoshik, WL Miller, and PJ O›Connor. 1993. Selecting individual or group interviews. In Successful focus groups: Advancing the state of the art, ed. DL Morgan. Thousands Oaks, CA: Sage.
12. Defra. 2007. An introductory guide to valuing ecosystem services. London: Department for Environment, Food and Rural Affairs.
13. Dietz, T, PC Stern, and A Dan. 2009. How deliberation affects stated willingness to pay for mitigation of carbon dioxide emissions: an experiment. Land Economics 85(2): 329–347.
14. Duncan, Marie T, and David L Morgan. 1994. Sharing the caring: family caregivers› views of their relationships with nursing home staff. The Gerontologist 34(2): 235–244.
15. European Environmental Agency, The. 2001. Late lessons from early warnings: the precautionary principle 1896–2000. In Environmental issue report No 22, ed. P Harremoës, D Gee, MG Malcolm, Stirling Andrew, J Keys, B Wynne, and S Guedes Vaz. Luxembourg: Office for Official Publications of the European Communities.
16. Foley, Naomi S, Tom M van Rensburg, and Claire W Armstrong. 2010. he ecological and economic value of cold-water coral ecosystems. Ocean &

Coastal Management 53(7): 313–326. doi:10.1016/j.ocecoaman.2010.04.009.

17. Fosså, JH, PB Mortensen, and DM Furevik. 2002. The deep water coral *Lophelia pertusa* in Norwegian waters: distribution and fishery impacts. Hydrobiologia 471: 1–12.
18. AM Freeman, AMyrick. 2003. The measurement of environmental and resource values: theory and methods. Washington, D.C.: Resources for the Future.
19. Freiwald, A, JH Fosså, A Grehan, T Koslow, and JM Roberts. 2004. Cold water coral reefs: out of sight - no longer out of mind. Camebridge, UK: UNEP-WCMC.
20. Garcia, SM, A Zerbi, C Aliaume, T Do Chi, and G Lasserre. 2003. The ecosystem approach to fisheries. Issues, terminology, principles, institutional foundaitons, implementation and outlook. In *FAO Fisheries Technical Paper*. Rome: FAO.
21. Gezelius, SS, and K Refsgaard. 2007. Barriers to rational decision-making in environmental planning. Land Use Policy 24(2): 338–348. doi:10.1016/j.landusepol.2006.04.002.
22. Glenn, H, P Wattage, S Mardle, T Van Rensburg, A Grehan, and N Foley. 2010. Marine protected areas-substantiating their worth. Marine Policy 34(3): 421–430.
23. Hadden, Susan G. 1991. Public Perception of Hazardous Waste. Risk Analysis 11(1): 47–57. doi:10.1111/j.1539-6924.1991.tb00568.x.
24. Howarth, RB, and MA Wilson. 2006. A theoretical approach to deliberative valuation: aggregation by mutual consent. Land Economics 82: 1–16.
25. Husebø, AL Nøttestad, JH Fosså, DM Furevik, SB Jørgensen. 2002. Distribution and abundance of fish in deep-sea coral habitats. Hydrobiologia 471:91–99.
26. Jackson, JBC, MX Kirby, WH Berger, KA Bjorndal, LW Botsford, BJ Bourque, RH Bradbury, R Cooke, J Erlandson, JA Estes, TP Hughes, S Kidwell, CB Lange, HS Lenihan, JM Pandolfi, CH Peterson, RS Steneck, MJ Tegner, and RR Warner. 2001. Historical overfishing and the recent collapse of coastal ecosystems. Science 293: 629–638.
27. Kahane, David, Kristjana Loptson, Jade Herriman, Max Hardy. 2013. Stakeholder and Citizen Roles in Public Deliberation. Journal of Public Deliberation 9 (2):Article 2.
28. Kumar, M, and P Kumar. 2008. Valuation of the ecosystem services: A psycho-cultural perspective. Ecological Economics 64(4): 808–819. doi:10.1016/j.ecolecon.2007.05.008.
29. LaRiviere, Jacob, Mikołaj Czajkowski, Nick Hanley, Margrethe Aanesen, Jannike Falk-Petersen, and Dugald Tinch. 2014. The Value of Familiarity: Effects of Knowledge and Objective Signals on Willingness to Pay for a Public Good. Journal of Environmental Economics and Management 68: 376–389.
30. Liu, S., R Costanza, S Farber, A. Troy. 2010. Valuing ecosystem services Theory, practice, and the need for a transdisciplinary synthesis. Annals of the New York Academy of Sciences, In Ecological Economics Reviews, 1185: 54–78.
31. Lo, Alex Y, and Clive L Spash. 2013. Deliberative monetary valuation: In search of a democratic and value plural approach to environmental policy. Journal of Economic Surveys 27(4): 768–789. doi:10.1111/j.1467-6419.2011.00718.x.
32. Lutringer, A, D Blamart, N Frank, and L Laberyrie. 2005. Paleotemperatures from deep-sea corals: scale effects. In Cold-water corals and ecosystems, ed. A Freiwald and J Roberts, 1097–1108. Berlin Heidelberg: Springer-Verlag.
33. MacIsaac, K, C Bourbannais, E Kenchington, D Gordon, and S Gass. 2001. Observations on the occurrence and habitat preference of corals in Atlantic Canada. In First International Symposium on Deep-Sea Corals, ed. JHM Willison, J Hall, SE Gass, ELR Kenchington, M Butler, and P Doherty, 58–75. Halifax, Nova Scotia: Ecology Action Centre.
34. Maier, Norman R. 1967. Assets and liabilities in group problem solving: the need for an integrative function. Psychological Review 74(4): 239–249. doi:10.1037/h0024737.
35. Mason, Claire, Paxton Gillian, Parr Joanna, and Boughen Naomi. 2010. Charting the territory: Exploring stakeholder reactions to the prospect of seafloor exploration and mining in Australia. Marine Policy 34(6): 1374–1380. doi:http://dx.doi.org/10.1016/j.marpol.2010.06.012.
36. MEA. 2005. Ecosystems and Human Well-being: Synthesis. Washington, DC: Millennium Ecosystem Assessment, Island Press.
37. Morgan, David L. 1996. Focus Groups. Annual Review of Sociology 22: 129–152. doi:10.2307/2083427.
38. Morris, J, A Wildavsky, W Beckerman, T Gilland, HI Miller, G Conko, CT Rubin, H Guldberg, B Durodie, B Yandle, IM Goklany, and R Matthews. 2000. Rethinking risk and the precautionary principle. Edited by J. Morris. Oxford: Butterworth-Heinemann.
39. Munda, Giuseppe. 2004. Social multi-criteria evaluation: Methodological foundations and operational consequences. European Journal of Operational Research 158(3): 662–677. doi:http://dx.doi.org/10.1016/S0377-2217(03)00369-2.
40. Myers, RA, and B Worm. 2003. Rapid worldwide depletion of predatory fish communities. Nature 423(6937): 280–283.
41. Nelson, Carl H. 2001. Risk Perception, Behavior, and Consumer Response to Genetically Modified Organisms: Toward Understanding American and European Public Reaction. American Behavioral Scientist 44(8): 1371–1388. doi:10.1177/00027640121956737.
42. Niemeyer, S, and CL Spash. 2001. Environmental valuation analysis, public deliberation, and their pragmatic syntheses: a critical appraisal. Environment and Planning C 19(4): 567–585. doi:citeulike-article-id:2749103.
43. Pearce, D, and D Moran. 1994. The economic value of

biodiversity. London: Earthscan Publications Limited.

44. Perman, R, Y Ma, M Common, D Maddison, and J McGilvray. 2011. Natural Resource and Environmental Economics. London: Pearson Education Limited.

45. Puglise, KA, RJ Brock, and JJ McDonough. 2005. Identifying critical information needs and developing institutional partnerships to further the understanding of Atlantic deep-sea coral ecosystems. In Cold-water corals and ecosystems, ed. A Freiwald and JM Roberts, 1129–1140. Berlin-Heidelberg: Springer-Verlag.

46. Sagoff, M. 1998. Aggregation and deliberation in valuing environmental public goods: A look beyond contingent pricing. Ecological Economics 24(2–3): 213–230. doi:10.1016/s0921-8009(97)00144-4.

47. Söderholm, Patrik. 2001. The Deliberative Approach in Environmental Valuation. Journal of Economic Issues 35(2): 487–495. doi:10.2307/4227681.

48. Soma, Katrine, and Arild Vatn. 2010. Is there anything like a citizen? A descriptive analysis of instituting a citizen's role to represent social values at the municipal level. Environmental Policy and Governance 20(1): 30–43. doi:10.1002/eet.529.

49. Spash, CL. 2006. Non-economic motivation for contingent values: Rights and attitudinal beliefs in the willingness to pay for environmental improvements. Land Economics 82(4): 602–622.

50. Tajfel, H. 1970. Experiments in intergroup discrimination. Scientific American 223: 96–102.

51. TEEB. 2010. The Economics of Ecosystems and Biodiversity: Mainstreaming the Economics of Nature: A synthesis of the approach, conclusions and recommendations of TEEB.

52. Tinch, Rob, Claire W, Armstrong, S, van den Hove. 2011. Report on policy demands for value evidence. In Deliverable 6.3 of HERMIONE (Hotspot Ecosystem Research and Man's Impact on European Seas) project. http://median-web.eu/IMG/pdf/hermione-policy_demands.pdf.

53. Turner, RK Morse-Jones, S, and Fisher B. 2010. Ecosystem valuation. Annals of the New York Academy of Sciences 1185(1): 79–101. doi:10.1111/j.1749-6632.2009.05280.x.

54. UNESCO. 2005. The Precautionary Principle. World: Commission on the Ethics of Scientific Knowledge and Technology.

55. van den Hove, S. 2000. Participatory approaches to environmental policy-making: the European Commission Climate Policy Process as a case study. Ecological Economics 33(3): 457–472. doi:10.1016/s0921-8009(99)00165-2.

56. van den Hove, Sybille. 2006. Between consensus and compromise: acknowledging the negotiation dimension in participatory approaches. Land Use Policy 23(1): 10–17. doi:http://dx.doi.org/10.1016/j.landusepol.2004.09.001.

57. Wattage, P, H Glenn, S Mardle, T Van Rensburg, A Grehan, and N Foley. 2011. Economic value of conserving deep-sea corals in Irish waters: A choice experiment study on marine protected areas. Fisheries Research 107(1–3): 59–67. doi:10.1016/j.fishres.2010.10.007.

58. Wilson, Matthew A, and Richard B Howarth. 2002. Discourse-based valuation of ecosystem services: establishing fair outcomes through group deliberation. Ecological Economics 41(3): 431–443. doi:10.1016/s0921-8009(02)00092-7.

CHAPTER 12 FISH TRADERS AS KEY ACTORS IN FISHERIES

Sara Fröcklin, Maricela de la Torre-Castro, Lars Lindström, Narriman S. Jiddawi

Keywords: Fish value chain, Fish market, Gender analysis, Middlemen, Small-scale fisheries, Zanzibar

INTRODUCTION

Women have an important role in world fisheries (Bennett 2005), but traditionally, fisheries have been associated with men (Davis and Gerrard 2000; Williams et al. 2005; Samuel 2007; Williams 2008; Choo et al. 2008) with focus primarily on capture fisheries, rather than women-dominated pre- and post-harvest activities, e.g., finance, processing, and marketing the catch (Weeratunge et al. 2010; Harper et al. 2013). Women also perform diverse unpaid tasks such as mending nets, collecting bait, preparing food for fishers, and keeping accounts, which are unacknowledged or undercounted as employment (Williams et al. 2005; FAO 2006; Williams2008). In addition, women's gleaning of invertebrates and near shore fishing is largely undervalued and almost invisible in management plans and fisheries statistics. In spite of women's central role, a number of factors reduce their access to fisheries resources and assets. These include traditional beliefs, norms and laws, which in turn confine them to the lowest end of fish value chains (FAO 2006; Porter 2006). Women's relatively limited participation in fish value chains further results in gender disparities in terms of income, and often in women being entrenched in poverty, as for example, in West African fisheries (FAO 2008). It is, however, important to remember that in most developing countries men also struggle to make a living, and fisheries have often been described as a last resort occupation (Allison and Ellis 2001). However, since women are associated with reproductive work, gender inequalities in access to fisheries resources affect not only the livelihoods of women, but also the entire household (Weeratunge et al. 2010).

The importance of including gender aspects in research and policy is recognized, but commonly overlooked in the fisheries and aquaculture sectors (Tietze 1995; Diamond et al. 2003; Fröcklin et al. 2012; Harper et al. 2013). Although there are valuable studies regarding women's role in fisheries, gender-disaggregated data on fishing activities are often absent (e.g., Weeratunge et al. 2010; Harper et al. 2013), which hinders robust comparisons and comprehensive gender analysis much needed for improved fisheries management and policy (de la Torre-Castro et al., unpublished). Further, the importance of gender aspects in adaptive fisheries management, i.e., an integrated, multidisciplinary approach based on learning-by-

Citation: Sara Fröcklin, Maricela de la Torre-Castro, Lars Lindström, Narriman S. Jiddawi, Fish Traders as Key Actors in Fisheries, DOI: 10.1007/s13280-013-0451-1.

doing (Holling 1978; Walters1986; Lee 1993; Gunderson et al. 1995), has not to our knowledge been addressed. One of the key elements in adaptive management is the iterative link between knowledge and action, and the need for responsive institutions. In an attempt to fill this knowledge gap we focus on the growing enterprise of fish trade in Zanzibar (Unguja Island), Tanzania, from a gender perspective and further link the findings to adaptive management aspects.

In much of Africa, women are actively involved in trade with both fish and other commodities (Walker 2001, 2002; ICSF 2002; Madanda 2003). However, in for example, the Lake Victoria fishery men tend to control the profitable large scale operations of high-value fish, while most women focus on the local market and low-value fish (Lwenya and Abila 2001). In Zanzibar, fish trade has traditionally been dominated by men but over the last years women traders (*Mama Karangas* in Swahili) have become a common sight at the local fish markets, which give rise to the following questions: What are the underlying reasons for the recent increase in women traders? Has the increase in women challenged traditional gender roles and led to women's empowerment such as improved decision making power in the household and/or in the society at large? Is the management system accounting for gender differences in Zanzibar? We further argue that in order for fisheries management to be adaptive, knowledge about both ecological and social spheres is needed. Thus, this study fills an important gap towards adaptive fisheries management in Zanzibar by acknowledging the role of both men *and* women traders, but also their needs and challenges in the fish value chain.

MATERIALS AND METHODS

The research was undertaken in Zanzibar (Unguja Island) in March and July 2012; during these periods semi-structured interviews with men ($n = 21$) and women ($n = 21$) traders, as well as observations, at ten different fish markets took place. In addition, we analyzed documents from the Ministry of Livestock and Fisheries, with its subdivisions: Department of Fisheries Development DFD and Department of Marine Resources DMR (the institution responsible for marine resource management), to investigate to what extent gender is considered in fisheries management. Also, price information on traded fish species was collected from the monitoring agents (*Bwana Dikos* in Swahili) directly at the auction at each of the study sites.

Study Area

Zanzibar is located in the Western Indian Ocean, about 40 km off the coast of mainland Tanzania (Fig. 1). It comprises a number of small islands and two large ones; Unguja and Pemba, and is inhabited by about 1 000 000 people (Ruitenbeek et al. 2005). This study focuses on Unguja Island, hereafter referred to as Zanzibar. Fishing activities have historically provided some of the most important livelihoods and assets on the Island, thus contributing to the formal and informal economic sectors. Fish is also a relatively cheap and accessible source of protein for Zanzibar's mainly low income households (Jiddawi and Öhman 2002), and a popular food item among tourists. Thus, the importance of fisheries based livelihoods has increased with more people engaged in fishing (Jiddawi and Khatib 2007) and fish trade-related activities.

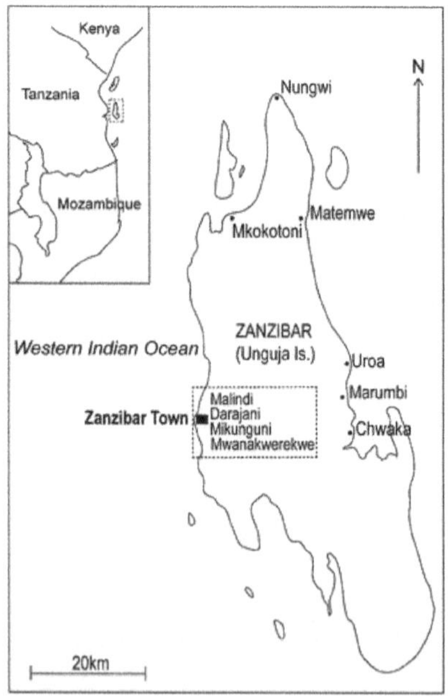

Figure 1: Map of Zanzibar (Unguja Island) 6°8′0″S, 39°19′0″E, with studied fish markets ($n = 10$). Village markets are shown by *small black dots* ($n = 6$) and Town markets are found within a *dotted line* ($n = 4$).

The fish value chain in Zanzibar is rather simple; most of the sector is artisanal and characterized by non-selective fishing and the use of traditional gears and equipment (Jiddawi2012). Fish catches are landed at sites officially selected by the DFD/DMR in Zanzibar, and monitored by an employee, the *Bwana Diko*, of the DFD (de la Torre-Castro 2006). The monitoring agents report the number of fish landed, weight and price. Fish are sold in a number of ways by the fishers, but often to traders via auctions held at the local fish markets (the local market place generally consists of brick buildings or open constructions that lack cold storage facilities except for the two main markets in Zanzibar Town) (de la Torre-Castro 2012; Thyresson et al. 2013). Traders then sell the fish, processed (salted, sun-dried, or smoked) or fresh (whole or filleted), to local customers, hotels and restaurants or via another auction. Lobsters and shrimps are, however, mainly traded directly between fishermen and men traders. Trading activities in Zanzibar involve a wide range of actors operating at different scales (Thyresson et al. 2011; Eriksson et al. 2012), and has by tradition been dominated by men or so-called "middlemen" (Crona et al. 2010), i.e., intermediaries buying the fish for further resell; they can operate by direct contact with the fishers or by buying the products at the local fish market. Middlemen may lease fishing equipment (e.g., boats, gears, and engines) (Jiddawi and Öhman 2002) or provide fishermen with credit. However, only a small proportion of Zanzibar's fish traders have the resources to supply fishermen with equipment, boats or credit and are thus not involved in these kinds of agreements. Instead, many traders buy and sell fish independently. In this study we focus on small-scale traders operating at the local markets; to acknowledge the gender perspective we use the more suitable and gender neutral term "fish trader" instead of "middlemen".

Interviews

Interviews were held with 21 women (age 20–55) and 21 men

(age 27–78) fish traders at ten different fish markets around the Island; Chwaka, Darajani, Malindi, Marumbi, Matemwe, Mikunguni, Mkokotoni, Mwanakwerekwe, Nungwi, and Uroa (Fig. 1). The sites were selected based on the local knowledge of scientists at the Institute of Marine Science (IMS), University of Dares Salaam (UDSM), to cover markets of different character (i.e., men/women-dominated, high- vs. low-value fish species, and tourism influences, Table 1). The value of each species was classified according to price per kg; high value >3 US dollar (USD), medium value ≈3 USD, and low value <3 USD. Fish markets per se were not set as a unit of analysis but rather as a way to capture general differences between men and women according to the components analyzed. The targeted respondents were fish traders who regularly purchased fish at the auction and then sold it further via another auction, fish market, directly to local customers, or to hotels and restaurants. Depending on market characteristics the number of women and men interviewed at each market varied, however, summing up to a total number of 21 men and 21 women respondents (Table 1). A semi-structured interview following Kvale (1994) and Denscombe (2007) was applied to solicit information about: (i) markets, customers, and mobility, (ii) material and economic resources including initial capital, daily income, and assets (e.g., equipment), (iii) traded species, (iv) contacts and organizations with focus on membership in fish trading organizations, fishing committees, cooperatives, and contact with DFD/DMR, and (v) perceptions and experiences to examine life improvements, possible effects on a community level, equality and capacity to cope with fluctuating markets. These components were developed and chosen based on the literature (e.g., Lwenya and Abila 2001; Medard et al. 2002; Weeratunge et al. 2010) and our own experiences working with coastal livelihoods in East Africa. The components provide a comprehensive picture of fish trade and help analyze the way in which fish trade and commodity flows are interwoven in wider social, cultural, economic, and environmental structures and processes. The interviewer introduced the theme and the respondents were able to speak freely on each topic. The semi-structured interview was chosen since it is one of the most powerful methods, allowing flexibility to probe for answers, follow-up the original questions and pursue new lines of questions (e.g., Kvale 1994; Denscombe 2007). Respondents were selected in consultation with the *Bwana Diko* at each fish market and according to availability and willingness to participate in the study. As there is no information on the total number of traders in Zanzibar the representativeness of the sample is difficult to assess. The interviews lasted from 40 min to 1 h and took place at the fish market or close to the respondent's house. All interviews were performed by the same researcher and trained assistant from IMS, the latter with Swahili as mother tongue, and all interviews were recorded for further analysis. About 6 to 10 days were spent at each fish market, which helped to establish a better understanding about the activity through observations and conversations with other fish traders, *Bwana Dikos* and local people.

Table 1: Market characteristics including men/women dominated, fish value (high, medium, low), relative closeness to hotels/restaurants and the number of men and women interviewed at each market

Market	Men/women-dominated	High-value fish Medium Low-value fish	Relative tourism influences (hotels/restaurants close to the market)	No. of men interviewed ($n = 21$)	No. of women interviewed ($n = 21$)
Chwaka	M	Mixed	Medium	2	1
Darajani	M	Mixed	High	2	1
Malindi	M	Mixed	High	2	2
Marumbi	W	Low to medium	Medium	1	2
Matemwe	M	Mixed	High	3	0
Mikunguni	W	Low to medium	Absent	0	4
Mkokotoni	M	Mixed	Absent	2	2
Mwanakwerekwe	W	Low to medium	Absent	1	4
Nungwi	M	Mixed	High	6	2
Uroa	M	Mixed	High	2	3

Data Analysis

The respondents' answers were organized, transcribed, and categorized by using the pre-determined components. The data were then analyzed according to the categories to provide a comprehensive picture of the gender dimensions of fish trade. Also, familiarity with the location and community enabled us to grasp wider social issues and further reduced the interviewer effect. Text analysis of the fisheries acts and management plans was done with focus on the inclusion of gender.

RESULTS

Increase in Women Traders—What is the Underlying Reason?

As mentioned previously fish trade in Zanzibar has traditionally been associated with men, which is reflected in the average number of years men and women traders had been active. Whereas the men had been involved in trading activities for on average 18 years, the number for women was nine. Based on observations and discussions with researchers at IMS, the number of women entering local fish markets has increased over the last years. The respondents reported that this increase was due to the lack of alternative economic activities and the need for all family members to contribute to household income.

Characteristics of Fish Traders and General Work Situation

The educational level among interviewed traders was generally quite low; however, more women than men had finished primary and secondary school and a complete lack of formal education was more prevalent among men. Although women in the present study showed higher educational levels, few women traders had managed to start-up the business themselves. Instead, the majority were initially introduced and trained by husbands, friends or relatives although once introduced most women traders continued on their own. The few men who received assistance during auctions did so as a way to increase the business, or to reduce work effort by hiring people to transport and sell the fish. However, it requires economic resources and did not occur on daily basis. Women were instead in need of support in order to obtain time to continue their reproductive work in addition to productive work, and not as a means to increase profit.

Although men and women traders spent an equal amount of time (about 8 h per day) on fish trade, women's responsibility for household duties and childcare resulted in much longer workdays. Men traders were generally present at the fish market hours before the auction started, and consequently used that "free time" to socialize with other traders, while most women showed up just in time for the auction and left immediately after it was finished. In addition to fish trade, 33 % of the women and 19 % of the men were engaged in other economic activities such as selling food and soap or agriculture to make ends meet.

Fish Markets, Customers, and Mobility

Both men and women generally bought fish at the same auction every day, which was reported to be due to familiarity, not only with the auction place, but also with other fish traders and fishers. Quality and availability of fish, price, and safety were reported to be less important and clearly not as important as having well established contacts. Some women mentioned the relative closeness between household and market as an important factor, which was again explained by commitment to household responsibilities. In terms of marketing, many women traders were operating by the road in a suburb to Zanzibar Town. However, they were obliged to pay a daily fee of 1 USD to a private landowner, apart from the mandatory 50 cents to the Municipality, just to access a small piece of land. Contrary to women traders men had more options in terms of fish marketing; some were selling from bicycles in Zanzibar Town which require no fees, while a few were involved in trade within the tourism industry. However, some men reported that hotels do not always pay traders at once thus only the traders with high financial capital are able to maintain a continuous trade while waiting for payment. In addition, many men were operating inside the main market buildings in Town, which offer a greater variety of customers and better profits due to higher prices (Figs. 2, 3a). Also, a typical strategy among men traders to increase the profit was to buy fish to a cheaper price at a village market, and then sell it to a higher price at one of the main markets in Zanzibar Town. The current price of renting a table inside the main market building was 50 cents per day, which is less than half the price compared to the women selling in the suburb. The main markets do not only offer a variety of customers but also cold storing facilities at a price of 20 cents per kilogram fish, which helps to preserve the fish longer and reduces a potential loss. Currently, no women rented tables inside the main markets. However, improved access to the main markets was requested by most women but regardless of free spots they reported to lack the right contacts within the Municipality and the market itself. Also, a common statement among men traders was that women should not interact with men inside the main markets, which was based on social and traditional norms and values about what a "respectable" woman should or should not do.

Figure 2: Photos showing **a** a men-dominated market in Zanzibar Town with **b** high-value fish species and **c** better access to infrastructure; **d** a women-dominated market in a suburb to Zanzibar Town with **e** low-value fish and **f** no or little access to infrastructure such as cold storing facilities. Photos: S Fröcklin.

To access the fish markets most traders used the local buses (*Dala dala* in Swahili). However, 11 of the 21 men owned a bicycle, one owned a car and another one a motorcycle (Fig. 3b). Having access to own transport means makes the traders more flexible on both a temporal and spatial scale. None of the women had access to any of these assets. To cope with current market changes women borrowed money, joined other traders so as to increase the total capital, engaged in other activities to increase income or simply bought less fish. Despite a potential decline in fish landings most women stated that they would or could not change market. Instead, they would return home, empty-handed, and do something else until the market improved. On the contrary, all men reported to have the possibility to change market (Fig. 3a), and most would do so if necessary. Only a few would borrow money, use savings, or engage in other activities to keep the business going. A few men said that they would rest at home since "no market is bad for long".

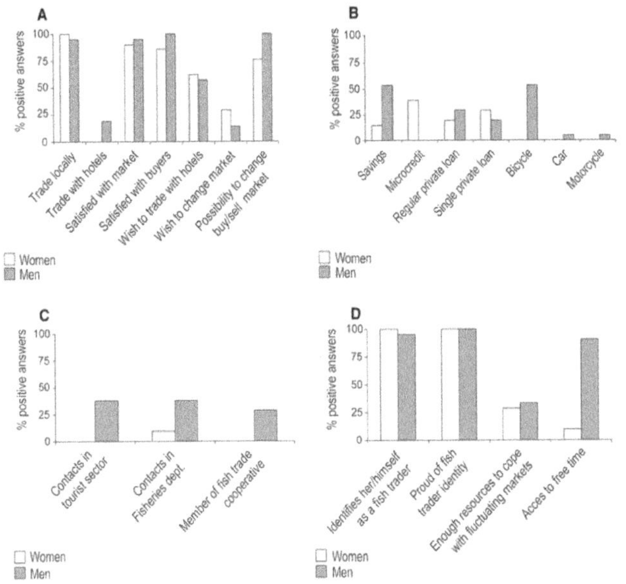

Figure 3: Summary of gender inequalities in **a** markets, customers, and mobility; **b** material and economic resources; **c** contacts and organizations; and **d** perceptions and experiences (percent positive answers based on 21 women and 21 men).

Material and Economic Resources

Capital and Income

To engage in fish trade a start-up capital is needed and fish traders were thus asked to state the source of their initial funding. The three main sources included microcredit, savings, and money lent (often interest-free) by friends or relatives. Most of the women traders (86 %) used borrowed money to start-up their business and a little less than half was involved in microcredit (mainly women from Zanzibar Town) or had taken a single loan from their husband or other relative, while the rest borrowed money on a daily basis from a wealthy man in their village. Only three of the 21 women used savings. In comparison, more than half of the men used savings, some took private loans on regular basis, and just a few had taken a single loan (Fig. 3b). The capital used for fish trade varied greatly between men and women. Men purchased fish for about 64 USD at a 13 USD profit per day (median value), whereas women used approximately 48 USD at a 6 USD profit per day (median value). Daily capital used was more or less constant for both men and women, i.e., money generated was always re-invested in a "special pot" used for fish trade only. The money left after removing additional expenses such as transport, frying oil, firewood and ice, and after re-investing in the pot, was considered as income.

Assets

There are a number of necessary and supplementary items linked to fish trading activities such as cell phones to communicate with buyers, fish traders, or the auctioneer, and freezers to preserve fish that is not sold. About half of the men and women traders had access to phones and among the remaining traders this item was highly requested. Less than half of the men and women owned a freezer. Thus, in an attempt to preserve fish most traders bought ice on daily basis. Hitherto, men possessed greater access to freezers due to their rented tables at the main markets.

Traded Species

All men interviewed sold fresh fish whereas many women sun-dried and/or smoked it before selling. The fish species reported to be of the highest economic importance for both men and women traders were Emperors (*Changu* in Swahili) (76 % of all men and 81 % of all women), which could be explained by a high demand but also by high natural fish abundances. Men tended to trade with a variety of medium- to high-value species such as tuna (71 % of all men), kingfish (67 %), octopus/squid (43 %), swordfish (43 %), barracuda (38 %), and trevally (33 %), while women's income was mainly based on various medium- to low-value species such as octopus/squid (48 %), goatfish (48 %), mackerel (38 %), and rabbitfish (38 %). No women were involved in trade with shark and lobster, which are the two single most expensive species on the Zanzibar fish market, compared to 19 and 14 %, respectively, of the men traders (Fig. 4). The reasons given for choosing one species over another were linked to financial capital, availability, customers and access to cold storing facilities.

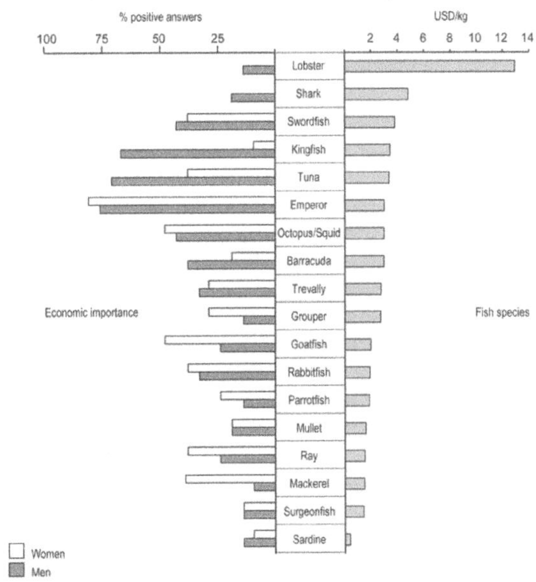

Figure 4: Average market prices for traded species (USD/kg) and economic importance for women (n = 21) and men (n = 21) traders, respectively (% positive answers).

Contacts and Organizations

Contacts

Both women and men expressed a wish to trade with hotels and restaurants but this requires contacts within the industry.

Whereas 38 % of the men traders had contacts in hotels and/or restaurants, not a single woman did (Fig. 3c). This inequality was reported to be due to women's lack of time to establish contacts, but also a result of the cultural setting and the common perception in Zanzibar that tourism-related work is unsuitable for women. However, apart from contacts within the tourism industry, about half of both the men and women traders had daily contact with other traders, fishers, or the auctioneer as a way to exchange information about, e.g., fish landings and demand. Further, all traders were asked about their relation to the former Department of Fisheries and Marine Resources, the institution responsible for fisheries management in Zanzibar. From the group, two women and eight men reported to have contact with the Department but most traders requested collaborations and information exchange.

Membership in Fish Trading Associations

Does an increasing number of women traders lead to a higher participation rate in formal organizations? This study found that there are no formal fish trading associations in Zanzibar. Instead, some men traders were members of the village's fisheries committee, but since fishermen and traders have different interests (capture fisheries problems vs. market interests) this was reported to be unsatisfying. In one of the main markets in Zanzibar Town, 25 men had formed a group where they discussed fish trade-related issues, shared experiences and borrowed money from each other. This kind of self-organized group was present among men traders in other places around the island as well. No women were involved in the official fisheries committee neither in any kind of self-organized group. Most women respondents had no "free time" to spend on organizing and many of them stated a lack of arenas for discussion. However, there was an overall wish for organizations with focus on their needs and interests, but help from the Government was requested since "traders are just grassroot people and cannot organize without assistance" (Interview with woman trader in Mikunguni 2012). A minority of the traders were reluctant to organizations since "people are not trustworthy" (Interview with woman trader in Mwanakwerekwe 2012) and "everybody should run their own business" (Interview with woman trader in Malindi 2012).

Perceptions and Experiences

Life Improvements and Equality

The majority of the respondents identified themselves as fish traders with high levels of job satisfaction (Fig. 3d). According to the respondents, society perceives fish trade as a high status activity; being a fish trader carries more prestige and greater economic status than being a fisher, seaweed farmer, or gleaner. The two main attributes for high job satisfaction were the sense of importance and recognition among people in their respective community. Another central factor was income. The majority of the interviewees said that fish trade has improved their living standards. The benefits included increased income and protein intake in the household, and many traders had been able to send their children to school, renovate houses and contribute to funerals, weddings, and loans. A majority of the traders also provided fish to people worse off than themselves, which increased their social status. In addition, most respondents reported gender equality at the auction sites. "What matters to succeed is a person's capital and not whether he or she was born a man or a woman". However, a large number of the men were worried about the increasing number of women traders and reported how women generally benefitted more from microcredit, which allowed them to buy higher quantities of fish. Women traders, on the other hand, found it difficult to compete with men on the auction site because men traders generally have higher capital, more experience and a broader contact net. From a vulnerability perspective, most traders reported not to have sufficient economic resources to cope with fluctuating markets (Fig. 3d). The women were mainly in need of credit and freezers to improve their business, withstand price hikes and cope with decreasing fish stocks. Similarly, most men requested credit but also a motorcycle to reduce time spent on local buses. Nevertheless, the overall attitude towards fish trade was very positive, especially among women who felt a sense of independence and more self-confident, due to an increased income and family well-being.

DISCUSSION

Feminization of Fish Trade?

The first question posed was whether or not men still dominate this activity or if fish trade is slowly turning into a typical women profession, similar to what has been observed in other activities in Zanzibar such as seaweed farming. For example, men were initially engaged in seaweed farming but as income turned out to be very low in relation to workload, most men left the activity and returned to fishing; today 90 % of Zanzibar's seaweed farmers are women (Fröcklin et al. 2012). Put in a fish trade context, decreasing stocks and smaller sizes of fish may open up space for women, whereas men would engage in other activities due to reduced profits. However, the results show that this does not seem to be the case in Zanzibar and the number of both men and women traders has increased, which is linked to the lack of livelihood options, but may also be a result of the financial crisis. However, the active involvement of women in an activity traditionally associated with men has the potential to challenge existing gender roles and empower women, and whether or not this is happening will be discussed below.

To What Degree Has Fish Trade Led to Women Empowerment?

The results show that women traders had a higher educational level but, regardless of education, the women surveyed shared less access to profitable markets and mobility, financial capital, contacts and organizations, and high-value fish species. However, trading with less valuable fish seems to have some advantages; it requires less capital and has a ready market due to high local consumption. This could probably explain why most men traded with both commercially valuable species and species of lower value. By diversifying the customer base, traders may be less vulnerable to fluctuating prices and demand. However, most traders in this study had regular customers and bought fish according to the orders whereas other agreed upon, from day to day, what species to buy so as to minimize competition over customers. Men's greater access to finance, profitable markets and customers, contacts, and commercially valuable fish resulted in income differentials. This shows that access to initial capital is a key factor for income enhancement, which in turn increases quality of life. However, since the women generally lacked own savings there are clear links to gender; women's access to start-up capital is very much restricted by traditional gender roles in which women are associated with household duties and childcare. Reproductive work is unpaid work and thus makes it hard for women to accumulate start-up capital and gain experience. Also, lack of free time reduces the chances to network and establish contacts much needed to access more profitable markets. In addition, microcredit and loans do not seem to provide money enough to trade with high-value species or higher fish volumes required for a substantial income increase. These types of gender inequalities have been observed in other places as well (e.g., Medard et al. 2002; Tindall and Holvoet 2008) and not only reduce women's access to fisheries resources, but also limit their bargaining power in case of price shocks or reduced fish landings. Nevertheless, due to the obviously commercial character of the activity, the average income for all fish traders irrespective of gender was quite high. A fish trader generally earns more than people involved in other coastal activities such as for example women invertebrate gleaners (0.5–0.6 USD per day) (Håkansson et al. 2012) and seaweed farmers (1 USD per day) (Fröcklin et al. 2012).

Despite of all problems and disadvantages illustrated above, women have still managed to enter a typical men-dominated arena. Hypothetically, a higher number of women could bring about change, not only by increasing the total household income and thus reducing poverty, but also women's social and economic status. However, women's greater involvement in productive work has not changed basic time-consuming activities associated to women's reproductive role; for example, no women were involved in any type of fisheries organization, which was largely due to household responsibilities but also the lack of organizational skills. Unlike Zanzibar, there are places, such as for example Ghana, where women do belong to fish trader associations (Walker2002), but membership in formal or informal fisheries associations and cooperatives is still more prevalent among men. Also, a higher number of women members may not per se increase women's status unless the existing norms and structures change to accommodate women's interests. Strong traditional norms and values in society may constrain women from actively participating in decision making. For example, in Mali many women traders are members of the traders' association but they are disadvantaged by the social setting (Tindall and Holvoet 2008). This seems to be a general trend in many developing countries. Also, in many parts of Africa women have been involved in fishing through ownership, processing, and fish trade (Geheb et al. 2008; Béné et al. 2009), but most fishing communities are still dominated by men (Rangan and Gilmartin 2002). Thus, women's efforts to improve their situation have placed them in men's action arena, without necessarily challenging traditional gender roles. There are also examples of how women have addressed the poverty situation through agency and social entrepreneurship, and by doing so, also male subordination (Onyango and Jentoft 2011). Clearly, there are ways for women to break out of their traditional spatial boundaries and overcome the socially constructed roles ascribed to women and men, but in Zanzibar this process seems to lag behind or be slower in character. This demonstrates how well-rooted gender roles can be in the social and cultural setting of a specific location. Nevertheless, despite women's limited power to economic and social resources, as well as decision making authority and organization, women traders reported high levels of job satisfaction, which is a common feature among fishermen (Pollnac et al. 2001; Allison and Horemans 2005), fisherwomen, fish traders, and processors (Appleton 2000). They further stated an increased level of self-esteem and awareness due to improved business skills. In conclusion, fish trade may have led to some level of empowerment but there is still a long way to go to diminish existing gender inequalities.

Gender Sensitivity and Adaptive Management

Due to an increasing number of traders, and particularly women, policy and management must address the gender dimension to include the interests and needs as well as environmental impacts of both men *and* women. We therefore suggest an adaptive fisheries management approach. The first step in adaptive management processes is to identify problems and desired goals, and call for the development of suitable policy. Next is the implementation of policy and monitoring of results. According to Walters (1986) this process is iterative and offering not only the potential to engage different groups but also the opportunity for them to learn from each other. At present, Zanzibar's fisheries management lack (i) research about the role of women in the fish value chain, (ii) gender-disaggregated data, and (iii) inclusion of these two in policy and management. The analysis of formal documents such as the Fisheries Act (Act 7 2010) (RGZ2010) showed that the focal point of the Department of Fisheries Development (DFD), and its sub-sections such as deep-sea fishing and artisanal fisheries, is primarily on men-dominated capture fisheries and export whereas very little attention is drawn to local domestic fish trade. This clearly answers the question whether or not Zanzibar's fisheries management and policy is gender sensitive. To overlook typical women activities like gleaning of invertebrates in the intertidal areas (Nordlund et al. 2010; Håkansson et al. 2012), or women traders' activities, results in problem and goal identification since information is based on some sources only. Further, the lack of information available on fish traders reveal that they operate on an open informal market, without clear rules, control or taxation. This type of informal trade is rather common in developing countries throughout Africa but also in other places such as Mexico (Pedroza 2013). At present Zanzibar's fisheries management seems not to be flexible and responsive enough to changing conditions, i.e., increase in traders and gender inequalities in the fish value chain, and there is no link between traders and formal institutions. Clearly, some of the most fundamental elements in adaptive management are missing.

Implications for Policy Interventions

This study raises some important issues and should act as a stepping stone towards policy interventions. So, what is the way forward? First, one essential step is to collect statistics about the number of traders, sex, finance, traded species, and so on. Second, the establishment of fish trading organizations is essential to gather information about fish traders' needs, problems, and interests. These types of organizations could be promoted by the Government and work as a platform for men and women to discuss common or different issues related to fish trade, and share knowledge and ideas while empowering women. Third, arenas that facilitate the formation of self-organized groups are needed and can act as a catalyst for better management and conservation. Arguably, a constant flow of fisheries resources should be in the traders' interest. The creation of such meeting arenas could also lead to collaborations and exchange of ideas in the development of policies. Four, to facilitate the implementation of the recommendations above one could use existing structures and institutions. For example, statistics could be collected by increasing the number of monitoring agents (*Bwana Dikos*) already present at each market. In addition there are some experiences already gained in trying to organize seaweed farmers. Five, and most importantly, gender *must* be mainstreamed in all steps of fisheries management and policy making. We argue that Zanzibar's fisheries management would benefit from an adaptive approach, which must be informed by extensive learning about the changing environment and people, both men and women, who depend upon it for daily survival, along with participation and an active dialogue throughout the process.

ACKNOWLEDGMENTS

The authors thank all respondents for participating in this study, as well as Muumin Hamad and the *Bwana Dikos* for interpretation work and field assistance. Thanks to Hampus Eriksson for assistance with illustrations and to Cindi Katz for input on an earlier version of the paper. Appreciation to the Institute of Marine Sciences and the Department of Fisheries and Marine Resources, Zanzibar. This research has been financed by the Swedish International Development Cooperation Agency (Sida) and The Swedish Research Council (Vetenskapsrådet, VR, diarienr. 344-2011-5448).

REFERENCES

1. Allison, E.H., and B. Horemans. 2005. Poverty alleviation, sustainable livelihoods and management in small-scale fisheries. Discussion Paper 17. FAO, Rome, Italy.
2. Allison, E., and F. Ellis. 2001. The livelihoods approach and management of small-scale fisheries. *Marine Policy* 25: 377–388.
3. Appleton, J. 2000. At my age I should be sitting under that tree: The impact of AIDS on Tanzanian lakeshore communities. *Gender and Development* 8: 19–27.
4. Béné, C., E. Steel, B.K. Luadia, and A. Gordon. 2009. Fish as the "bank in the water"—Evidence from chronic-poor communities in Congo. *Food Policy* 34: 108–118.
5. Bennett, E. 2005. Gender, fisheries and development. *Marine Policy* 29: 451–459.
6. Choo, P.S., B.S. Nowak, K. Kusakabe, and M.J. Williams. 2008. Guest editorial: Gender and fisheries. *Development* 51: 176–179.
7. Crona, B., M. Nyström, C. Folke, and N. Jiddawi. 2010. Middlemen, a critical social–ecological link in coastal communities of Kenya and Zanzibar. *Marine Policy* 34: 761–771.
8. Davis, D., and S. Gerrard. 2000. Introduction: Gender and resource crisis in the North Atlantic fisheries. *Women's Studies International Forum* 23: 279–286.
9. de la Torre-Castro, M. 2006. Humans and seagrasses in East Africa—A social–ecological systems approach. PhD Thesis. Stockholm, Sweden: Stockholm University.
10. de la Torre-Castro, M. 2012. Governance for Sustainability: Insights from Marine Resource Use in a Tropical Setting in the Western Indian Ocean. *Coastal Management* 40: 612–633.
11. Denscombe, M. 2007. *The Good Research Guide for small-scale social research projects*. Maidenhead: Open University Press.
12. Diamond, N.K., L. Squillante, and L.Z. Hale. 2003. Cross currents navigating gender and population linkages for integrated coastal management. *Marine Policy* 27: 325–331.
13. Eriksson, H., M. de la Torre-Castro, and P. Olsson. 2012. Mobility, Expansion and Management of a Multi-Species Scuba Diving fishery in East Africa. *PLoS ONE* 7: 1–11.
14. FAO. 2006. Gender policies for responsible fisheries. FAO, Rome, Italy, 12 pp.
15. FAO. 2008. Achieving poverty reduction through responsible fisheries. Lessons from West and Central Africa. FAO, Rome, Italy, 190 pp.
16. Fröcklin, S., M. de la Torre-Castro, L. Lindström, N.S. Jiddawi, and F.E. Msuya. 2012. Mariculture as a development project in Zanzibar, East Africa: A price too high to pay?*Aquaculture* 356–357: 30–39.
17. Geheb, K., S. Kalloch, M. Medard, A.T. Nyapendi, C. Lwenya, and M. Kyangwa. 2008. Nile perch and the hungry of Lake Victoria: Gender, status and food in an East African fishery. *Food Policy* 33: 85–98.
18. Gunderson, L.H., C. Holling, and S. Light. 1995. *Barriers and bridges for the renewal of ecosystems and institutions*. New York: Columbia University Press.
19. Håkansson, E., S. Fröcklin, and M. de la Torre-Castro. 2012. Invertebrate Harvesting in Chwaka village: Resilience, gender and sustainability. In *People, nature and research in Chwaka Bay, Zanzibar, Tanzania*, ed. M. de la Torre-Castro, and T.J. Lyimo, 255–264. Zanzibar Town: WIOMSA. ISBN: 978-9987-9559-1-6.
20. Harper, S., D. Zeller, M. Hauzer, D. Pauly, and R.U. Sumaila. 2013. Women and fisheries: Contribution to food security and local economies. *Marine Policy* 39: 56–63.
21. Holling, C.S. 1978. *Adaptive environmental assessment and management*. London: Wiley.
22. ICSF (International Collective in Support of Fishworkers). 2002. Report of the study on the problems and prospects of artisanal fish trade in Africa. ICSF, Chennai, India, 92 pp.
23. Jiddawi, N.S., and H. Khatib. 2007. Zanzibar fisheries frame survey 2007. Census Report. Department of Fisheries and Marine Resources, Ministry of Agriculture, Livestock and Environment 2008. Zanzibar, Tanzania, 61 pp.
24. Jiddawi, N.S. 2012. Artisanal fisheries and other marine

resources in Chwaka Bay. In *People, Nature and Research in Chwaka Bay, Zanzibar, Tanzania*, ed. M. de la Torre-Castro, and T.J. Lyimo, 193-212. Zanzibar Town: WIOMSA. ISBN: 978-9987-9559-1-6.

25. Jiddawi, N.S., and M.C. Öhman. 2002. Marine fisheries in Tanzania. *AMBIO* 31: 518–527.
26. Kvale, S. 1994. *An introduction to qualitative research interviewing*. Lund: Studentlitteratur.
27. Lee, K.N. 1993. *Compass and gyroscope*. Washington, DC: Island Press.
28. Lwenya, C.A., and R.O. Abila. 2001. The role of women in artisanal fish processing and trading on Lake Victoria (Kenya). *African Journal of Tropical Hydrobiology and Fisheries* 10: 1–2.
29. Madanda, A. 2003. Commercialization and gender roles among Lake Victoria shore fishing communities of Uganda. Department of Women and Gender Studies, Makerere University, Kampala, Uganda, 66 pp.
30. Medard, M., F. Sobo, T. Ngatunga, and S. Chirwa. 2002. Women and gender participation in the fisheries sector in Lake Victoria. In *Global Symposium on Women in Fisheries*, ed. M.J.Williams, N.H. Chao, P.S. Choo, K. Matics, M.C. Nandeesha, M. Shariff, I. Siason, and J.M.C. Wong, 155–168. Kaohsiung: Worldfish Center.
31. Nordlund, L., J. Erlandsson, M. de la Torre-Castro, and N.S. Jiddawi. 2010. Changes in an East African social–ecological seagrass system: Invertebrate harvesting affecting species composition and local livelihood. *Aquatic Living Resources* 23: 399–416.
32. Onyango, P.O., and S. Jentoft. 2011. Climbing the hill: Poverty alleviation, gender relationships, and women's social entrepreneurship in Lake Victoria, Tanzania. *MAST* 10: 117–140.
33. Pedroza, C. 2013. Middlemen, informal trading and its linkages with IUU fishing activities in the port of Progreso, Mexico. *Marine Policy* 39: 135–143.
34. Pollnac, R.B., R.S. Pomeroy, and I.H.T. Harkes. 2001. Fishery policy and job satisfaction in three southeast Asian fisheries. *Ocean and Coastal Management* 44: 531–544.
35. Porter, M. 2006. Gender and fisheries: A global perspective. Paper Presented at Global Coasts: Gender, Fisheries and Contemporary Issues, International Symposium, University of Tromso, Norway, 16 pp.
36. Rangan, H., and M. Gilmartin. 2002. Gender, traditional authority, and the politics of rural reform in South Africa. *Development and Change* 33: 633–658.
37. RGZ. 2010. *Fisheries Act*. Zanzibar, Tanzania; Printed by the Government Press.
38. Ruitenbeek, J., I. Hewawasam, and M. Ngoile. 2005. *Blueprint 2050: Sustaining the marine environment in mainland Tanzania and Zanzibar*. New York: The World Bank.
39. Samuel, L. 2007. Women, work and fishing: An examination of the lives of fisherwomen in Kerala. *South Asia Research* 27: 205–227.
40. Thyresson, M., M. Nyström, and B. Crona. 2011. Trading with resilience: Parrotfish trade and the overexploitation of key-ecosystem processes in coral reefs. *Coastal Management* 39: 396–411.
41. Thyresson, M., B. Crona, M. Nyström, M. de la Torre-Castro, and N. Jiddawi. 2013. Tracing value chains to understand effects of trade on coral reef fish in Zanzibar, Tanzania. *Marine Policy* 38: 246–256.
42. Tietze, U. 1995. FAO's role and experiences with improving the social and economic status of women in fishing communities in Asia and the Pacific. FAO, Rome, London, 9 pp.
43. Tindall, C., and K. Holvoet. 2008. From the lake to the plate: Assessing gender vulnerabilities throughout the fisheries chain. *Development* 51: 205–211.
44. Walker, B.L.E. 2001. Sisterhood and seine-nets: Engendering development and conservation in Ghana's marine fishery. *Professional Geographer* 53: 160–177.
45. Walker, B.L.E. 2002. Engendering Ghana's seascape: Fanti fish traders and marine property in colonial history. *Society and Natural Resources* 15: 389–407.
46. Walters, C. 1986. *Adaptive management of renewable resources*. New York: McGraw Hill.
47. Weeratunge, N., K.A. Snyder, and C.P. Sze. 2010. Gleaner, fisher, trader, processor: Understanding gendered employment in fisheries and aquaculture. *Fish and Fisheries* 11: 405–420.
48. Williams, S.B., M. Hochet-Kinbongui, and C.E. Nauen. 2005. Gender, fisheries and aquaculture: Social capital and knowledge for the transition towards sustainable use of aquatic ecosystems. EU, Brussels, Belgium, 32 pp.
49. Williams, M.J. 2008. Why look at fisheries through a gender lens? *Development* 51: 180–185.

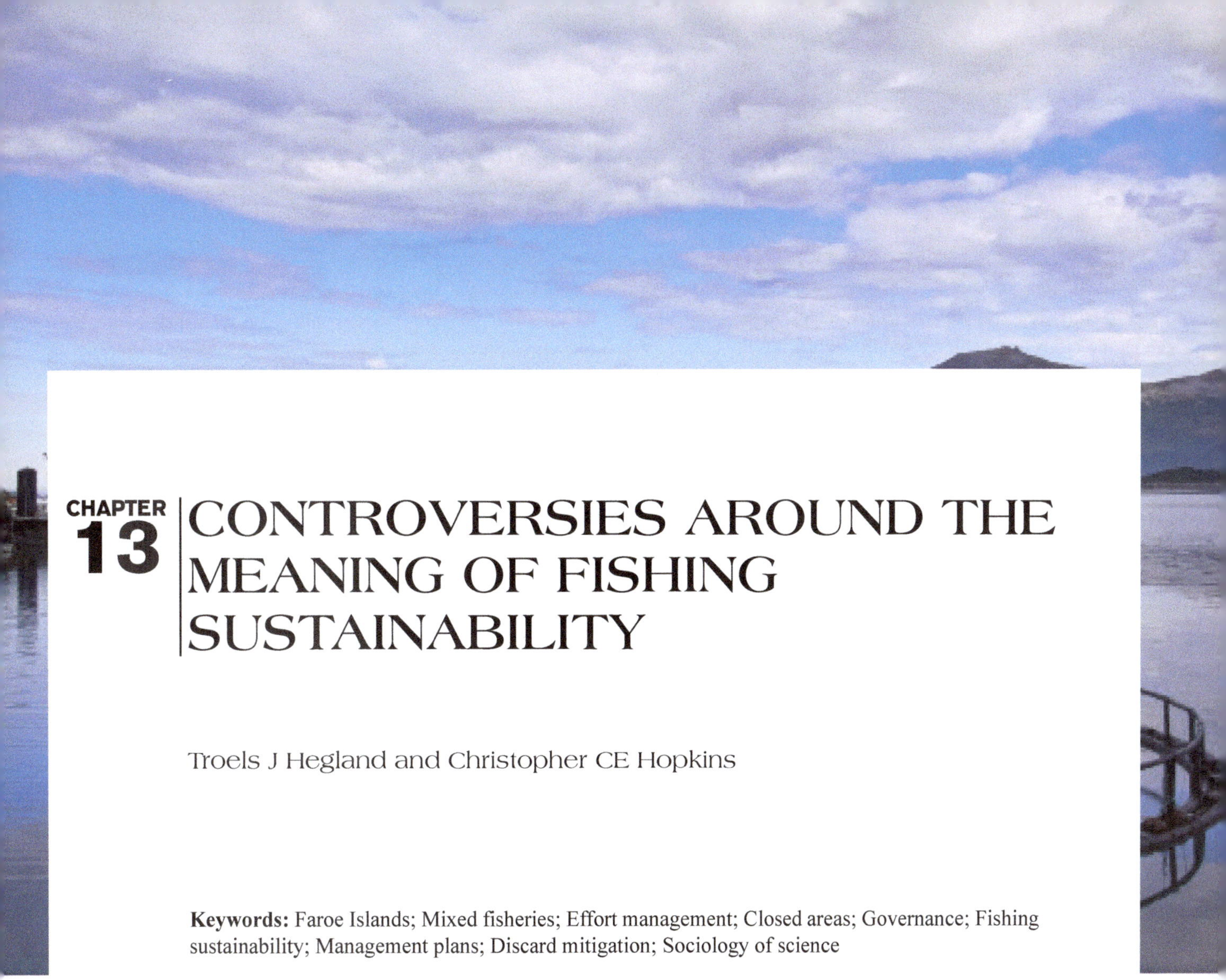

CHAPTER 13

CONTROVERSIES AROUND THE MEANING OF FISHING SUSTAINABILITY

Troels J Hegland and Christopher CE Hopkins

Keywords: Faroe Islands; Mixed fisheries; Effort management; Closed areas; Governance; Fishing sustainability; Management plans; Discard mitigation; Sociology of science

INTRODUCTION

Fishing and fish processing have been the main sources of income for the Faroe Islands since the 1920s. The fisheries industry — including the catching and processing sectors, and a comparatively smaller but growing aquaculture sector — contributes roughly 20 per cent of the gross domestic product (MFNR [2008]). Additionally, seafood related products represented over 94 per cent of the Faroese merchandise exports in 2012 (MFNR [2013a]). Close to being a `mono-product' economy, the Faroese national economy to a large extent stands and falls with the fisheries industry[a] and is highly vulnerable to the volatility of catches and prices of the most important fish stocks (Búskaparráðið [2010]). Populated by only about 48,000 people, the Faroe Islands are a self-governing territory within the Realm of Denmark with its own Faroese Home Government (FHG) and parliament (*Løgtingið*). Under the framework of the Home Rule Act of the Faroe Islands, powers are divided between the Faroe Islands and Denmark. The Faroe Islands decided not to join the European Union (EU) when Denmark did so in 1973. Thus, as fisheries management within the Faroese exclusive economic zone (EEZ) under the Home Rule Act remains the responsibility of the FHG, the Faroe Islands are not subject to the EU's Common Fisheries Policy (CFP) (Statsministeriet [1948]; MFNR [2008]; SFI [2012]).

Traditionally, the most important fishery conducted in Faroese waters is the mixed demersal fishery targeting cod (*Gadus morhua*), haddock (*Melanogrammus aeglifinus*) and saithe (*Pollachius virens*) primarily on the Faroe Plateau and Faroe Bank. Between 1996 and about 2002, the mixed demersal fishery thrived and the Faroese gained a reputation for being near the forefront of fisheries management and sustainable fishing, due to a large extent to the performance of a new fisheries management system established in 1996 (Chuenpagdee and Alder [2001]). Contrasting to the approach in most other European countries, the system implemented aims at controlling fishing effort (i.e. input control) rather than directly controlling the amounts of fish caught or landed (i.e. output control). Major advantages of the Faroe Islands' effort-based system included resolving (by a landing requirement for all fish caught) the previously prevalent at-sea discarding of the

Citation: Troels J Hegland, Christopher CE Hopkins, Controversies around the meaning of fishing sustainability, DOI: 10.1186/s40152-014-0012-7.

unwanted bycatch of fish which was disliked by the fisheries industry, the relative simplicity of management administration of the new effort system compared with the previous output-based system and the major role of the catching sector in co-developing the new system (Jákupsstovu et al. [2007]). However, in the last decade, various concerns have emerged concerning weaknesses in the Faroese system, including scientific advice not taken into account properly when deciding on the effort and fishing mortality applied to the target stocks; failure to set up a system for effectively monitoring fishing effort; lack of implemented fishery management plans (FMPs)/ harvest control rules (HCRs); overcapacity, poor economic performance and profitability (Jákupsstovu et al. [2007]; Christensen et al. [2009]; Baudron et al. [2010]; Búskaparráðið [2010]; ICES ACOM [2012a]; Nielsen et al. [2012]). The system is currently under heavy pressure, both nationally and internationally, to reform.

Under the EU's CFP, fish stocks are managed primarily by annual catch limits, i.e. output as total allowable catch (TAC) together with technical measures (Holden [1994]). EU fisheries management, however, has been particularly impaired when using single-species TACs in mixed fisheries situations due to associated discarding at sea (Holden [1994]; Daan [1997]). Thus, TACs restrict the official landings but not the catch, such that the discrepancy between the two output measures may be substantial and difficult to determine accurately. In January 2014, the reformed CFP came into force, including a ban on discarding of fish together with a requirement for full catch accountability, as well as a legally binding commitment to fish sustainably according to the principle of maximum sustainable yield (MSY, i.e. the largest yield/catch that can be taken from a species' stock over an indefinite period) in the context of multiannual FMPs (EU [2013]).

Having given a short introduction to the Faroe Islands' context, we present the Faroese effort management system. We then discuss the system's ability (or lack thereof) to sustain a level of fishing effort that produces long-term resource sustainability and the associated controversies and uncertainties around this. Finally, we explore how the Faroe Islands are attempting to move towards a fisheries management system along the lines of current international standards and provide some ideas on how the Faroe Islands can move forward.

The current paper is primarily concerned with describing and diagnosing the Faroese fisheries management system with the aim of being able to offer advice on how the system could be reformed. However, the article could also be read in the context of some more generic debates in fisheries science.

One such debate relates to the role of science in fisheries management and how science is fed into decision-making. In a relatively recent paper by several of the important scholars in the field in Europe, calls are made *"for a more interactive system of producing a common knowledge base"*, thereby changing the role of scientists from traditional 'experts' to 'transparency experts' assisting *"stakeholders in trying to build an accurate common picture of the marine environment"* (Schwach et al. [2007]: 803). The Faroese system offers a case of a system that adheres to a traditional approach that separates science and stakeholders (with the implications hereof) and is therefore a useful reference. In addition to the above, the Faroese case is also an example of an effort-based system. Effort-based systems are interesting not only because they are rarer than their TAC/quota-based counterparts but also because they are considered less bureaucratically cumbersome, which is of increasing interest as the economic importance of capture fisheries sector decreases. Furthermore, effort-based systems present an attractive fix to the notorious discards-problem, which has for instance been a major issue under the CFP. Currently, the EU is implementing a discard-ban in the TAC/ quota-approach and there are few discussions on moving towards effort based management. However, in connection with the 2002/03 reform of the CFP effort management was considered as a serious alternative to the current TAC/quota-approach (e.g. Shepherd [2003]). Similarly, only a few years ago the North Sea Regional Advisory Council was actively promoting an experiment with effort-management in the Kattegat. This was welcomed by the European Commission but never materialised (Fisheries Secretariat [2014]). In exploring effort-management and the features that can possibly inspire TAC/quota-based approaches, the Faroe Islands remains the only 'local' inspiration for EU decision-makers.

MATERIALS AND METHODS

Publications on the Faroese effort management system were collected as part of a comprehensive literature review. A systematic approach was applied to conduct an electronic search of catalogues, bibliographies, and discriminating use of internet search engines. Relevant material sources (articles in peer-reviewed scientific journals or books, as well as grey literature) were read and new references identified through the citations.

Having familiarised ourselves with Faroese fisheries management in general, and the effort management system in particular, we conducted a series of semi-structured interviews (average about 1½hours) predominantly with fisheries industry representatives (catching, processing, and fishery unions), fisheries managers and ecologists/biologists and social scientists associated with the mixed demersal fishery. Regarding the fisheries industry, we predominantly interviewed representatives from the catching sector, which is the sector most concerned with and vocal about fisheries management. It was not possible to identify any relevant representatives of environmental non-governmental organisations (ENGOs). A total of 17 individuals[b] were interviewed (in 11 interviews) during a field trip to the Faroe Islands in August 2012. The interviews followed a list of standardised questions/topics, prepared in advance, and took into account our literature review and our wish to learn more about the effort management system and its challenges.

The interviews, which were recorded and subsequently transcribed[c] and organised by means of textual analysis software, served a dual purpose. On one hand the interviews filled in gaps in the factual knowledge on the Faroese effort management system obtained through the literature review. Equally important, however, the interviews uncovered controversies and different perceptions on issues related to the Faroese system. The respondents were granted anonymity and will only be identified by their affiliation with one of the stakeholder categories.

For comparisons of the levels of fishing mortality (F) exerted on stocks of cod, haddock and saithe between 1987 and 2011, we used the Multiple Sample Comparison (MSC) procedure in the Statgraphics Centurion XVI Professional package (StatPoint Inc.). Tests were run to determine whether or not there are significant differences between the means, variances, and/or medians of the F values pertaining to these three stocks. In the MSC procedure, an analysis of means (ANOM) plot (Ott et al. [2005]) displayed the sample means for each of the three stocks on a chart in order to easily determine which means are significantly different from the `grand' mean for all stocks.

BACKGROUND

While most governments around the North Atlantic adopted fisheries management based on TACs following the establishment of extended national fisheries jurisdictions in the mid-1970s, the Faroe Islands continued to manage the demersal fishery in their EEZ by means of traditional technical regulations, including application of minimum mesh sizes and closed areas. Thus, the Faroese regulations had the objective of controlling the catch composition, for instance by reducing the catch of juvenile fish, rather than constraining the volume of the catch (Gezelius [2008a]). It is not clear why the Faroese continued to apply the somewhat traditional approach while most other states moved `forward'. It is likely that a number of factors contributed to this.

A licensing system for the demersal fishery was first implemented in 1987, thereby providing a means to regulate catch levels (Gezelius [2008a]). Thereafter, but before 1994, the demersal fishery in the Faroese EEZ was managed by a combination of license limits (curbing the number of fishing vessels), area restrictions for trawl fisheries (e.g. closed areas and seasons), minimum mesh sizes, and measures to protect juvenile fish (Maguire [2001]; Jákupsstovu et al. [2007]). To reduce F and recover declining stocks in the wake of a near collapse of the Faroese economy in 1992, a new TAC-based management system, implemented by means of individual transferrable quotas (ITQs), coupled with a discard ban, was introduced in 1994 when the Faroe Islands constructed its first, comprehensive legal framework for fisheries management, the Commercial Fisheries Act of 1994 (Løgtingið [1994]). However, the ITQ system was abandoned after only two years, not only due to the substantial costs and administrative effort necessary for this management form, but also because it received considerable criticism and resistance from the catching sector (Gaard et al. [2002]). It has also been argued that the TAC-system was in part forced upon the Faroe Islands by the Danish government (at least so it was perceived) as part of a package in which the Danish government helped solve the Faroese financial problems in 1992/93 (Gezelius [2008b]). This did not contribute to the perceived legitimacy of the new system.

It is widely recognised that management by TACs of species caught in mixed fisheries — as the Faroese mixed demersal fishery — is problematic because the quota of different species may be exhausted at different rates. Thus, fishers face a dilemma when the quota for one species is exhausted: stop fishing and underutilise the quota for other species, or continue fishing and discard or illegally land over-quota fish (Daan [1997]). With the latter option, the F level dictated by the TAC will be exceeded, and the scientific basis for stock assessment and future management advice will be compromised if the stock assessment is based only on official landings data, assuming that landings equal the catches (Kraak et al. [2008]).

In the Faroe Islands, the TAC-based system resulted in extensive discarding as well as under- and misreporting of substantial parts of the catch, related to bycatch problems that were increasingly difficult for fishers to handle, particularly when single species quotas had been used up (Maguire [2001]; Jákupsstovu et al. [2007]; Gezelius [2008a]; Johnsen and Eliasen [2011]). The situation was exacerbated also by unusually low primary production on the Faroe Plateau ecosystem during the early 1990s, which subsequently had a devastating effect on the Faroese demersal fishery and society in the mid-1990s (Gaard et al. [2002]).

In response to the wide-reaching criticism of the TAC-based system, a general desire emerged to establish a new system that was easy to administer and enforce, and should remove incentives for discarding incidental catches, misreporting and black landings (Gezelius [2008a]). The chosen solution was a distinct Faroese management system focusing on the regulation of fishing effort rather than catches.

ELEMENTS OF THE FAROESE EFFORT MANAGEMENT SYSTEM

The Faroese effort management system was implemented in 1996 after a short, joint design-effort between managers, scientists and key fisheries sector representatives (Jákupsstovu et al. [2007]). This system has evolved somewhat over the years, but the main elements are still basically the same today.

A fundamental objective of the Faroese effort management system is to regulate fishing effort so that the annual catch of the three most important demersal stocks (cod, haddock and saithe) does not exceed 33 per cent of the stocks, corresponding to controlling F at ≤ 0.45 on each of the three component stocks (Jákupsstovu et al. [2007]).

Beneath, based on available literature (Løkkegaard et al. [2007]; Jákupsstovu et al. [2007]; Gezelius [2008a]; Gezelius [2008b]; Christensen et al. [2009]; Zableckis et al. 2009; Johnsen and Eliasen [2011]; Løgtingið [1994]; ICES NWWG [2012]; DNV [2013]), as well as our fact-finding fieldtrip to the Faroe Islands, the most important elements of the Faroese effort management system are described.

Fleet segmentation

Fleet segmentation is a central element in controlling and dispersing fishing effort and the fishing pattern (e.g. where and when fishing takes place, and by which vessels), and in limiting conflicts between segments and active and passive gears. Different regulations apply to different fleet groups (segments), but within a given group the same rules apply. In addition, the objective of this intricate system of fleet segmentation and group-specific regulations, set out in the Commercial Fisheries Act, is to ensure that different fleet segments — including trawlers, longliners, coastal vessels over 15 gross registered tonnes (GRT), coastal vessels under 15 GRT, and others — as far as possible each catch a pre-defined share of respectively cod, haddock, saithe and redfish (*Sebastes marinus*).[d] As a general rule, holders of fishing days who fish outside a so-called `ring' line, in waters deeper than 200 m, can fish for three days for each day allocated inside the `ring'. Illustrative examples of specific regulations for various fleet groups include:

- One fishing day by longliners under 110 GRT is considered equivalent to two fishing days for jiggers in the same gear category. Thus, longliners under 110 GRT could double their fishing days' entitlement by converting to jigging.[e]
- Trawlers are generally not allowed to fish within the 12 nautical miles (nm) limit. In addition several other areas are closed to trawling either throughout the year or parts of it.
- Single trawlers under 400 horse power (HP) are given special licenses to target flatfish inside 12 nm with a bycatch allocation of 30 per cent cod and 10 per cent haddock. In addition, they must use sorting devices in their trawls in order to minimize bycatches.

Capacity regulation

Capacity regulation aims to maintain fleet capacity at the 1997 level within each fleet group. It is taken for granted that the fleet capacity (fishing effort) is sustainable by not resulting in overfishing of the targeted stocks in the long-term. There are rules for allowing vessel transfers between groups (e.g. vessel replacement) and merging of capacity. The capacity policy is based on vessel licenses: a `harvesting licence' attached to a specific vessel > 15 GRT, and a `fishing licence' which allocates a certain number of fishing days in the EEZ and tonnes of fish outside the EEZ. The transferability of fishing days is restricted, and they cannot be transferred freely between gear and vessel categories.

Effort regulation and tradability of effort

Maximum total effort expressed as fishing days (i.e. days-at-sea) is fixed annually for the coming fishing year (running from 1 September in one year to 31 August the next year) for each of the fleet groups and sub-groups. Excepting the artisanal fleet groups, the total effort is then allocated equally between individual vessels in each of the fleet groups. For the coastal fishery, 60 per cent of the total effort is allocated to full-time fishers who receive individual, equal-sized effort quotas. However, artisanal part-time fishers receive a common effort quota, i.e. not individually allocated, and their fishery is closed when the quota has been used. Fishing days may be traded within fleet groups and, with some restrictions, between groups. Fishing days can leased out for one year or sold permanently. Official effort conversion keys are used when trading effort between fleet groups in order to account for differences in fishing capacity across vessel sizes, engine power and gear types. Individual vessels can meet restrictions from effort limitation regulations by purchasing days-at-sea from other vessels. Thus, the effort management system effectively allocates individual transferrable fishing effort rights. The effort regulation is maintained through the fishing license system.

Area closures

An important characteristic of the Faroese regulatory regime is the comprehensive use of areas closed in certain periods of the year, as well as areas closed for certain types of fishing, in particular trawling. Such areas have been used since the establishment of the 200 nm Faroese EEZ in 1977. Initially the main purpose was to avoid gear conflicts between longline and trawl fishers, but it is now mainly used to regulate access rights to fishing areas. Additional uses of area closures — for the protection of aggregating adult fish on spawning grounds and juvenile fish in nursery and feeding grounds, limitation of bycatch, or the protection of vulnerable species and habitats — have played an increasing role in the case of temporary and seasonal closures. Up to about 1993, the extent and number of area closures in the EEZ gradually increased. Since about 1996 the closed areas have remained basically unchanged, with the exception of three new areas closed to trawling in 2005 to protect cold-water corals and additional area closures in 2011 within the 6 nm boundary.

Additional technical measures

Additional technical measures span regulations regarding mesh sizes, obligatory, permitted or banned fishing gear, and bycatches. Specific minimum mesh size regulation and use of sorting grids are applied to particular fisheries. Use of beam trawls and Danish seine is banned. The general discard ban includes real-time rules for changing fishing areas when bycatch occurs, with obligations for reporting when bycatch levels reach 30 per cent of the catch under a certain size limit, and a system of minimum landing sizes for target species. In the context of discarding, all fish of commercial interest caught, including bycatch, must be retained onboard, landed and registered.

Governancef and decision-making

Although the effort management system is a complex combination of rules and regulations intended to achieve the target catch as well as accommodating the interests of different fleet segments, the central decision-making with regards to the Faroese effort management system concerns the setting of the overall number of days-at-sea (an effort quota comparable to the overall catch quota in an output management system). This annual exercise starts when the Ministry of Fisheries and Natural Resources (MFNR) receives the annual report from the Faroe Marine Research Institute (FAMRI). This report is passed on to the Committee on Fishing Days (*Fiskidaganevndin*), appointed by the Minister, consisting of representatives from the catching sector as well as a chairman who is not from the sector. This committee drafts its own report based on the catching sector's views and experiences. Building on FAMRI's report and the committee's report, the Minister drafts a proposal for the coming year's fishing effort concerning the number of days-at-sea. The Minister's proposal has tended to lie closer to the views of the committee than to those of FAMRI. The Minister's proposed bill is evaluated by the Fisheries Advisory Council (*Fiskivinnurádid,* a council under the MFNR, which includes broader fisheries industry representation), concerning any potentially associated changes that may be needed to fisheries policy and legislation. The Minister then presents the proposed bill to the Parliament, which makes the final decision and amends the Commercial Fisheries Act, see Figure1.

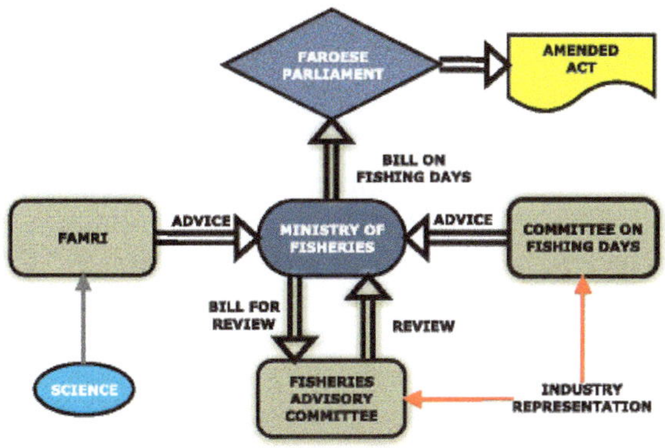

Figure 1: Flow chart for setting of the number of days-at-sea (Adapted from MFNR[2013b]).

With the exception of FAMRI scientists, prior to the proposal's submission to Parliament all stakeholders taking part formally in the process are from the fisheries industry. Thus, commercial fishing interests — especially from the catching sector — have substantial influence on the decision-making process. The fisheries industry is organised into special interest associations (e.g. Ship-owners Association, Fishermen's Union, Fish Processors Union, Association of Coastal Fishermen) and representation remains within these groups. ENGOs are absent from the process and even from the public debate, as indicated by our failure to find a representative to interview during our fieldtrip. On the other hand, the diverse fleet structure of segments and métiers results in a large variety and complexity of frequently differing interests in the catching-related representation. This complexity makes it difficult to gain agreement on effort cuts across different fleet fractions.

As mentioned in several interviews, the process sketched above can be viewed as suboptimal as it provides little incentive or support for the integration of diverse views between scientists, managers and industry representatives. In particular, it `plays off' the scientists against the industry and does not provide an effective forum for building understanding, trust and unification between these key actors. This leaves the politicians, in particular the Minister of Fisheries and Natural Resources, relatively free to determine the final decision between the frequently conflicting positions of the scientists and fisheries industry representatives. Whereas substantial progress has been made in the USA and in the EU in detaching politics from the fishery decision-making arena, due to the emergence of agreed rules-based FMPs (see later) and integrative stakeholder-based consultative forums (USA's regional management councils; EU's regional advisory councils) (Hartley and Robertson [2006]; Ounanian and Hegland [2012]), Faroese decision-making remains entrenched in a politicised, adversarial system. Notably, the absence of ENGO activity on the Faroe Islands may have contributed to the situation by pushing the scientists (more than their colleagues in for instance mainland Europe) towards a role as `issue advocates' (i.e. protector of the fish) making them appear more as political actors than as `honest brokers' (Pielke [2007]).

EVALUATING THE SUSTAINABILITY OF THE EFFORT MANAGEMENT SYSTEM

In the current section we explore the extent to which the Faroese management system can be considered biologically sustainable. Consequently, we are not looking into other measures of sustainability, i.e. social or economic.

The Faroese effort management system enjoys considerable support from both the fisheries industry, particularly the catching sector, and wide parts of the general public (Jákupsstovu et al. [2007]; Christensen et al. [2009]). Criticism comes mainly from the scientific community, which argues that the level of fishing effort has for a long time been too high, thus exerting excessive F on the target stocks. The criticism is basically divided into two arguments: 1) the implementation of the effort management system is in breach of the relevant legislation, and 2) the legislation behind the effort management system is out of step with major international agreements and standards of best practice.

The core argument related to the *implementation of the effort system* is that it produces F levels that are not in accord with the requirements. As mentioned, the Commercial Fisheries Act has a major objective of regulating fishing effort so that the annual catch of each of the three species does not exceed 33 per cent of the stocks in numbers, corresponding to an average annual F = 0.45 (Jákupsstovu et al. [2007]).

In Figure 2 the scientifically estimated F-levels for 1987 to 2011 for the three main demersal species from the International Council for the Exploration of the Sea (ICES) are presented.

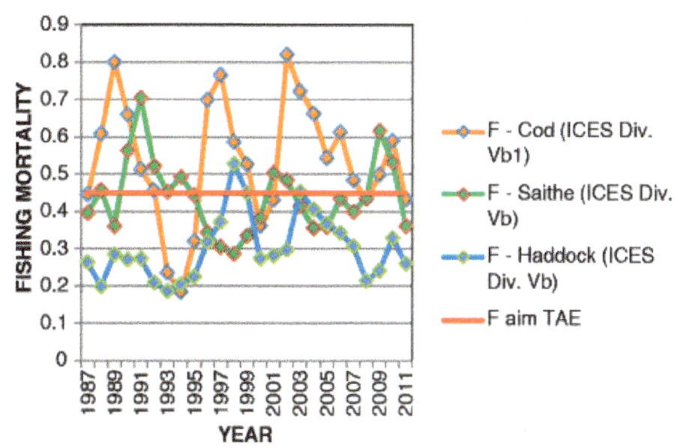

Figure 2: Variation in fishing mortality (F) for cod, saithe and haddock (Data from ICES ACOM[2012a]).

The F levels of all the stocks have varied substantially both before and after the implementation of the effort management system in 1996 (Figure 2). Although allowing for occasional overshoots is implicit in the present system, the F-levels for cod, in particular, have been high compared to the 0.45 target.

An analysis of means (AOM) plot (Figure 3) shows which means of the F values for the individual stocks are significantly different than the `grand mean' of the F values for all the samples (i.e. all three stocks together). The grand mean for all three gadoids was 0.43 and cod (mean 0.54) and haddock (mean 0.30) had significantly different means than the grand mean, whereas saithe (mean 0.44) was not significantly different. Given that the Faroese effort management aims at an average (mean) F = 0.45, it is notable that there is no statistical difference between 0.45 and the grand mean (0.43) of F for the three gadoids in the period 1987 to 2011. This also applies for the period following 1996 when the effort management system was operational. Thus, the mean F for the `basket' of gadoids (0.43) essentially has been in accord with the 0.45 aim of the effort management system as the 0.45 value falls within the upper and lower 95 per cent confidence limits of the plot. Of the `basket' of gadoids, saithe falls within the 95 per cent confidence limits and haddock falls below the 95 per cent confidence limits thereby fulfilling the desire of the effort management system that F ≤ 0.45. As cod falls above the 95 per cent confidence limits, its F has significantly exceeded the management aim.

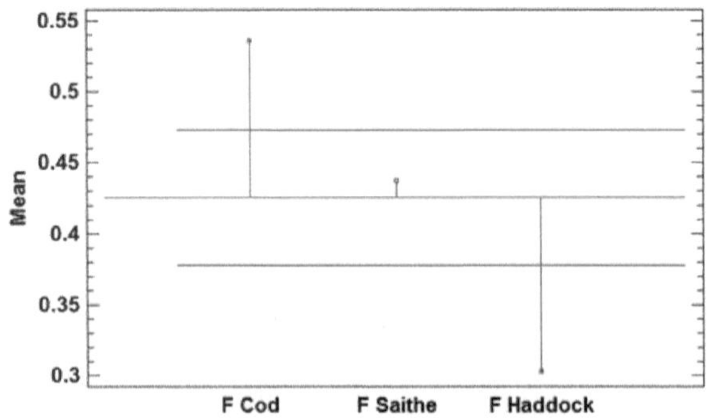

Figure 3: Analysis of means plot for fishing mortality (F) levels of cod, saithe and haddock in the Faroese mixed fishery from 1987 to 2011. The central line shows the grand mean (0.43) for all three stocks, and upper and lower decision (95 per cent confidence) limits are shown at 0.47 and 0.38 respectively.

The F levels in Figure 2 support Jákupsstovu et al.'s ([2007]) conclusion that the objective of constraining the exploitation of the major demersal stocks by the effort management system, so that F ≤ 0.45 on *each* of the three component gadoids, has not been achieved. Thus, Faroese vessels have been fishing cod more intensively than the legislation stipulates.

The question of whether *the legislation behind the effort management system is out of step with major international agreements and standards of best practice*, the second strain of criticism, relates to the absence of operational Faroese FMPs designed to ensure that the catches (yields) from these stocks are sustainable. Influential international protocols and agreements setting such standards and relating to the requirement for FMPs, aiming to produce maximum sustainable yield (MSY) from fish stocks have emanated from the United Nations (UN) system. These instruments are primarily:

- The 1995 Code of Conduct on Responsible Fisheries (FAO [1995]), especially Article 7.3.3 "*Long-term management objectives should be translated into management actions, formulated as a fishery management plan or other management framework*";
- The UN [1995] Fish Stocks Agreement (UN [1995]), especially Article 5(b) emphasising that measures should be "*...designed to maintain or restore stocks at levels capable of producing maximum sustainable yield,...*", and Article 6 specifying that "*a stock should be kept at a sustainable level by keeping it above a minimum biomass benchmark and by keeping the fishing mortality below a maximum fishing rate benchmark*"; and
- The 2002 Johannesburg Declaration of the World Summit on Sustainable Development (FAO [2003]), especially Article IV.31(a) of the Implementation Plan calls for actions to "*maintain or restore stocks to levels that can produce the maximum sustainable yield....*".

The reiteration of the MSY approach from the Johannesburg Summit represented an extension to, rather than an abrupt change in, most national fishery policies (Lassen et al. [2014]).

These instruments call *inter alia* for stock sustainability, in the form of a MSY-based management and FMPs, consisting of stock and fishing intensity related reference points (benchmarks) and HCRs. In this regard, effective fishery management requires an understanding of how the fishery is performing relative to stock-related reference points; the most commonly used reference points relate to the stock size (biomass, B) and the levels of F that will result in changes in stock sizes, given various relationships between the stock, recruitment, natural mortality, and growth (Beddington et al. [2007]).

The FMP's core contains the policy-related goal for fish stock management, expressed in terms of fishing mortality/effort and targeted stock size. The FMP presents an indication of the preferred balance between stock-related conservation concerns and yield (Lassen et al. [2014]). Such FMPs encompass precautionary approach (PA) standards, related to the above UN policy instruments, in the form of `target' and `limit' reference points — which are designed to account for risk/uncertainty and for keeping fishing within sustainable limits — and recovery plans for stocks depleted beyond specified levels. The Guidelines in Annex 2 of the 1995 UN Fish Stocks Agreement outline *inter alia* that: a) a limit reference point (LRP) indicates a state of a fish stock/fishery that is considered undesirable, and a target reference point (TRP) indicates a state that is considered desirable as a management objective; b) a fish stock/fishery is anticipated to approach or fluctuate around a TRP, and should have a very low statistical probability of exceeding a LRP; c) TRPs are set to maximize yields (MSY) as a management objective, and should not be exceeded `on average'; d) It is anticipated that decision rules, such as HCRs, are established so that such reference points (i.e. LRPs, TRPs) shall be used to trigger pre-agreed conservation and management action; and e) recovery measures should be established for restoring stocks depleted beyond specified levels.

The fundamental reference points for fisheries referred to in the above-mentioned international instruments are the biomass (B_{MSY}) that can produce MSY resulting from fishing at F_{MSY} over a long time. It is clear that allowable catches can deviate from MSY due to environmental and economic factors, interdependence of stocks, *etc.*, but catch rates above MSY (i.e. > F_{MSY}) are likely to result in declines below B_{MSY} and lower catches in the long term (Froese et al. [2010]). Thus, HCRs should adaptively and periodically (i.e. autonomously) adjust fishing effort (E), and thereby F, to sustainable levels reflecting the changing stock biomass. Having biologically-based MSY as a policy objective has an intuitive appeal because MSY implies efficiency and maximized revenues, and associated socio-economic benefits (Larkin et al. [2011]). FMPs not only promote stock sustainability, but also form a key means for engaging diverse stakeholder participation in the knowledge production and dialogue process, which underpins FMP formulation (Hartley and Robertson [2006]; Mackinson et al. [2011]). In the USA, FMP development involves a standard federally-mandated public participation process involving scoping, public comment periods, hearings, etc. (Hartley and Robertson [2006]). In the USA and the EU, regional fishery management councils and regional advisory councils, respectively, provide forums for stakeholder consultations and agreements on FMPs (Hartley and Robertson [2006]; Hegland and Wilson [2009]). Furthermore, the adoption of MSY-based FMPs, which are consistent with the PA, constrain the fishery to pre-agreed parameters, which anticipate that catch limits do not exceed the level specified in formal scientific advice. With catch limits `locked in' empirically, there are sparse prospects to embark on politically-driven `decision-overfishing' (*sensu* Aps and Lassen [2010]) whereby fishing opportunities are set systematically higher than the scientific advice based on sustainable exploitation. Because FMPs should be robust and foster stock stability and yield, they are also called multiannual or long-term management plans (LTMPs).

ICES is the intergovernmental marine science organisation, which provides fishery advice in the context of international agreements and addressing the policy and legal needs of ICES 20 member countries and the regional fishery management organizations (RFMOs), which they have established, around the North Atlantic and adjacent regional seas (Rozwadowski [2002]; Lassen et al. [2014]). Due to the influence of the UN fishery instruments, in 2009 ICES developed the MSY and PA framework for giving its future advice connected with LTMPs and HCRs (Lassen et al. [2014]). The LTMPs, and their HCRs, result from political decisions at the national political and RFMO levels, and ICES currently provides advice according to such plans where they are found to be consistent with PA standards (Lassen et al. [2014]). The advice has international legitimacy as it has been agreed in a committee of scientists which includes scientists appointed by all governments of the ICES member countries. It is notable that whereas Denmark, as an EU member state, has signed up to MSY-based management, the Faroe Islands have not. This reflects the latter's fishery autonomy from Denmark and it not being within the EU. However, the Faroe Islands take part in ICES activities, including the ICES advisory process, and may request scientific advice from ICES based on an agreement between Denmark, the FHG and ICES (ICES [1989]).

ICES has estimated F_{MSY} for Faroese cod (ICES Subdivision Vb1, the Faroe Plateau), haddock and saithe (in Division Vb comprising both the Faroe Plateau and the Faroe Bank) at 0.32, 0.25, and 0.28, respectively (ICES ACOM [2012a]). Thus, the F = 0.45 target for these stocks, as set out in the Faroese Commercial Fisheries Act, is much too high to fulfil the above-mentioned international standards. For many years, ICES has advised a number of different actions with regards to these Faroese stocks, all with the aim of substantially reducing the F-levels.

The concerns of ICES and FAMRI regarding the excessive F exerted by Faroese vessels were relatively easy to ignore for the half dozen years of the effort management system while trends for landings as well as the spawning stock biomass of the three stocks were generally increasing. However, the trend has reversed over the last decade (ICES ACOM [2012a]). In particular the landings of haddock have plummeted, together with its biomass level falling below the LRP and recruitment being close to 40-year minimums, such that ICES has recommended instigating a recovery plan in 2013 with F = 0.15. The cod's current status is not good either. The biomass of cod on the Faroe Bank is so low that fishing has essentially been stopped since 2009. The biomass of the Faroe Plateau cod has been below or near the LRP since 2006 and recruitment is close to a 40-year minimum, such that ICES has recommended instigating a recovery plan in 2013 with F = 0.20. Currently saithe is the only stock whose biomass is consistent with the PA as its spawning stock biomass is above the precautionary reference point (B_{pa}), but ICES considers nonetheless that it is harvested unsustainably ($F > F_{MSY}$).

THE CONTROVERSIES

The deteriorating situation of the Faroese demersal stocks has, over the last years, fuelled a debate on the reasons for this and even more importantly what to do to rectify it. At its core are various disagreements on the mechanisms of the effort management system.

As noted earlier, the two opposing camps are — in the absence of ENGOs and in the presence of a politicised, adversarial decision-making system - in the Faroese context often the scientists and the catching sector. The role of scientists and the relationship and interface between science and management in fisheries is a topic that has attracted considerable scholarly attention lately, not least regarding studies of ICES and the CFP (e.g. Schwach et al. [2007] and Wilson [2009]). Consequently, in addition to providing an insight into various controversies specific to the Faroe Islands, the following is also a study of the relationship between scientific rationalities and other rationalities.

The disagreement on the need for continuous adjustment of the number of fishing days

As described earlier, the number of fishing days for the coming fishing year is decided annually at the political level, in principle based on input from the scientists and the fisheries industry, mainly the catching sector in the latter case. Interestingly, the two advisory parties strongly disagree on the basic point of this exercise.

The catching sector's dominating perspective is that reductions in fishing days should, as a matter of principle, not be used as a response to fluctuations in fish stocks. Rather, cutting days should only happen in response to increased efficiency of the fleet (i.e. effort `creep'). The sector's argument is that the Faroese fleet has traditionally always shifted its effort towards where the greatest fishing opportunities were. So if there were few cod, vessels targeted other, more abundant species, which is perfectly possible because of the absence of output quotas.

The catching sector argues that the `automatic' reallocation of effort towards the most abundant species, together with the large areas that are either completely closed for commercial fishing or closed for certain types of fishing, safeguards the stocks being driven to depletion: "*We believe that we in this system have regulated so that we will never catch so large a share of the stocks that they will collapse*" (Quote 1, fisheries industry, translated).

Besides protecting particular areas and the fish within them, the catching sector argues that the closed areas reduce the effective fishing pressure that can be exerted on a fishing day (e.g. by preventing access to the best fishing spots) to an extent that has to be accounted for in the scientific advice, something which the sector argues is not the case:

The recommendation from FAMRI builds on intervening directly in the number of fishing days, while what we are saying is that when you close a giant area for longliners [in this case inside 3 nm] *then this will—all things being equal—have an effect on effort. Because you might very well still have a 100 days but you cannot go to the best spots to fish. Clearly, this must have an effect.* (Quote 2, fisheries industry, translated)

Thus, the current low level of cod is, for significant parts of the fisheries industry, explained not by reference to overfishing but by changes in the natural environment. The catching sector finds support in the fact that massive cuts in fishing days have not resulted in improvements for the cod stock.

Similarly, other fisheries industry respondents emphasise that there has been more or less consistent overfishing of the saithe stock when measured against the current PA and MSY reference points from ICES without this having led to a collapse of the stock; rather the industry representatives emphasize that the stock continues to be resilient and healthy.

The above, combined with a much referred to failure by Faroese biologists to predict a swift recovery of the cod stock in the mid-1990s, has created a situation where the catching sector generally considers that the system is already essentially sustainable. Or at least that shifting to a system that is more directly based on stock assessments would not be considerably more biologically sustainable.

As already mentioned, the catching sector does, however, generally accept that it is in principle necessary to continuously cut capacity to make up for effort `creep', traditionally estimated to be about two to four per cent per year (Pauly and Palomares [2010]). Effort regulation incentivises fishers to increase their catchability because they are limited by the time they are allowed to fish (Nielsen et al. [2006]). So called `technological creep' coupled with improvement of knowledge regarding best fishing practice to maximise potential catches over time, increases the fishing efficiency of fishing vessels, and thereby increased fishing capacity regarding catch levels per fishing day (Jákupsstovu et al. [2007]; Nielsen et al. [2012]). However, the catching sector argues from the assumption that the number of fishing days allocated from the outset of the system was appropriate and sustainable:

We clearly accept that if there has been an increase in efficiency, there has to be a reduction of days. That is what the system should be doing. From the perspective of the biologists, you look at how the fish stock is doing; the traditional ICES approach. Then you say: `Well, cod is not doing so well, therefore we have to cut fishing days.' We are sort of not on the same page. That is the problem. (Quote 3, fisheries industry, translated)

Jákupsstovu et al. ([2007]) emphasised that there was no evidence that fleets switch target species according to the relative abundance of all three species, which was one of the underlying assumptions of the effort management system. ICES also questions the mechanism of shifting effort, and has emphasised that for the Faroese effort system to work properly in a mixed fisheries context, the *relative* effort directed towards each species has to remain nearly constant, to ensure that the catches of each species are adapted to the abundance of each stock (ICES ACOM [2008a], ICES ACOM [2008b]). ICES noted additionally that the recent decline in the cod stock was associated with a marked increase in fishing mortality on cod, indicating that the management system has not functioned properly in that respect.

The scientific community considers that the failure to set up a credible system for monitoring changes in effective fishing effort is a major flaw (ICES ACOM [2012a]). Furthermore, there are big differences regarding the potential and actual efficiency gains across the different catch segments. This makes cutting fishing days across the board evermore illegitimate for certain parts of the fleet — in particular among traditional, coastal vessels, which do not see themselves as able to increase effort:

The problem is that whenever you have to cut days, you have the lawnmower-model and that is difficult to use. My personal opinion is that as long as you do not have another model to take days away, then it will be complicated to cut them. (Quote 4, fisheries administration, translated)

In contrast to the catching sector, the ICES-related scientific community believes that the number of allocated fishing days (effort) must be linked to how stocks are fluctuating via the proposed FMP, which reflects international best practice. The basis for the linkage is to `hard-wire' into the FMP the empirical relationship between effort and F so as to ensure that the effort level would be in accord with the principle of MSY (i.e. keeping the actual $F \leq F_{MSY}$ and $B \geq B_{MSY}$). This reflects the scientific conviction that fishing pressure is a contributor to stock fluctuations, or at least that too high fishing pressure can have detrimental effects at low stock levels. This is, as mentioned above, not readily accepted by the catching sector, which highlights the closed areas in this regard and emphasises the uncertainty of nature and the complex interactions that the demersal stocks are perceived to be part of.

The disagreement on the sustainability of the initial level of fishing days agreed

When the effort management system was originally set up, one of the key issues was to decide on the initial number of fishing days. Since shortly after the start there has, however, been disagreement on whether the number first set was sustainable.

From the scientific community's perspective, one of the fundamental problems of the effort management system is that the number of days originally allocated was too high (Jákupsstovu et al. [2007]; ICES NWWG [2012]); too many days were distributed in order to get everybody onboard. However, in the following years, when the number of days should have been cut to better reflect the resource situation, there was no will for this. This created a conflict between FAMRI, which considered that the industry was violating a joint agreement, and the catching sector, which accused FAMRI for turning its back on the system it had originally partly designed.

From the outset of the system—when it was set in place—there were far too many days in the system. And that gave a tremendous throwback; because the unanimous decision to take the system on led people to believe that the system was self-regulatory, which of course it is not. So when [after the first year] *we advised from the fisheries research side to reduce the days, so the number of days at sea agreed with what the law said, this was met with opposition and although the days have been reduced gradually over the years, they have not been reduced to the extent to which we thought was necessary.* (Quote 5, fisheries scientist, translated)

According to one industry respondent, the FAMRI scientists found themselves squeezed between loyalty to the agreement on the effort management system and loyalty to international colleagues in the ICES system. The latter were purportedly unable to understand/accept the Faroese effort system, which fundamentally differs from the TAC systems that ICES usually caters for. Based on our interviews with scientists, it appears that as ICES advice is basically geared towards TAC systems this has created a complicated situation for FAMRI scientists. They have to balance between a) the ICES system, which is not geared up to give targeted advice regarding the Faroese system, and b) domestic expectations which lack guidance on how to do this due to a dearth of clear domestic policy objectives and procedures.

The disagreement on how to interpret the presence of unused fishing days

Each year fishing days remain unutilised by Faroese vessels. When exploitation rates are clearly under 100 per cent, which has been more common than not in several fleet segments (Zableckis et al. [2009]; ICES NWWG [2012]), one can infer that insufficient fish are available to sustain economically justifiable levels of catch per unit effort. So much the scientists and the catching sector agree on.

However, the above disagreements on the need to link fishing days closer to the stock status, as well as on the sustainability of the initial level of fishing days, spill over into competing interpretations of what the presence of unused fishing days really means. To the scientists, the presence of unused fishing days demonstrates that the initial effort allocation was too high to constrain F, particularly for cod, and that too many available days remain in the system. So, fishing days should be removed to bring the effort level down to match the actual fishing opportunities. To the catching sector, the unused days are contrarily seen as indicating that the system is working: *"This is the core of the system. If there are few fish, then we use few days; if there are many fish, we use many days"* (Quote 6, fisheries industry).

The catch sector recognises that unutilized fishing days contribute to the system possibly appearing unsustainable if considering single years, where fishing pressure on one stock might surpass the F = 0.45 threshold, or any valid F level considered sustainable. However, in the long-term, the sector argues that the system is in equilibrium where stocks are not in a risk of collapse, at least not due to fishing pressure. From the sector's perspective, the excess capacity, rather than being the problem, is the feature that will allow the fleet to fish on the 'tops' whenever stocks are in good condition:

> *We are back to this issue of how you look at things. The biologists would say that when stocks decline you should reduce capacity; however, the fishermen's logic is: 'we need to have enough capacity to fish the stocks, when they are at their maximum, we cannot decrease capacity just to suddenly need it two years later; that is no good'.* (Quote 7, outside observer, translated)

In addition, the catching sector has little hope that the number of days will be increased again after having been cut. That situation, of course, might result in more stable stock and catch levels, but stability does not seem to be a crucial objective of the sector when it fears that stability will be at a considerably lower level than the present average.

A SELF-REGULATING SYSTEM?

The scientific community and the catching sector start from very different perspectives concerning the above disagreements. In current MSY-related fisheries management, self-regulation of fishing effort (E) occurs via the HCR which, as already noted, should adaptively and periodically (i.e. autonomously) adjust E, and thereby F, to the level necessary to achieve MSY as guided by scientific assessments informing on the status of the stock relative to established reference points. Thus, scientists see the Faroese effort management system as being unbalanced from the outset (i.e. excessive E and F due to too many fishing days) and furthermore unable to bring itself into balance due to the absence of dynamic self-regulation, i.e. the lack of a FMP containing a dynamically reactive HCR relating effort to stock fluctuations. However, the catching sector, views the system as being in balance from the start and also as able to deal, in a semi-automatic way, with imbalances that may arise. In our interviews, there were often references to this disagreement regarding the extent to which the effort management system can be considered partially or fully *self-regulating*. Meaning that it is an in-built feature that the system's design prevents an application of a level of fishing pressure, on a particular stock, high enough to drive it to depletion or conversely that allows a stock at a low biomass level, due to human or naturally induced impacts, to recover to desired levels (not to be confused with the ongoing discussions of governance models resting on self-management/self-regulation by the fisheries industry; for examples, see Townsend et al. [2008]).

Compared with traditional TAC management systems, the Faroese industry is proud of operating what they perceive as a system in which there is little need to interfere as most adaptations (except for effort 'creep') occur automatically. As one respondent explained: *"Even though you could say that if all fishing days were utilized, then the fishing pressure would be very high, it is not the case that all days are used. That is what you call a self-regulating system."* (Quote 8, fisheries administration, translated). Consequently, the catching sector generally argues that the presence of closed areas, which protect spawning areas and function as a stock refuge, in combination with economic incentives to shift effort towards the most abundant species in a system that is inherently in balance, ensures that stocks will not be driven to depletion.

To the scientists the notion of the self-regulating system has been used primarily as a fig leaf by the catching sector to avoid having to cut days sufficiently (c.f. Quote 5). An example of the weakness of the current notion of the fully self-regulating system is the fact that it is not merely the state of the stock that determines whether it is economically viable to continue fishing:

> *This is where they say that it is self-regulating, because you would not go out unless it is economically viable. But the problem is just that this is not just determined by the stock; it is also dependent on oil prices and other things, so it is possible to fish the stock down for the wrong reasons.* (Quote 9, outside observers, translated)

Consequently, although there are self-regulating mechanisms in the system, it is not clear how strong these are and this causes fundamental disagreement. Furthermore, the differences in interpretation of the sustainability of the original level of fishing days are important to have in mind when trying to understand the present situation. Finally, to understand the differences in perspectives it is also necessary to recall that the standard scientific models of ICES are generally not accepted across the board in the catching sector. The latter is generally skeptical towards the scientific community and specifically question whether the models are appropriate in the context of the Faroe Islands and the unique effort management system.

A feature in the current setup of the effort management system, which seems to have reinforced the differences described above, is that the final decision on fishing days is taken at the political level, based on separate input from two sides: the biologists and the catching sector. As earlier noted, the decision-making system provides little incentive for integration of views (c.f. Figure1). Thus, the two parties position themselves solidly in opposing camps. As one respondent put it: *"It sort of becomes a trade-off. If you give in too much you risk losing ground"* (Quote 10, outside observers, translated).

ADAPTING THE FAROESE EFFORT MANAGEMENT SYSTEM TO INTERNATIONAL BEST PRACTICE

As highlighted earlier, the Faroese effort management system could benefit from aligning itself with the previously mentioned UN policy instruments including adopting MSY-related management via LTMPs and HCRs in a PA approach (i.e. encompassing limit and precautionary reference points for fishing levels and stock biomass). However, the main issue regarding fisheries management in the Faroe Islands is not the effort management system itself, but rather its inability to adjust to scientific recommendations (emanating from ICES as the custodian of best practice) and to variability and trends in catchability (Baudron et al. [2010]). This in turn is linked to the conviction that the initial effort was set too high by the Faroese decision-makers, and that it was not decreased sufficiently thereafter.

It has been increasingly apparent in recent years that a major deficiency in the Faroese effort management system is the lack of a proper FMP, and especially the absence of related HCRs, for the demersal stocks (see also Anon. [2011a]). In response to this deficiency, as well as pressure to adopt MSY-related management by 2015 stemming from broad-based international accord from the 2002 Johannesburg Declaration (FAO [2003]), in October 2011 a group[g] appointed by the Faroese Minister of Fisheries and Natural Resources formulated a draft FMP, based on general MSY principles developed by ICES, for cod, haddock and saithe including HCRs (Anon. [2011b]; ICES NWWG [2012]). The outcome currently awaits discussion and necessary approval at the political level (ICES ACOM [2012b]).

The draft Faroese FMP (Anon. [2011b]) includes a proposal for a stepwise reduction of the fishing mortality of each of the three stocks to F_{MSY} in 2015 and a recovery plan if the SSB declines below a `triggering' level. Key MSY and PA parameters for the three stocks are mutually outlined by ICES (ICES ACOM [2012b]) and by the draft Faroese FMP (Anon. [2011b]). These plans in draft emerged as three, independent single species management plans, rather than a multispecies management plan. The multispecies management plan approach is necessitated because in mixed fisheries the single species objectives, concerning MSY and the PA, are difficult to meet simultaneously due to differences in dynamics and initial state of the various stocks. The approach in a mixed fishery is to manage on a `basket' approach (e.g. the suite of targeted gadoids) where it is acknowledged that not all target stocks will be able to achieve either their own or a `pooled' MSY target at the same time. In reality, the mixed fishery will be managed such that the various stocks which are in the `basket' oscillate within a band as close as possible to the MSY target with respect to the exerted fishing effort, while ensuring that the biomass of individual stocks do not fall below their particular precautionary or limit reference points. Effort levels must be monitored credibly and periodically adjusted to compensate for effort `creep' and achieve necessary optimization. A properly functioning management system should contribute towards autonomous regulation, i.e. adjusting effort and F to reflect changing stock status in MSY-related management, and ensuring that the market for trading fishing days works properly. For the Faroese mixed fishery, this would ideally entail hard-wiring the need for periodic increases or decreases in effort levels to the FMP, i.e. relating effort to fishing mortality and stock biomass as ascertained by stock assessments. The rights-based, tradable nature of the fishing days (i.e. an individual effort quota) provides for some autonomous adjustment, but the Faroese Council of Economic Advisers considers that restrictions on tradability limit its efficacy (Búskaparráðið [2010]). The existence of excess effort/capacity in the Faroese mixed fishery results in the economic revenue generated being lower than when fishing sustainably (i.e. $F \leq F_{MSY}$, $B \geq B_{MSY}$). It is notable that the Australian Northern Prawn Fishery, which is an effort-based mixed fishery for several prawn species, systematically monitors and regulates the effort needed to achieve maximum economic yield (i.e. net economic returns) by adjusting the gear units (headrope length) (Kompas et al. [2010]). Thus, effort in the Faroese context potentially may reflect not only fishing days but technical measures connected to fishing capacity such as hook numbers for long-liners and trawl openings – possibly modifying the extent of closed areas could enter the equation as well, reflecting the Faroese tradition for this.

The fishing effort levels set for the `basket' of stocks depends on how management wants to optimise the relative catch of the particular stocks. Agreeing on this, in the context of multispecies management plans, opens up for a complex debate not only on stock conservation but also on matters related to societal choice in a wider socio-economic context. Thus, there will be options and potential trade-offs in fishing opportunities as highlighted, for example, by recent ICES advice on mixed fisheries in the North Sea ICES ACOM [2012c]) underpinned by the Fleet Fishery Forecast framework (Ulrich et al. [2011]). Informed decision-making depends on detailed and disaggregated knowledge of fleet/métier fishing behaviour in time and space, a feature currently monitored well in real time by the Faroese authorities.

The Faroese FMP is currently in limbo in the political system and has not been embraced in the recent annual revision of the regulations. This is despite mounting pressure to reform the current system. It is likely that the Faroese reluctance to move forward with the plan(s) is due to the controversies which we have described.

CONCLUSIONS AND WAYS TO MOVE FORWARD

The Faroese effort management system does not currently live up to international best practice in fisheries management. In particular, the current system does not include a FMP, with associated HCRs, promoting autonomous effort management and thereby long-term stock sustainability in the form of MSY. Furthermore, the target F for cod, haddock and saithe enshrined in the legislation significantly exceeds that necessary to achieve MSY.

Whether or not the above is a contributory factor to the current overall state of the Faroese demersal stocks is a matter of dispute between the industry and scientists. But scientists argue that the associated overfishing constitutes part of the explanation. Nonetheless, it is clear that — irrespective of its shortcomings — the system has several very appealing elements. As examples, the effort management system has curtailed the discards problem which blights many mixed fisheries. In addition, the system enjoys considerable support from the industry, which is essential in any fisheries management system. Further, the system exhibits great ingenuity in using targeted effort allocations and extensive use of closed areas to both balance and steer catches, and cater for the needs of diverse fleet segments.

Thus, the exercise that the Faroese face, insofar that they wish to move towards updating their management system and reclaim their position at the forefront of fisheries management, is to adopt reform without jeopardising the positive elements of the system.

A major problem regarding actually moving forward with a Faroese FMP in line with current scientific knowledge is the enduring and substantive incongruity between the perspectives of the catching sector and the scientific community. Based on our analysis, we therefor suggest that a necessary first step should be to work on reconciling the two camps. The following represent our suggestions on connected remedial actions which could be taken.

Integration of the advice process

As described earlier, there is currently little formalised integration between scientist and industry viewpoints. This reinforces the existence of divergent perspectives. Consequently, creating a more integrated (or `interactive' (Schwach et al. [2007])) advisory process and forums to consult and do business in could diminish this gap and thereby lead to increased collaboration and trust between the primary actors. This may already be on the way due to generational change and industry-scientist research cooperation. For inspiration, examples of such integration in the EU include the presentation of ICES advice by visiting scientists in the CFP's stakeholder-led regional advisory councils, and these councils acting as the primary stakeholder forum for promoting agreement on FMPs, as well as the opening up of most ICES meetings to observers from for instance the industry.

ICES to commence work specifically on the Faroese effort management system

The Faroese authorities should officially request that ICES starts working on producing tailor-made advice for the Faroese management system. This would release the Faroese scientists from being stuck between `a rock and a hard place' as `translators' of ICES advice. This would in turn require full openness and transparency of all appropriate data necessary for monitoring and analyzing fishing effort. The context of the effort management system suggests that the Faroe Islands could become a valuable laboratory for advancing the management of mixed fisheries by using effort management and closed areas. For the Faroe Islands this concerns few stocks and there is a high degree of data availability on the stocks and the fleets.

Start to set up a system for monitoring capacity development

It is essential to develop a credible approach to measure and fully analyse fishing effort, including `effort creep', in order to measure how much fishing effort/mortality changes over time in terms of fishing days. This should be possible due to the close monitoring of Faroese vessels and their relatively modest numbers. Peer reviewed data and conclusions on the status and trends of fishing effort deployment are essential, not least for developing models to relate fishing effort to fishing mortality. The latter relationship should eventually be `hard wired' into the FMPs.

Improved understanding of the role and efficacy of closed areas

Faroese scientists, working via appropriate ICES expert groups, should – in cooperation with the industry - intensify efforts to document and quantify the usefulness of various closed areas for particular purposes. This lack of documentation concerns the catching sector, which otherwise generally supports using closed areas in combination with effort management.

Although the above actions would not in themselves make a reform of the core parts of the effort management system uncontroversial or easy, we believe that this would facilitate making the difficult decisions that lie ahead, not least in relation to the inevitable reduction of fishing pressure on the stocks that is required to achieve MSY-related best practices. The latter furthers both the long-term sustainability of the stocks and the fisheries industry. Taking action on the above, challenges scientists and the industry to collaborate in balancing conservation concerns and yield.

ENDNOTES

[a]In the following, the fisheries industry will generally be defined as consisting of the catching sector and the processing sector and not the aquaculture sector, although the latter is often included especially for statistical purposes.

[b]More specifically four scientists, seven industry representatives, three managers and three `outside observers'; the latter including persons who had been vocal in the debate on the effort management system but did not fit the other categories.

[c]Most of the Faroe Islands interviews were conducted in Danish/Norwegian/Swedish. Whenever necessary (and indicated), the original quote has been translated to English by the authors.

[d]Although managed through the effort management system,

redfish are not usually caught in a mixed fishery with the remaining species.

[e]These conversion possibilities provide fishers with flexibility in selecting gear type to use/how to deploy their effort.

[f]We define fishery governance as the sum of the legal, social, economic and political arrangements used to manage the fishery. The establishment of institutions, legislation/policies, management plans and processes through which management may be realized at appropriate levels (e.g. national and local) is fundamental to effective fishery governance. Governance also concerns how power, knowledge and decision-making are shared amongst the key stakeholders.

[g]The group consisted of scientists from FAMRI, and one representative each from the MFNR, the University of the Faroe Islands, and the catching sector (trawlers).

ACKNOWLEDGEMENTS

The research leading to these results received funding from the European Union Seventh Framework Programme (FP7/2007-2013) under grant agreement No. 289257. The funded project, MYFISH, deals with the issue of how to integrate the concept of MSY into the EU's CFP (www.myfishproject.eu), and the research on the Faroe Islands was conducted as part of an effort to collect experiences from outside the EU. We are very grateful to our contacts on the Faroe Islands who took time to talk to us and inform us very openly about their fisheries system. In addition, three anonymous reviewers provided constructive proposals for improving the manuscript. We dedicate this paper to Douglas C. Kongshøj Wilson (IFM), our deceased friend and collaborator. He played a key role in proposing and designing the studies of best practices in fishery governance from which this work derives.

REFERENCES

1. Anon. 2011a. ICES, rådgivning og beslutningsprocesserne - med vægt på Færøerne som case [ICES, advisory and decision-making processes – with weight on the Faroe Islands as a case]. Report No. ANP 2010:780. Nordic Council of Ministers. . webpage accessed 8 June 2012., [http://www.norden.org/da/publikationer/publikationer/2010-780]
2. *Umsitingarætlan fyri toska-, hýsu- og upsastovnarnar í føroyskum sjógvi. [Management plan cod, haddock and saithe in Faroese EFZ]*. 2011b.
3. Aps R, Lassen H: Recovery of depleted Baltic Sea fish stocks: a review. *ICES Journal of Marine Science* 2010, 67: 1856–1860. 10.1093/icesjms/fsq118CrossRef
4. Baudron A, Ulrich C, Nielsen JR, Boje J: Comparative evaluation of a mixed-fisheries effort-management system based on the Faroe Islands example. *ICES Journal of Marine Science* 2010, 67: 1036–1050. 10.1093/icesjms/fsp284CrossRef
5. Beddington JR, Agnew DJ, Clark CW: Current problems in the management of marine fisheries. *Science* 2007, 316: 1713–1716. 10.1126/science.1137362CrossRef
6. *Economic Report, Spring 2010*. The Council of Economic Advisers, Thorshavn; 2010.
7. Christensen AS, Aranda M, McCay B, McLay HA, Rova C, da Silva AL, Wolff F: Understanding social robustness in selected European fisheries management systems. In*Comparative Evaluations in Innovative Fisheries Management - Global Experiences and European Prospects*. Edited by: Hauge KH, Wilson DC. Springer, Dordrecht; 2009:163–189. 10.1007/978-90-481-2663-7_8CrossRef
8. Chuenpagdee R, Alder J: A sustainability ranking of North Atlantic fisheries. In *Fisheries Impacts on North Atlantic Ecosystems: Evaluation and Policy Exploration* Edited by: Pitcher T, Sumaila UR, Pauly D. 2001. Fisheries Centre Research Reports 9(5):49–54 Fisheries Centre Research Reports 9(5):49-54
9. Daan N: TAC management in North Sea flatfish fisheries. *Journal of Sea Research* 1997, 37: 321–341. 10.1016/S1385-1101(97)00026-9CrossRef
10. *MSC Fishery Full Assessment Report, Faroe Islands Saithe Fishery: Clients P/F JFK and P/F Faroe Origin. Public comment Draft Report. Report No. 2012–017, Revision No. 4*. Det Norske Veritas Certification AS, Høvik, Norway; 2013.
11. EU. 2013. Regulation (EU) No 1380/2013 of the European Parliament and the Council of 11 December 2013 on the Common Fisheries Policy, amending Council Regulations (EC) No 1954/2003 and (EC) No 1224/2009 and repealing Council Regulations (EC) No 2371/2002 and (EC) No 639/2004 and Council Decision 2004/585/EC.
12. *Code of Conduct for Responsible Fisheries*. Food and Agriculture Organization, Rome; 1995.
13. *World Summit on Sustainable Development 2002 and its implications for fisheries. Committee on Fisheries, COFI/2003/Inf. 14*. 2003.
14. *Effort System*. 2014.
15. Froese R, Branch TA, Proelss A, Quaas M, Sainsbury K, Zimmermann C: Generic harvest control rules for European fisheries. *Fish and Fisheries* 2010, 12(3):340–351. 10.1111/j.1467-2979.2010.00387.xCrossRef
16. Gaard E, Hansen B, Reinert J: Ecological features and recent trends in the physical environment, plankton, fish stocks, and sea birds in the Faroe Shelf ecosystem. In *Large Marine Ecosystems of the North Atlantic*. Edited by: Sherman K, Skjoldal H-R. Elsevier, Amsterdam; 2002:245–265.
17. Gezelius SS: Management responses to the problem of incidental catch in fishing: A comparative analysis of the EU, Norway, and the Faroe Islands. *Marine Policy* 2008, 32: 360–368. 10.1016/j.marpol.2007.08.008CrossRef
18. Gezelius SS: From catch quotas to effort regulation: Politics and implementation in the Faroes fisheries. In *Making fisheries management work*. Edited by: Gezelius SS, Raakjær J. Springer, New York; 2008b:99–129. 10.1007/978-1-4020-8628-1_4CrossRef
19. Hartley TW, Robertson RA: Stakeholder engagement, cooperative fisheries research and democratic science: The case of the Northeast Consortium. *Human Ecology Review* 2006, 13(2):161–171.
20. Hegland TJ, Wilson DC: Participatory Modelling in EU

Fisheries Management: Western Horse Mackerel and the Pelagic RAC. *Maritime Studies* 2009, 8(1):75–96.

21. Holden M: *The Common Fisheries Policy. Origin, Evaluation and Future*. The Buckland Foundation, London; 1994.
22. *Memorandum of Understanding Between the Government of Denmark and the Homes Governments of the Faroe Islands and of Greenland on the One Hand and the International Council for the Exploration of the Sea on the Other Hand*. International Council for the Exploration of the Sea, Copenhagen; 1989.
23. *Report of the ICES Advisory Committee 2008, Book 4. Section 4: The Faroe Plateau Ecosystem*. International Council for the Exploration of the Sea, Copenhagen; 2008.
24. *Report of the ICES Advisory Committee 2008, Book 4. Section 4.3 Assessments and advice*. International Council for the Exploration of the Sea, Copenhagen; 2008.
25. *Report of the ICES Advisory Committee 2012, Book 4. ICES Advice for 2013, Faroe Plateau Ecosystem, Advice for 2013 Sections 4.4.1–4.4.4*. International Council for the Exploration of the Sea, Copenhagen; 2012.
26. *Report of the ICES Advisory Committee 2012, Book 1. Section 1.2 General context of ICES advice*. International Council for the Exploration of the Sea, Copenhagen; 2012.
27. *Report of the ICES Advisory Committee 2012. Section 6.4.25 Ecoregion North Sea, stock mixed-fisheries advice*. International Council for the Exploration of the Sea, Copenhagen; 2012.
28. *Report of the North-Western Working Group (NWWG), 26 April - 3 May 2012*. ICES Headquarters, Copenhagen; 2012.
29. Jákupsstovu SH, Cruz R, Maguire J-J, Reinert J: Effort regulation of the demersal fisheries at the Faroe Islands: a 10-year appraisal. *ICES Journal of Marine Science* 2007, 64: 730–737. 10.1093/icesjms/fsm057CrossRef
30. Johnsen JP, Eliasen S: Solving complex fisheries management problems: What the EU can learn from the Nordic experiences of reduction of discards. *Marine Policy* 2011, 35: 130–139. 10.1016/j.marpol.2010.08.011CrossRef
31. Kompas T, Dichmont CM, Punt AE, Deng A, Che TN, Bishop J, Gooday P, Ye Y, Zhou S: Maximizing profits and conserving stocks in the Australian Northern Prawn Fishery.*Australian Journal of Agriculture and Resource Economics* 2010, 54: 281–299. 10.1111/j.1467-8489.2010.00493.xCrossRef
32. Kraak SBM, Buisman FC, Dickey-Collas M, Poos JJ, Pastoors MA, Smit JGP, van Oostenbrugge JAE, Daan N: The effect of management choices on the sustainability and economic performance of a mixed-fishery: a simulation study. *ICES Journal of Marine Science* 2008, 65: 697–712. 10.1093/icesjms/fsn045CrossRef
33. Larkin SL, Alvarez S, Sylvia G, Harte M: *Practical Considerations in Using Bioeconomic Modelling for Rebuilding Fisheries*. 2011.CrossRef
34. Lassen, H, C Kelly, and M Sissenwine. 2014. ICES advisory framework 1977–2012: from F_{max} to precautionary approach and beyond. ICES Journal of Marine Science 71:166–172.CrossRef
35. *Løgtingslóg nr. 28 frá 10. mars 1994 um vinnuligan fiskiskap, sum seinast broytt við Løgtingslóg nr. 89 frá 17 august 2012*. 1994.
36. Løkkegaard J, Andersen J, Boje J, Frost H, Hovgaard H: *Report on the Faroese Fisheries regulation—the Faroe Model*. Report of the Institute of Food and Resource Economics (FOI), University of Copenhagen; 2007.
37. Mackinson S, Wilson DC, Galiay P, Deas B: Engaging stakeholders in fisheries and marine research. *Marine Policy* 2011, 35: 18–24. 10.1016/j.marpol.2010.07.003CrossRef
38. Maguire J-J: Fisheries science and management in the North Atlantic. In *Fisheries impacts on North Atlantic Ecosystems: evaluation and policy exploration* Edited by: Pitcher T, Sumaila UR, Pauly D. 2001. Vol. 9(5), University of British Columbia: 36–48 Vol. 9(5), University of British Columbia: 36-48
39. *Faroe Islands fisheries and aquaculture: Responsible management for a sustainable future*. 2008.
40. *Fishing*. 2013a.
41. *The fishing days system*. 2013b.
42. Nielsen JR, Sparre PJ, Hovgaard H, Frost H, Tserpes G: Effort and Capacity Based Fisheries Management. In *The Knowledge Base for Fisheries Management*. Edited by: Motos L, Wilson D. Elsevier, Amsterdam; 2006:163–216. Vol. 36 Vol. 36 10.1016/S0167-9309(06)80010-4CrossRef
43. Nielsen M, Flaaten O, Waldo S: Management of and Economic Returns from Selected 65 Fisheries in the Nordic Countries. *Marine Resource Economics* 2012, 27: 65–88. 10.5950/0738-1360-27.1.65CrossRef
44. Ott ER, Schilling EG, Neubauer DV: *Process Quality Control: Troubleshooting and Interpretation of Data*. American Society for Quality, Quality Press, Milwaukee, Wisconsin; 2005.
45. Ounanian K, Hegland TJ: The regional advisory councils' current capacities and unforeseen benefits. *Maritime Studies* 2012, 10.
46. Pauly D, Palomares MLD: *An empirical equation to predict annual increases in fishing efficiency. Fisheries Centre Working Paper #2010–07*. University of British Columbia, Vancouver; 2010.
47. Pielke RA Jr: *The Honest Broker: Making Sense of Science in Policy and Politics*. Cambridge University Press, Cambridge; 2007.CrossRef
48. Rozwadowski HM: *The Sea Knows No Boundaries: A Century of Marine Science Under ICES*. International Council for the Exploration of the Sea and University of Washington Press, Seattle and London; 2002.
49. Schwach V, Bailly D, Christensen A-S, Delaney AE, Degnbol P, van Densen WLT, Holm P, McLay HA, Nolde Nielsen K, Pastoors MA, Reeves SA, Wilson DC: Policy and knowledge in fisheries management: a policy

brief. *ICES Journal of Marine Science* 2007, 64: 798–803. 10.1093/icesjms/fsm020CrossRef

50. *Faroe Islands in figures 2012*. 2012.
51. Shepherd JG: Fishing effort control: could it work under the common fisheries policy? Viewpoint in. *Fisheries Research* 2003, 63: 149–153. 10.1016/S0165-7836(03)00125-5CrossRef
52. *Lov om Færøernes Hjemmestyre*. 1948.
53. *Case studies in fisheries self-governance. FAO Fisheries Technical Paper. No. 504*. FAO, Rome; 2008.
54. Ulrich C, Reeves SA, Vermard Y, Holmes SJ, Vanhee W: Reconciling single-species TACs in the North Sea demersal fisheries using the Fcube mixed-fisheries advice framework. *ICES Journal of Marine Science* 2011, 68: 1535–1547. 10.1093/icesjms/fsr060CrossRef
55. UN. 1995. Agreement for the Implementation of the Provisions of the United Nations Convention on the Law of the Sea of 10 December 1982 relating to the Conservation and Management of Straddling Fish Stocks and Highly Migratory Fish Stocks [United Nations Fish Stocks Agreement].
56. Wilson DC: *The Paradoxes of Transparency – Science and the Ecosystem Approach to Fisheries Management in Europe*. Amsterdam University Press, Amsterdam; 2009.CrossRef
57. Zableckis S, Raid T, Arneson R, Murillas A, Eliasen S, Sverdrup-Jensen S, Kuzebski E:*Costs of management in selected fisheries*. 2009.

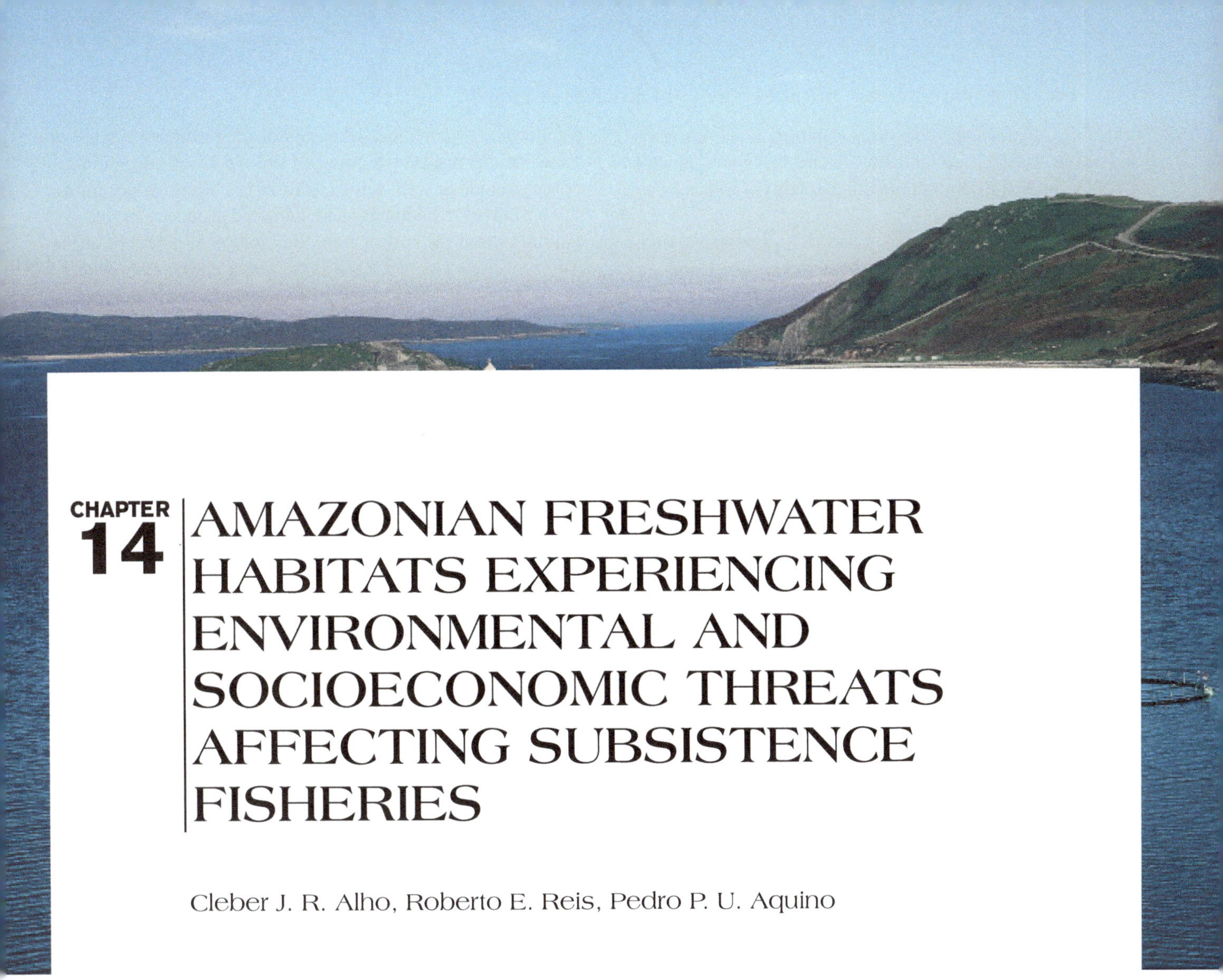

CHAPTER 14

AMAZONIAN FRESHWATER HABITATS EXPERIENCING ENVIRONMENTAL AND SOCIOECONOMIC THREATS AFFECTING SUBSISTENCE FISHERIES

Cleber J. R. Alho, Roberto E. Reis, Pedro P. U. Aquino

Keywords: Amazon basin, Biodiversity, Fishery resources, Freshwater habitats, Environmental and socioeconomic threats, Seasonal flooding

INTRODUCTION

Human uses of natural resources have affected the freshwater ecosystems and the surrounding terrestrial environment during the recent decades of land use along all the large rivers on the planet. There has been worldwide concern over aquatic ecosystems, with freshwater biodiversity being the overriding conservation priority. River fisheries experts, meeting at the Symposium on the Management of Large Rivers for Fisheries, held in *Phnom Penh*, *Cambodia* in 2003, expressed alarm that freshwater fish stocks have declined in many of the world's largest rivers (Dudgeon et al. 2006). Further evidence of this concern was the initiative of the United Nations General Assembly, which proclaimed the period from 2005 to 2015 as the International Decade for Action—"Water for Life." The resolution calls for a greater focus on water issues to achieve freshwater conservation. The document "Living Planet Report 2014—Species and spaces, people and places" (http://wwf.panda.org/about_our_earth/all_publications/living_planet_report/) organized by WWF International, Zoological Society of London, Global Footprint Network, and Water Footprint Network, points out that although human beings are a product of the natural world, we have become the dominant force that shapes ecological and biophysical systems. In doing so, we are not only threatening our health, prosperity, and well-being, but our very future. Human well-being depends on natural resources such as water and fish, as well as ecosystem services.

Furthermore, biodiversity contributes to safeguarding subsistence use for traditional human communities (Millennium Ecosystem Assessment 2005). The document of the Intergovernmental Panel on Climate Change (IPCC) held in *Yokohama*, *Japan*, March 25–29, 2014, stated that a large fraction of both terrestrial and freshwater species face increased extinction risk under projected climate change during and beyond the twenty-first century, especially as climate change interacts with other stressors, such as habitat modification, over-exploitation, pollution, and invasive species (IPCC 2014).

From a biodiversity perspective, the Amazon basin is unequaled, with its aquatic habitats that are home to the world's richest assemblages of freshwater flora and fauna, including reptiles (*Caiman crocodilus*, *Melanosuchus niger* among other

Citation: Cleber J. R. Alho, Roberto E. Reis, Pedro P. U. Aquino, Amazonian freshwater habitats experiencing environmental and socioeconomic threats affecting subsistence fisheries, DOI: 10.1007/s13280-014-0610-z.

crocodiles, *Podocnemis expansa*, *P.unifilis*, *P. sextuberculata,* and other turtles); amphibians; aquatic habitat-dependent birds, with good indicators of habitat quality (*Harpia harpyja*, *Morphnus guianensis*, *Neomorphussquamiger*, *Penelope pileata*, *Simoxenops ucayalae*, *Synallaxis cherriei*); mammals (otters such as *Pteronura brasiliensis* and *Lontra longicaudis*; river dolphins such as *Inia geoffrensis and Sotalia fluviatilis*, and manatee *Trichechus inunguis*); and more than 2200 fish species (Albert and Reis 2011; Alho 2011).

Other mammal species life history of which is connected with water are marsupials such as*Chironectes minimus*, *Micoureus demerarae*, *Marmosops noctivagus*, and *Marmosa murina.*Wild rodents strongly linked to habitats close to water are the semiaquatic *Nectomys squamipes*and the "toró" arboreal bamboo rat *Dactylomys dactylinus* usually seen vocalizing on branches over the water of rivers and seasonal lakes. Large mammals also exhibit dependence on aquatic habitats, like the semiaquatic capybara *Hydrochoerus hydrochaeris* and the tapir *Tapirus terrestris.*

The number of Amazonian aquatic species is a clear underestimation, because a significant portion of biodiversity is yet to be discovered and described, particularly considering amphibians and fish. The magnitude of the region is unique: the Amazon basin covers an area of approximately 7 000 000 km^2, of which about 58 % (4 100 000 km^2) is located in Brazil. These species play a fundamental role in the aquatic ecosystems because of their high diversity, and the ecological and evolutionary patterns they exhibit. For example, the Amazonian freshwater fish is very old and has distinct historical origins. Publications like "Check List of the Freshwater Fishes of South and Central America" (Reis et al. 2003), and Böhlke et al. (1978) point out the steeply ascending curve of new species being described, suggesting that the Amazonian number may exceed 3200 species.

The Amazon freshwater ecosystems are packed with the highest diversities of fish to be found anywhere on the planet (Lévêque et al. 2008; Albert and Reis 2011). The traditional socioeconomic activities such as fishing and extraction of other live natural resources have been affected by the latest increase in human migration, caused by new possibilities of access to land or driven by new infrastructure projects, livestock, agriculture, mining, commercial fishing, or other motivations. More than 24 million people live in Brazilian Amazonia (Santos et al. 2014).

In comparison with the *Mekong* River in Southeast Asia, one of the largest inland rivers and fisheries in the world, the total diversity of fish species—estimated at 1000 species—is significantly lower than that of the Amazon—estimated at more than 2200 (Albert and Reis2011). The *Mekong* basin has been subjected to hydrological alterations, overfishing, and environmental degradation, and its fishery has been declining considerably over the recent years (Baran and Myschowoda 2008; Dugan et al. 2010).

Taking water discharge as a parameter, the Amazon River discharges 219 000 m^3/s, while another of the world's large rivers, the *Congo* River in Africa, discharges 41 800 m^3/s (Reis2013). The Amazon River is the largest single source of freshwater runoff on Earth, representing some 15–20 % of global river flow (Salati and Vose 1984). Amazonian biodiversity is not only a source of biological and scientific wonder and fascination. The wealth of biological resources found in the Amazon biome has proven to be of enormous benefit to human well-being (Alho 2012). From wood, foods, and beverages to medicines and industrial products derived, for example, from the rubber tree, the biological diversity of the Amazon forest has had a profound effect on the development of human society. Many of these biodiversity secrets have been revealed to Science by indigenous peoples, whose livelihoods and cultures depend on their environment and surrounding biodiversity.

The Amazonian biodiversity represents a critical socioeconomic resource for local people who have long relied on fish resources as their major source of food and income (Ruffino 2004, 2005,2008; Batista et al. 2012). Most notably, three freshwater turtle species are ecologically and socioeconomically important in the Brazilian Amazon due to traditional human consumption: the giant turtle *Podocnemis expansa*, the yellow spotted turtle or "tracajá" *Podocnemis unifilis*, and the six-tubercled turtle or "pitiú" *Podocnemis sextuberculata* (Alho 1985, 2011).

The "várzea" floodplain appears as the first focus of settlement of the Amazon region, with evidence of the presence of lowland indigenous peoples for about 2000 years. These indigenous groups sought the environments of rivers and streams (as a source of food and water) to settle their villages and spread through the region. However, the hydrographic network of the region not only conditioned the process of occupation by indigenous tribes and, later, by settlers, but also guided the path which the regional economy would take (Madaleno 2011). The different levels of landscape transformation, throughout the historical human occupation, are synthesized from landscape diversification to simplification and suppression (Lui and Molina 2009).

The aim of this paper is to identify and evaluate environmental and socioeconomic threats to Amazonian aquatic ecosystems to pursue the conservation and sustainable use of its freshwater biodiversity, taking fish diversity and fisheries as factors for analysis. Two primary concerns are established: (1) the protection of natural aquatic habitats and (2) the relationship of biodiversity resources, mainly fish and river turtles, to local people's welfare.

MATERIALS AND METHODS

Field studies were carried out in three designated Amazonian threatened areas, rich in biodiversity, which are known as hotspots, providing a good representation of the different natural aquatic habitats: the Upper Xingu River, the Lower Tocantins River, and the Mid Negro River, highlighting the aquatic biodiversity, with emphasis on the diversity of fish and fisheries in their various forms in different kinds of aquatic habitats (Fig. 1). Ten days were spent working at each site, visiting different habitats, interviewing riverine inhabitants, fishermen and other fisheries professionals, and covering a sampling path of about 300 km in each hotspot. The interviews were performed to detect the riverine fishermen's perception of the environmental threats and fishing conflicts under a qualitative or descriptive focus rather than a quantitative ones. The people targeted by the interviews were the leaders of the organized fishermen's institutions and other actors in the fishing productive web, such as artisanal fishermen; the owners of ice boats who purchase the fish to transport to consumer centers and other intermediate traders; the leaders of the fishing colonies and cooperatives; and others. Among the interview topics were fishing arts and artifacts, including the mesh size of fishing nets; the use of underwater harpoons; period of closed fishing season and government financial support to fishermen during the closed season; the role of legislation and fishing regulation enforcement; conflicts between resident artisan fishermen and larger fishing boats; conflicts between riverine fishermen and sport fishing activities, as well as perceptions of different fishing stressors such as riparian deforestation; advance of agricultural fields and cattle pastures; effects of infrastructure such as hydroelectric dams, roads, mining, boat traffic; and water pollutants. We conducted the interviews to assess the attitudes and perceptions of fishers and other actors toward the environmental threats and fishing conflicts. Surveys were based on informal group discussions conducted with fishers and other respondents. Participants within the fishing areas were chosen opportunistically. We surveyed fishers belonging to both traditional subsistence and commercial categories. We qualitatively analyzed positive perceptions, oral histories, and convictions, related to fishing activities, documented in survey responses to open questions. We also examined the fish catch harvested across the study sites.

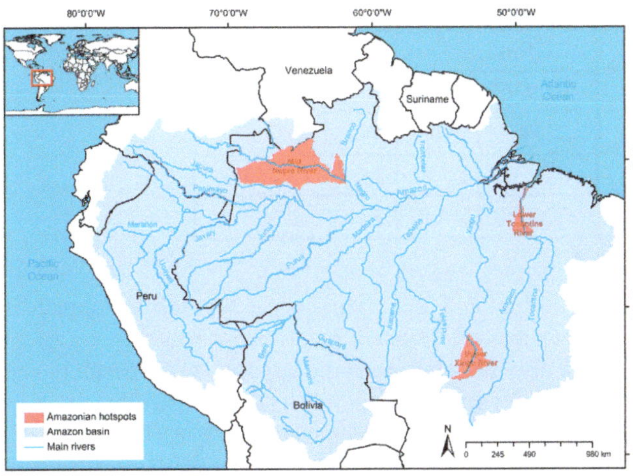

Figure 1: Study regions (upper Xingu River, Lower Tocantins River, and mid Negro River) Amazonian threatened areas, rich in aquatic biodiversity focus of the fieldwork.

Study sites

The Xingu River is a major tributary of the Amazon basin and drains the Brazilian Shield, along with the Tocantins, Araguaia, Tapajós, and part of the Madeira rivers. The Xingu River basin forms the Aquatic Ecoregion 322 of the freshwater ecoregions map of the world (Abell et al.2008). In April 2013, 300 km were covered in a motorboat throughout the headwater region of the *Xingu* River, sampling the selected areas (Table 1).

Table 1: Study sites in the Upper Xingu River region informing geographic position and altitude of each site

Study sites	Latitude	Longitude	Altitude (m)
Headwater of the *Xingu* River	−12.880338°	−52.814270°	300
Jointure of *Culuene* and *Sete de Setembro*rivers	−12.924960°	−52.825969°	302
Upper *Culuene* River	−13.077444°	−52.885847°	310
Coronel *Vanick* River	−13.126701°	−52.689272°	304
Upper Coronel *Vanick* River	−13.237053°	−52.629633°	317
Upper *Sete de Setembro* River	−13.239636°	−52.551104°	311
Upper *Xingu* River	−14.045409°	−52.344515°	355

This region is an ecotone zone, transition between the savanna (Cerrado) biome of central Brazil and the Amazon basin. This study area was chosen for three main reasons: (1) the Xingu River represents the tributaries of the right-hand side of the Amazon, which have their headwaters on the Brazilian Shield; (2) it is an endemic region for fish species: out of 142 fish species recorded for the Xingu basin, 36 species are endemic (Albert and Reis 2011; Buckup et al. 2011), and most of this endemism occurs in the upper Xingu basin; and (3) this region has been under strong threat of deforestation.

The second selected study site was the Lower Tocantins River, inside ecoregion 324—Tocantins–Araguaia—(Abell et al. 2008). In early June 2013, during the fieldwork, approximately 300 km were covered along the Tocantins River, sampling the selected sites (Table 2).

Table 2: Study sites in the Lower Tocantins River region informing geographic position and altitude of each site

Study sites	Latitude	Longitude	Altitude (m)
Baião village	-2.792377°	-49.674500°	10
Joana Peres village	-3.016875°	-49.748470°	12
Ituquara village	-3.031125°	-49.646270°	13
Vila Nova village	-3.450295°	-49.608875°	13
Moru village	-3.558638°	-49.612157°	14
Tucuruí city	-3.756088°	-49.664309°	16
Tucuruí dam	-3.832021°	-49.643278°	70
Porto do Onze	-3.848169°	-49.680821°	70
Porto Novo	-4.414191°	-49.387704°	73

Santa Rosa village	−4.522811°	−49.408602°	73

The Lower Tocantins River was selected for field study, considering the following reasons: (1) it is a river strongly impacted by a huge hydroelectric dam and reservoir (Mérona et al. 2010); (2) it presents high endemism of fish species; and (3) it is an important Amazonian river, extending up to 2750 km in length.

Finally, the third study site selected was the Mid Negro River, included in ecoregions 314 and 315 (Abell et al. 2008). In late July 2013, 310 km along the Negro River were covered, throughout the selected sampling areas adjacent to the town of Barcelos (Table 3).

Table 3: Study sites in the Mid Negro River region informing geographic position and altitude of each site

Study sites	Latitude	Longitude	Altitude (m)
Fishing area 1	−0.856116°	−62.765446°	35
Barcelos city	−0.966978°	−62.928499°	30
Bacabal village	−0.483100°	−62.921512°	29
Daracuá village	−0.506514°	−63.214250°	27
Ponta da Terra village	−0.770953°	−63.142164°	25
Aracá River	−0.771188°	−62.937692°	28
Cubá Stream	−0.924075°	−62.917655°	29
Mamulé Stream	−0.808618°	−63.237575°	27
Murumuru Stream	−0.467140°	−62.928506°	39
Fishing area 2 (*Calibuque*)	−0.629722°	−62.868497°	33
Fishing area 3	−0.464315°	−63.154313°	28
Fishing area 4	−0.466774°	−62.936246°	40
Demini River	−0.414927°	−62.903415°	44
Fishing area 5	−0.833065°	−63.231232°	35
Fishing area 6 (*Zamula*)	−0.865862°	−62.774625°	32

The Negro River was chosen because of the following reasons: (1) it is entirely contained inside the lowland forested area of the Amazon basin, being the major black-water river on the left bank of the Amazon River, and it is the largest black-water river in the world; (2) its black water is poor in mineral nutrients but presents fish endemism; and (3) traditionally, the region of the Mid Negro River supports an extractive activity of ornamental aquarium fish with high socioeconomic relevance.

The identification, characterization, and evaluation of environmental threats to aquatic ecosystems are important tools in planning strategic conservation actions. These aspects of threats were therefore involved in finding ways to implement policies and environmental management, taking into account any particular environmental modifier that potentially threatens the Amazonian aquatic environment. Some of these threats to aquatic ecosystems are of a scientific and technical nature, such as those that influence the structure and function of ecosystems. Others are of a political and administrative nature, and can often be anticipated and avoided, such as the legal fishery standards that must be implemented and followed.

Identifying and qualifying the effect of the threat is somewhat subjective, since it involves not only the ecological focus, but also the socioeconomic point of view. Therefore, the environmental variable threat translates into the integration of physical and chemical components (soil, water quality, etc.), as well as the biotic component (flooded vegetation, aquatic biodiversity, etc.) and socioeconomic elements (fishing, hunting, capture of turtles and their eggs, extraction of other natural products, social organization of fishermen, etc.).

A wide range of environmental threats, detected during the fieldwork, interact with natural biological and physical components of the Amazon freshwater ecosystem to change biodiversity productivity and ecological community structure.

RESULTS

While conducting fieldwork in the three designated river basins (Upper Xingu River, Lower Tocantins River, and Mid Negro River) some environmental threats to the aquatic ecosystems and their biodiversity were identified (Table 4).

Table 4: Identification and description of environmental and socioeconomic threats in the three Amazonian river basins sampled during the fieldwork: Upper Xingu River, Lower Tocantins River, and Mid Negro

Threat	Description of threat	Upper *Xingu*River	Lower*Tocantins*River	Mid *Negro*River
Alteration and loss of fish habitats due to deforestation of riparian vegetation	Habitat conversion of riparian communities, from expansion of agriculture, cattle ranching and urbanization in floodplains	Intensively verified	Moderately verified	Not yet relevant
Effect of reduced river flow due to deforestation of headwater areas	Changes in upland areas (deforestation, expanding cattle ranching, urbanization) resulting indirectly in greater sediment loads and contaminants such as fertilizers and pesticide from run-off	Observed conversion of natural vegetation into huge areas of soybean crop fields	Moderately verified	Not yet relevant
Effect of environmental contaminants on the water quality	Direct contamination of rivers from increased dumping of organic and solid waste into rivers from expanding urban areas and from activities such as intensive agriculture and mining	Potential contamination of waters from agricultural fertilizers and pesticides	Urban sewage pollutants	Urban sewage pollutants
Effect of hydroelectric reservoirs on diversity and fish communities	Transformation of a lotic environment into a lake eradicating or reducing populations of rheophilic fish species, and providing favorable conditions for lentic species to proliferate	Moderate and specifically located effect from small hydroelectric plants	*Tucuruí* dam and plant operation: overall drastic effects on natural habitats and fish communities	Not relevant
Effect of infrastructure on diversity and fish communities	Changes in hydrological regimes through construction of infrastructure such as roads, ports and navigation channels	Moderate impact	Severe impact	Not relevant
Threats to turtles and freshwater mammals	Illegal hunting and commerce of wildlife	Moderate impact in the upper region but very severe in the lower portion (*Belo Monte* region)	Severe impact	Moderate impact

Effects of fishing conflicts	Detected growing number of conflicts among natural resource users	Moderate conflict between sport and subsistence fishing	Severe conflict among commercial fishing interests	Moderate conflict between ornamental and sport fishing
Decline of ornamental aquarium fishing	Capture of ornamental fish species for aquarium international commerce	Not relevant in the upper region but important in the lower area of Volta Grande do*Xingu*	Not relevant here	Negative socioeconomic effect due to decline in demand; fishermen are now starting to seek their livelihoods in other activities, such as agriculture
Effects of deficient implementation of fishing regulation	Lack of organizational and institutional capacity to deal with fishing in a participatory and integrated manner	Sport fishing observed here to be organized to take advantage of immediate income of fishing opportunity	Several fishing colonies exist but conflicts are still severe due to the lack of fishing regulation and implementation	Weakness of organization due to present decline in ornamental fishing
Effects of deficient dialogue between participant actors of fishing social organizations	Social organizations play a significant role in enabling fishery management to ensure sustainability	Strong organization among owners of sport fishing accommodation but low concern for sustainability	Conflicts among subsistence, commercial and sport fishing	Drastic decline in dialogue due to decline in demand for ornamental fishing
Effects of overfishing	Overexploitation of fishery resources is observed when fish stocks are suppressed to a level where fishing in a given region is no longer sustainable, that is, there is a need to increase fishing effort and stocks are not replenished naturally	Competition between sport fishing and artisanal subsistence fishing. Smaller fish sizes.	Observed strong pressure on fishing stocks to meet demand	Sport fishing in increasing demand and ornamental fishing in decline
Effects of global climate change on aquatic environments	The published literature indicates that it appears highly possible that the predicted climate change over the next decades may well cause additional damage to Amazon aquatic ecosystems	Regionally subject to drying change due to intense and extended deforestation	Regionally subject to change due to drastic environmental change—*Tucuruí* dam	Forest still well preserved

The identified environmental and socioeconomic threats imply that in addition to the decline in quantity and quality of fishery resources, based mainly on field observations, there is also a reduction in the size of fish being captured, and this trend has a strong linkage with regional human demographic growth in Amazonia. As a consequence, there is an increasing demand for fish and other natural resources, and conflicts have arisen between commercial and subsistence fisheries.

The Upper Xingu River: Sport fishing and the threat of soybean fields

The Xingu basin is the fourth largest watershed of the Amazon, comprising seven percent of the region. The higher altitudes of the headwaters are located in the state of Mato Grosso at 800 m elevation. The Xingu River runs on crystalline rocks, carrying a low amount of sediments, with clear water. The waters start to rise in September–October, reaching the maximum level in March–April.

The upper Xingu region has been under strong pressure of deforestation. Local inspection followed by analysis of satellite images has shown immense areas adjacent to visited rivers (Culuene, Sete de Setembro, Xingu and others) converted into arable fields and cattle pasture, as observed during the fieldwork. The advance of agriculture has been driven by government programs which encouraged crops in the region. In this area, the Brazilian Savanna biome had already been highly converted by 2008, according to the deforestation monitoring program PRODES (INPE 2008). Local observations show that crop fields have now advanced to the edge of the rivers, altering or eliminating the riparian forest. This radical deforestation violates Brazilian legislation (Brazilian Forest Code, Law 12 651/2012), which protects a strip of the riparian forest on both sides of the river as an area of permanent preservation for large rivers. The seasonally inundated riparian forest constitutes important feeding and reproductive habitats for fish. In addition, water contamination through the intensive use of fertilizers and pesticides, which are usually spread by airplane on the crop fields, potentially affects the freshwater wildlife.

The region also attracts sport fishing, and several fishing resorts were observed along the visited rivers. Local people interviewed reported a decline in fish quantity during the last 10–15 years. They also claim that more recently the size of fish collected by sport fishing has decreased.

The Lower Tocantins River: Infrastructure plants and change in the fish ecological community

The Tocantins River is a major tributary of the Amazon basin and drains, for the most part, from the Brazilian Shield. The catchment basin of the Tocantins River drains an area of 767 000 km^2 (Mérona et al. 2010). The last portion of these environments is now submerged by the Tucuruí hydroelectric reservoir.

The Tocantins basin is the third largest sub-basin of the Amazon River, with an average annual discharge of 11 000 m^3 per second. Its waters are clear and nutrient-poor due to draining away the crystalline formations of the Brazilian Shield. In this basin occurs a high endemism of fish species: "piranhas" such as *Serrasalmus geryi* and *S. eigenmanni*, and other species like "aracus" *Leporinus affinis* and *L. taeniatus*, the ray *Potamo trygonhenley,* and others (Mérona et al. 2010).

In 1984, the *Tucuruí* hydroelectric plant was built in the lower portion of the river. The dam created a 200 km-long reservoir with an area of 2875 km^2 (Mérona et al. 2010). Within the upper dam, to the north of the Tocantins River basin, still freely connected to the rest of the Amazon basin, some fish species occur that are absent or rare in today's portion under the influence of the dam. The Tucuruí dam has caused a decline in fish diversity living in the reservoir (Mérona et al. 2010). The change from a lotic environment in the reservoir eradicated or significantly reduced populations of the rheophilic fish species and provided the conditions for lentic species to proliferate, thereby altering the composition of the local fish community. Detritivorous, herbivorous and insectivorous fish decreased in abundance in the reservoir. At the same time predators, omnivorous and planktivorous species, were favored and increased in abundance (Mérona et al. 2010). Surveys at local fish markets during the fieldwork showed that some abundant commercial species caught in the reservoir are more adapted to a lentic environment, such as "mapará" *Hypophthalmus marginatus*, "tucunarés" *Cichla monoculus*and *Cichla* sp., and "corvina" *Plagioscion squamosissimus*.

Downstream from the dam, the decrease in abundance of fish species is mainly due to the regulation of river flow, which greatly disturbs the annual feeding and reproductive cycles of fish. The dam also interrupts migratory routes, since migratory fish cannot cross the barrier created by the dam, which has no fish ladders, to spawn upriver. Threats to fish and their aquatic habitats were identified during the fieldwork. Fishermen from fishing colonies were interviewed on the lower river region of Tucuruí: Colony Z-53 Breu Branco, Colony Z-43 Porto Novo Jacundá, both within the reservoir, and Colony Z-34 Baião downstream from the reservoir. Interviewees stated that there has been a significant decrease in fish abundance, both in the lake and downstream, especially in the last 10–15 years.

Local fishermen interviewed condemned the use of underwater harpoon fishing, a practice that has become commonplace in the last decade, as it allows for the capture of "tucunarés"—when adults feed little with restricted movement during the period in which they are defending spawning territories or taking care of offspring, and for this reason are not captured by the usual gear, like gillnets and hooks. The capture of these individuals displaying territorial reproductive behavior affects all offspring and thus has a great predatory effect. Most respondents complain of lack of legislation and fishing regulation enforcement, reporting that government officials rarely appear throughout the year or during the period of the closed fishing season. Because of this, interviewees stated that the closed season is not observed by most fishermen in the region.

According to the majority of the fishermen, however, the primary responsibility for the sharp decline in fish abundance after closing the Tucuruí dam is the daily downstream fluctuation of the water level and, especially, the weekly variation in river level as a function of the flow of water in turbines to attend different hours of energy production. This area was also visited during the fieldwork. The greatest demand for electricity in the early evening of each day, as well as lower demand by the industries on weekends, leads to a regulation of the amount of water passing through the turbines that is released into the river, causing a small increase in river level every night, and a reasonable decrease in level every weekend. These changes in water level disrupt the natural cycles of fish feeding and reproductive behavior, being especially destructive in the spawning season (October–March). Many of these areas covered during the fieldwork are shallow, and small variations in river level, such as those produced by the retention of water on the weekends, are sufficient to expose to the air and thus kill juvenile fish and eggs laid in the mass of vegetation.

The Tocantins River already has several hydropower plants in operation, and several others planned to be built. The existing threats here identified will then be intensified.

The Mid Negro River: Buy a fish and save a tree

The Negro River is entirely contained in the Amazon sedimentary basin, being the largest tributary of the left bank of the Amazon River. Most of this basin is on a lowland floodplain, rising to about 100–250 m altitude in the west and down to about 50 m altitude near its mouth.

The black water of the Negro River is extremely low in mineral content. It is 2230 km long, with a catchment area of 696 000 km^2, and an average flow of 28 400 m^3 per second, which represents 14 % of the annual average flow of the Amazon basin (Filizola et al. 2010). It is estimated that an area of 30 000 km^2 of the Negro basin is flooded seasonally between four and eight months of the year.

The current estimate of fish species richness in the Negro River basin exceeds 750 described species. Over 90 species are endemic to the basin, as well as some monotypic genera (*Tucanoichthys*, *Ptychocharax*, *Atopomesus*, *Leptobrycon*, *Niobichthys, and Stauroglanis*) currently found only in this basin (Petry and Hales 2013). An example of endemic species is the small catfish *Denticetopsis sauli* found only in the headwaters of the Negro River. The fish community composition differs from other parts of the Amazon basin primarily in its ecological constraints imposed by nutrient-poor black waters and low pH.

The iridescence of some species, such as the cardinal-tetra *Paracheirodon axelrodi*, may be an adaptive trait to live in the black water of the Negro River. Interesting ecological phenomena include large migrations of some fish species. The "jaraqui" (*Semaprochilodus insignis*), for example, migrates from the black waters of the Negro River to the white-water rivers to spawn. There are also unique assemblages of species in deposits of leaves and miniaturized habitats. Relictual species in this region include the arapaima *Arapaima gigas*, one of the largest freshwater fish in the world, of more than two meters in length, the "aruanã" *Osteoglossum bicirrhosum*, and the endemic black "aruanã" *O. ferreirai* (Petry and Hales 2013).

This region is distinguished by the large number of ornamental fish species captured to supply the international demand for ornamental aquarium fish. The interviewees revealed that there are over 100 commercially exploited aquarium species, and the small cardinal-tetra contributes over 80 % of individuals captured. To compare the local inspection to satellite images, it was confirmed that the forests of the mid Negro region are still very well preserved and, only recently, with the sharp decline in ornamental fishery, have areas begun to open up for subsistence agriculture. One human community visited, Daraquá, located in a flooded "igapó" area, had more than 20 families about 10 years ago, all relying exclusively on ornamental fishing. With the recent decline in the ornamental fishing trade, families were gradually abandoning the community and moving to land areas where other economic activities are possible. Therefore, in these new areas, a considerable portion of upland forest was removed to accommodate plantations of cassava, maize, and other crops. Currently, only a father and son from one family remain in Daraquá, where they rely on the capture of ornamental fish.

The "Projeto Piaba" (http://opefe.com/piaba.html) is a community-based interdisciplinary project established to understand the ecological and sociocultural systems of the *Negro* basin, in order to conserve and maintain the live ornamental fishery and other renewable resources at a commercially feasible and ecologically sustainable level. According to the interviewees, until around 10 or 12 years ago, the town of *Barcelos* had about 600 families relying on capture and commerce of ornamental fish. That figure has now fallen to less than ten percent of the previous number, and the main economic activities have become agriculture and guiding tourists for sport fishing, especially for "tucunarés." Indeed, nowadays the sport fishing season, which occurs between August and February, is what drives the city, both socially and economically.

Ornamental fishery used to be the main economic activity of the city of Barcelos and other communities of the Mid Negro, having started in the mid-twentieth century. It is estimated that, at the time, about 20 million ornamental fish were exported every year from Barcelos. Different stakeholders interviewed informed us that during the last 10–15 years, the entire production chain of ornamental fish in the Mid Negro River had severely declined. The direct and immediate cause of this decline is a sharp drop in demand for ornamental fish. Indirect causes underlying decrease in demand, however, are less clear and have multiple origins. Apparently, competition from new and growing export markets for ornamental fish in Peru, Venezuela, and especially Colombia, charging even lower prices than Brazil, have increased difficulties for Brazilian fish exporters. In addition, several commonly exported fish species from the Negro River can now reproduce in captivity, in the USA and some Asian countries such as Malaysia and Singapore, at reduced costs. Also, the cardinal-tetra, the main species captured in the Negro River, is now under management control to reproduce in captivity in the USA and the Czech Republic, albeit on a small scale.

The decline in ornamental fishery is believed to have had a negative impact on regional forest conservation. For example, the "Piaba Project," created by ichthyologist Dr. Labbish Chao, from the University of Amazonas at Manaus, implies that unlike mining, logging, agriculture, and other activities, ornamental fishing is not a destructive activity. This perception is very clearly expressed in the slogan of "Projeto Piaba": buy a fish and save a tree.

DISCUSSION

The Amazon's aquatic ecosystems linked to all biodiversity resources as well as human riverine communities, including in many places indigenous peoples who depend on natural resources, are increasingly at risk from the identified threats here described. Changes in fish species composition, alteration of the relative abundances of species and their source of food, and changes in reproduction and productivities of populations are some of the consequences of the environmental threats and unsustainable use of fishery resources by humans.

Natural changes are part of any aquatic ecosystem. However, present human demographic growth and land use in Amazonia, combined with rapid increases in fishery consumer demands, are bringing new pressures to damage fish resources and driving more profound changes, more rapidly than any natural impact has ever done.

The identified threats on the effects of reduced river flow due to deforestation and infrastructure (Table 4) affect feeding and reproductive strategies of fish. In fact, the seasonal dispersion of fish to perform feeding and reproduction strategies is one of the most striking behavioral complexes of the Amazonian ichthyological fauna. The majority of species move between the river and the adjacent flooded forest, including the well-known species such as "curimatã"*Prochilodus nigricans*, "jaraquis" *Semaprochilodus* spp., and "pacu" *Myleus* spp. (Batista et al.2012). The floodplains support organic production of the aquatic ecosystem and are directly responsible for fisheries' productivity, guaranteeing fishing yields and the maintenance of biodiversity. In addition to fish-feeding strategy as a function of the flooding regime, fish-reproductive strategy depends mainly on the flooded season, since the majority of Amazonian fish rely on inundated habitats to reproduce. The productivity of the aquatic ecosystem depends on two factors: (1) the amount of nutrients washed down by water from the headwaters, from the Andes, the Brazilian, and the Guyana Shields; and (2) from the extent of the floodplain (lowland flooded forest). Important representatives of fish in the seasonally flooded aquatic ecosystems are detritivorous species that feed on decaying plant debris, followed by species of herbivores, and frugivores.

Heterogeneous types of forested riparian vegetation cover about 30 % of the Amazon basin (Junk et al. 2014). The edges or banks of the rivers form a different complex of aquatic habitats, depending on the season of the hydrological cycle. Also the marginal lakes change according to the season of the year, since they have depressions or passages linking to the main course of the river.

Seasonal flooding is one of the key elements that control productivity and food webs. The flooded areas create new habitats and the movements of waters carry nutrients that supply the food webs. These seasonal forces favor productivity, promoting availability of new aquatic habitats to supply feeding and reproductive niches for several species. The biogeochemical cycle is an important process controlling the productivity of the aquatic ecosystem, combining physical and biological components of the natural system. While river dolphins, turtles and fish are able to respond promptly to the physical component of seasonal flooding, they are also integrated in the biological complex of competition and predation, as another component of trophic linkages.

During the beginning of the dry season, when resources are becoming scarce, many fish migrate into the river channels and change feeding strategy. Therefore, the feeding niches fluctuate depending on water flow. The evolutionary strategies of various fish species vary depending on the food supply: the predator piscivorous species such as the "tucunarés" (*Cichla* spp.) and the "pirarucu" *Arapaima giga*; the carnivores such as the "acará" *Geophagus altifrons* that feed on crustaceans, molluscs, bryozoans and insects; the herbivores that feed on aquatic macrophytes, such as "aracus" *Rhytiodus* spp.; the detritivorous species which are those that consume decaying organic matter, such as "jaraquis" *Semaprochilodus* spp.; and the omnivores that use various sources of food, such as "tambaqui" *Colossoma macropomum*.

There is a synchrony between fruit production in the inundated forest and the flooding season. Frugivore fish disperse to the flooded forest to feed, such as species of the genera *Colossoma*,*Bryconops*, *Tocantinsia*, *Leporinus*, *Tometes*, *Myleus* and *Triportheus*. In addition, these species play an important role in dispersing these forest seeds, contributing to the maintenance of the ecosystem. At low tide, the waters return to the main river channel, carrying decaying organic matter, which also contributes to the productivity of the aquatic ecosystem. Some fish species, such as *Myleus rhomboidalis*, change their feeding strategy depending on the season. During high water they feed on fruits in the flooded forest. During the dry season, they return to the river and feed on small prey such as invertebrates (Batista et al. 2012).

The ecological interactions among all the abiotic and biotic components of the freshwater ecosystem, including river types, flood regimes, distinct riparian forest, and many other elements of local biodiversity, are responsible for a complex mosaic of aquatic habitats. An important aspect to emphasize is that, depending on the specific water composition, the amount and quality of sediment and nutrients influence primary productivity, fish diversity, and community structure. These ecosystem components influence water visibility and electric conductivity, in consequence affecting the behavioral ability of fish to detect visual signals (characiform fish) or electric stimuli (gymnotiform fish) to swim, forage, escape from predators, and reproduce.

However, there is a group of sedentary fish species that are able to stay in the same habitat all year round, tolerating a low level of oxygen. This group includes "tucunarés" *Cichla* spp., "acarás" *Geophagus* spp., "aruanã" *Osteoglossum bicirrhosum*, "pirarucu" *Arapaima gigas*, various species of "piranhas" *Serrasalmus* spp., and *Pygocentrus* spp., among others (Fabré and Barthem 2005).

The Imazon Deforestation Alert System detected a deforestation area of 838 km^2 in August and September 2014, in Amazonia, which represents an increase of 191 % in relation to 288 km^2 of August–September 2013 (www.imazon.org.br/publication/forest-transparency/deforestation-report-for-the-brazilian-amazon-august-2014-sad). The impacts on the biogeochemical cycle, as consequences of clearing and burning of the Amazon forest, are being raised as the causes of extreme droughts in southeastern Brazil, with climate change on broader scales (Cavalcanti2012). Flooding or droughts affect the population and several sectors of the economy.

The effect of hydroelectric dams has a strong influence on fish movement and community structure (Table 4). Some species that exhibit long-distance migration along rivers, swimming thousands of kilometers upriver to lay eggs, such as the catfish "piramutaba"*Brachyplatystoma vaillanti* and "dourada" *B. flavicans,* find the dam an obstacle in their migration route (Fabré and Barthem 2005). The environmental stimulus that triggers this behavior is a complex set of elements that include water level and turbidity as well as available seasonal habitat structure. Lotic environment of the reservoir reduces rheophilic fish species and favors lentic species such as the phytophagous strategists that filter plankton like the "maparás"*Hypophthalmus* spp., found in abundance in the Tucuruí reservoir, as has been observed in our field studies. The Lower Tocantins River suffers from change in the fish community ecology, since the Tucuruí dam has caused the decline in fish diversity living in this area.

Besides the Tucuruí dam, examined in this study, there has been increasing construction of large hydroelectric dams in the Amazon, such as Jirau and Santo Antônio on the Madeira River and Belo Monte on the Xingu River. In addition, there has been a proliferation of hydroelectric dams in the Andean Amazon, with potential ecological impacts in terms of river connectivity and forest loss (Finer and Jenkins 2012). There are plans for

151 new dams greater than 2 MW over the next 20 years, more than a 300 % increase. The reservoir formed in an area previously without that body of water now has an influence on the phenological rhythms of the surrounding vegetation as well as altering the seasonal phases of the hydrological cycle (flood phase, drying phase, drought, and flooding phase). The movements of fish, aquatic mammals, and freshwater turtles fit in with this hydrological cycle, in function of food availability and appropriate reproductive habitats (Alho 2011).

Our survey shows that hydrological disruptions, together with a fast growing human population and increase in deforestation, provide major challenges in the way fisheries resources are sustainably used in the region.

The threat to turtle species has an impact on biological as well as on socioeconomic components (Table 4). Traditionally, in the Amazon, there is consumption of turtle meat and eggs, mainly by traditional peoples, for whom this is an ingrained cultural habit. Besides fish, turtles are important inhabitants of aquatic ecosystems. During the nesting season, the Amazon turtle*Podocnemis expansa* migrates from lakes connected to the rivers to selected areas of river beaches to lay eggs (Alho and Pádua 1982). This movement is synchronized with the ebbing regime and lowest water.

The threat of the decline of ornamental aquarium fishing (Table 4) showed that the Mid Negro River area, like other regions in the Amazon, is distinguished by the large number of ornamental fish species captured to supply the international demand for ornamental aquarium fish, but recently it has suffered a decline in the ornamental fishing trade, and people have abandoned the area, looking for other sources of income. In the Negro River, there are also unique assemblages of species in deposits of leaves and miniaturized habitats.

The growing number of conflicts detected among natural resource users (Table 4) revealed disagreements between commercial and subsistence people. The fishers reported a decline in their potential catch and livelihoods, leading to conflict. In general, fisheries operate under a system of open access to fish. It has been difficult for resident communities to exclude commercial fishermen from their traditional fishing territory. When profit is the goal, they tend to catch as many fish as possible. Capture and transport of commercial fish across the Amazonian floodplain is still mainly done by fisherman's canoe, which is evidence of artisanal fishery. The fishing boat has its role in storing fish in ice-boxes and transporting the fish caught by artisanal fishermen to the consumer centers. Fish are the main source of protein for much of the population in the Amazon River basin, exceeding 400 g of fish per person per day. Fishing is one of the most traditional activities for the region, with high socioeconomic importance. It is a source of food and income, as well as employment and leisure, for a large number of people (Batista et al. 2012).

Our fieldwork detected overexploitation of fishery resources and consequent decline in fisheries (Table 4). One piece of evidence that could well illustrate the trend for overfishing, with socioeconomic consequences, is the status of the "tambaqui" fish *C. macropomum*, which is widely appreciated and much sought after by the market. Fishing for "tambaqui" has been in decline due to overfishing since 1970, a time when this species was abundantly landed in the city of Manaus (Batista et al. 2012). The same trend occurs in relation to the other well-known Amazonian fish such as the giant "pirarucu" *Arapaima gigans*, "tucunaré" *Cichla* spp., or "jaraqui" *Prochilodus brama*, and also to regional freshwater turtles such as *Podocnemis expansa* and *P. unifilis* (Alho and Pádua 1982, Alho 1985, 2011).

The environmental threats identified in this study have strong links with fish ecology, negatively impacting their biological cycles. Amazon fish food originates in the synthesis of two sources: an autochthonous source, which depends on the productivity of the aquatic ecosystem itself (phytoplankton, zooplankton, aquatic macro-invertebrates, insect larvae, etc.) and an allochthonous source, which depends on the features near water, mainly the alluvial floodplain forest (fruit and seeds fallen into the water, etc.).

Water chemistry imposes additional constraints on distribution and abundance of Amazonian biodiversity. Direct contamination of water, as a result of the identified threat from environmental contaminants, alters fish abundance and modifies ecological community. Within the Amazon basin, there are rivers with water of different colors, including three major types: (1) Rivers which have their headwaters in uplands such as the Amazon River with its springs on the Andes, sediment-rich water with nutrients from rocky terrains. This is also the case of the Juruá, Purus, Madeira, Marañón, Meta, Napo, and Ucayali rivers. (2) Rivers with black waters due to natural dissolved organic matter such as tannins, including the Negro, Trombetas, and Tefé rivers, with very acidic waters that are nearly devoid of sediments and nutrients. These rivers rise in the Amazon forest and carry forest sediments, mainly decomposing fragments of leaves and other organic matter. (3) Rivers with clear water, relatively nutrient-poor that can range from alkaline to acidic waters, such as the Tapajós and Xingu that run throughout areas of low inundation, mainly the crystalline soils of the central Brazilian plateau, covered by savannah vegetation (Cerrado biome), before reaching the lowland seasonally flooded Amazonian "várzea." The Upper Xingu River suffers from serious problems with deforestation, sport fishing decline, and the threat of soybean fields, and this area suffers from the intensive use of fertilizers and pesticides, contaminating the freshwater wildlife.

The scientific information available encapsulates our understanding of how the Amazon biome operates as an ecological system and how the biological resources, including fishery resources, are regulated. The highly diverse range of Amazonian biodiversity lives in a number of different aquatic habitats, including large rivers, small streams, extensive wetlands and seasonally flooded areas, rivers and lakes in elevated areas, etc. These main types of aquatic habitats are based on altitude, land gradient, rainfall, temperature, vegetation cover, and soil type. The large river channels are of immense dimensions in the Amazon, such as the Amazon River itself or the Negro River near the city of Manaus. They have sinuous courses, forming islands, and flood large portions of the forest forming the "várzea" and "igapó" seasonally inundated ecosystems.

It has been recognized that conflicts over biodiversity resources can be diminished and resources better managed when stakeholders are more involved in management, and when access rights are distributed more effectively and equitably (Ruffino 2004, 2005, 2008; Batista et al. 2012).

The suggestion to build a management strategy for Amazon fisheries resources should include an integrated bank of information, with analysis of existing data on fish biology and fisheries. It has to consider options for planning, consultation, decision-making, distribution of tasks and resources, and formulation and implementation of rules for managing all fishing activities. This management aims to ensure environmental sustainability, the continued productivity of fisheries resources, and the socioeconomic compatibility of the actors involved.

Besides the integrated approach among all actors involved in fisheries, management should also emphasize conservation of the aquatic ecosystems, considering the environmental threats analyzed in this study.

Several authors studying fisheries in the Amazon indicate some models of participatory community management of fisheries as one of the successful strategies to achieve sustainable use (Ruffino 2004, 2005, 2008; Batista et al. 2012). Artisanal fishing is still crucial for riverine populations in the Amazon. On the other hand, the increased demand for fisheries has created conflicts between commercial fishermen and the inhabitants of rivers, streams and lakes who attempt to prohibit the entry of commercial fishermen in their traditional fishing areas, faced with the new situation of declining fish stocks.

CONCLUSION

In conclusion, this study shows that the conservation and management of the Amazonian aquatic ecosystems and their fishery resources can be enhanced if there is governmental and political motivation to implement management through participatory activities, integrating all actors involved, to enforce legislation and specific fishery norms. This is necessary to put into practice the laws and regulations, fishing calendar, environmental education, to minimize fishing conflicts, and to develop specific monitoring programs to mitigate the environmental threats to aquatic ecosystems. The lessons learned with some successful Amazon fishery projects are a positive direction to be taken and improved on, considering this new information on conservation and management of Amazonian freshwater ecosystems.

Sustainable fishing strives for the balance between the losses caused by human and natural growth, and the gains due to reproduction and growth, in such a way that the abundance of a species in a given area remains constant through time. Conservation and management are tools to encompass ecosystem interconnections, linking human use such as fisheries to scientific knowledge.

The efforts of Amazon human communities to reassume control of lakes and other traditional fishing territories have added the participation of riverine people, fishermen's associations, fishing associations, governmental and non-governmental organizations, religious organizations, municipalities, and other actors involved. The idea is to implement measures to regulate fishing in certain areas, by means of a fishing agreement (acordo de pesca), for the socioeconomic benefit of local people. Consistent with this line of action, some successful projects have supported some fishing agreements in the Brazilian Amazon, and the "ProVárzea" Project has played an important role with government support. The Mamirauá Institute, located near the city of Tefé, state of Amazonas, Brazil, has been implementing participatory community-based management of "pirarucu" (*Arapaima gigas*) in the Mamirauá Reserve of Sustainable Development for over a decade, successfully both in the application of the management model and with respect to socioeconomic aspects to increase the income of local fishermen. Fishermen involved in this program recognize that resource conflicts can be diminished and better managed when they and other stakeholders are more committed to management, and when access rights are distributed more efficiently and equitably. There has been a commitment from the Brazilian government (Ibama Institute, through the "ProVárzea" program) to implement policies regarding decentralization and community-based management. Also the Fishing Colonies and Fishing Zones are organizations that can better address the relationship of fisheries resources to local fishermen's welfare and the conservation of the resources by implementing an agreed management plan, including all the relevant attributes of legislation.

ACKNOWLEDGMENTS

We appreciate the many fishermen and other Amazon riverine people who helped during the fieldwork and research for this study. We especially thank Rubens Ghillardi Jr. from Eletronorte for providing logistic support for the fieldwork. Partial funding for this project came from the Amazon Cooperation Treaty Organization (ACTO) through the component of aquatic ecosystems of the project "Integrated and Sustainable Management of Transboundary Water Resources in the Amazon River Basin Considering Climate Variability and Climate Change" with support from the Global Environment Facility (GEF) and United Nations Environment Programme (UNEP). Celina Alho assisted in the preparation of the manuscript.

REFERENCES

1. Abell, R., M.L. Thieme, C. Revenga, M. Bryer, M. Kottelat, N. Bogutskaya, B. Coad, N. Mandrak, et al. 2008. Freshwater ecoregions of the world: A new map of biogeographic units for freshwater biodiversity conservation. *BioScience* 58: 403–414.
2. Albert, J.S., and R.E. Reis. 2011. Introduction to Neotropical freshwaters, 3–19. In *Historical biogeography of neotropical freshwater fishes*, ed. J.S. Albert and R.E. Reis. Berkeley, LA: University of California Press.
3. Alho, C.J.R. 1985. Conservation and management strategies for commonly exploited Amazonian turtles. *Biological Conservation* 32: 291–298.
4. Alho, C.J.R. 2011. Environmental effects of hydropower reservoirs on wild mammals and freshwater turtles in Amazonia: A review. *Oecologia Australis* 15: 616–627.
5. Alho, C.J.R. 2012. Importance of biodiversity for the human health: an ecological perspective.*Estudos Avançados* 26: 151–165. University of São Paulo (in Portuguese, English summary).
6. Alho, C.J.R., and L.F.M. Pádua. 1982. Reproductive parameters and nesting behaviour of the Amazon turtle *Podocnemis expansa* (Testudinata, Pelomedusidae) in Brazil. *Canadian Journal of Zoology* 60: 97–103.

7. Baran, E., and C. Myschowoda. 2008. Have fish catches been declining in the *Mekong* river basin? In *Modern myths of the Mekong: A critical review of water and development concepts,principles and policies.* ed. M. Kummu, M. Keskinen, and O. Varis, 55–64. Espoo: Helsinki University of Technology.

8. Batista, V.S., V.J. Isaac, N.N. Fabré, J.C.A. Gonzalez, O.T. Almeida, S. Rivero, J.N. Oliveira Jr., M.L. Ruffino, et al. 2012. *Fish and fisheries in the Solimões River: An integrated evaluation*, ed. Instituto Brasileiro do Meio Ambiente e dos Recursos Naturais renováveis. Ibama-MMA. Brasília: Projeto Manejo dos Recursos Naturais da Várzea (in Portuguese).

9. Böhlke, J.E., S.H. Weitzman, and N.A. Menezes. 1978. Current status of the systematics of South American freshwater fish. *Acta Amazonica* 8: 657–677. (in Portuguese, English summary).

10. Buckup, P.A., M.R. Britto, J.R. Gomes, J.L.O. Birindelli, F.C.T. Lima, J.A. Maldonado-Ocampo, C.H. Zawadzki, F.R. Carvalho, et al. 2011. Ichthyofauna inventory of the *Xingu-Tapajós* aquatic ecoregion. In *Xingu-Tapajós aquatic ecoregion*, ed. Z.C. Castilhos, P.A. Buckup, 163–174. Rio de Janeiro: CETEM-MCT (in Portuguese).

11. Cavalcanti, I.F.A. 2012. Large scale and synoptic features associated with extreme precipitation over South America: A review and case studies for the first decade of the 21st century.*Atmospheric Research* 118: 27–40.

12. Dudgeon, D., A.H. Arthington, M.O. Gessner, Z.I. Kawabata, D.J. Knowler, C. Lévêque, R.J. Naiman, A.H. Prieur-Richard, et al. 2006. Freshwater biodiversity: Importance, threats, status and conservation challenges. *Biological Reviews* 81: 163–182.

13. Dugan, P.J., C. Barlow, A.A. Agostinho, E. Baran, G.F. Cada, D. Chen, I.G. Cowx, J.W. Ferguson, et al. 2010. Fish migration, dams and loss of ecosystems services in the *Mekong* basin. *AMBIO*39: 344–348.

14. Fabré, N.N., and R.B. Barthem. 2005. *Fishery Management of the migrant giant catfishes piramutaba and dourada in the Solimões-Amazonas axis*, org. Manaus: PróVárzea-Ibama/Coleção Documentos Técnicos: Estudos Estratégicos (in Portuguese).

15. Filizola, N., N. Spíndola, W. Arruda, F. Seyler, S. Calmant, and J. Silva. 2010. The *Rio Negro*and *Rio Solimões* confluence point—hydrometric observation during the 2006/2007 cycle. In*River, coastal, and estuarine morphodynamics.* RCEM 2009, ed. Vionnet et al., 1003–1006 London: Taylor and Francis Group. ISBN: 978-0-415-55426-8.

16. Finer, M., and C.N. Jenkins. 2012. Proliferation of hydroelectric dams in the Andean Amazon and implications for Andes-Amazon connectivity. *PLoS One*. doi:10.1371/journal.pone.0035126.

17. INPE—National Institute for Space Research. 2008. *Amazon forest cover monitored by satellites*. Brasília: Ministry of Science and Technology (in Portuguese).

18. IPCC—Intergovernmental Panel on Climate Change. 2014. Climate Change 2014. Retrieved April 29, 2014, from www.ipcc.ch.

19. Junk, W.J., M.T.F. Piedade, R. Lourival, F. Wittmann, P. Kandus, L.D. Lacerda, R.L. Bozelli, F.A. Esteves, et al. 2014. Brazilian wetlands: their definition, delineation, and classification for research, sustainable management, and protection. *Aquatic Conservation Marine and Freshwater Ecosystems* 24: 5–22.

20. Lévêque, C., T. Oberdorff, D. Paugy, M.L.J. Stiassny, and P.A. Tedesco. 2008. Global diversity of fish (Pisces) in freshwater. *Hydrobiologia* 595: 545–567.

21. Lui, G.H., and S.M.G. Molina. 2009. Human settlement and landscape transformation in the Brazilian Amazon. *Amazonia* 1: 200–228. (in Portuguese, English summary).

22. Madaleno, I.M. 2011. Develop the Amazon? History of the human occupation of the Brazilian Amazon. *Espaço & Geografia (Space & Geography)* 14: 331–360. (in Portuguese, English summary).

23. Mérona, B., A.A. Juras, G.M. Santos, and I.H.A. Cintra. 2010. *Fishes and fisheries in the lower Tocantins River: Twenty years after the Tucuruí hydroelectric dam.* Brasília: Eletrobrás-Eletronorte. (in Portuguese).

24. Millennium Ecosystem Assessment, Ecosystems and Human Well-Being: Health Synthesis. 2005. *The Millennium Ecosystem Assessment series: A framework for assessment; current state and trends*, Vol. 1: Scenarios, Vol. 2: Policy responses, Vol. 3: Multiscale assessments, Vol. 4: Our human planet: Summary for decision-makers; Synthesis report: Ecosystems and Human Well-Being: Synthesis, biodiversity synthesis, diversification synthesis, human health synthesis, wetland and water synthesis, opportunities and challenges for business and industry. Washington, DC: Island Press; The Center for Resource Economics; World Resources Institute.

25. Petry, P. and J. Hales. 2013. Freshwater Ecoregions of the world; *Ecoregion* 314: Rio Negro. Retrieved May 22, 2014, from http://www.feow.org/ecoregions/details/314.

26. Reis, R.E. 2013. Conserving the freshwater fishes of South America. *International Zoo Yearbook*47: 65–70.

27. Reis, R.E., S.O. Kullander, and C.J. Ferraris, Jr. 2003. *Check list of the freshwater fishes of South and Central America*, ed. Editora da Pontifícia. Porto Alegre: Universidade Católica do Rio Grande do Sul.

28. Ruffino, M.L. 2004. *Fishery and its resources in the Brazilian Amazon*, ed. Ibama/ProVárzea/Edua, Manaus, Brasil (in Portuguese).

29. Ruffino, M.L. 2005. *Fishery resources use management in the Amazon*. Manaus : PróVárzea. (in Portuguese).

30. Ruffino, M.L. 2008. Integrated fishery statistic system for the Amazon. *Pan-American Journal of Aquatic Sciences* 3: 193–204. (in Portuguese, English summary).

31. Salati, E., and P.B. Vose. 1984. Amazon basin: A system in equilibrium. *Science* 225: 129–138.

32. Santos, D., D. Centano, J.G.A. Aranibar, and A. Veríssimo. 2014. *Social Progress Index for the Brazilian Amazon*. Instituto do Homem e Meio Ambiente da Amazônia (IMAZON): Belém.

CHAPTER 15 FISHERIES CO-MANAGEMENT: THE CASE OF JARAQUIS SEMAPROCHILODUS SPP. (CHARACIFORMES, PROCHILODONTIDAE)

Vandick S Batista and Liane G Lima

BACKGROUND

Fishing is a traditional activity in the Amazon region performed by Indians and local people for direct feeding, generally called subsistence fishery. Commercial activity became more intensive from the end of the decade of the 1960's with the introduction of the monofilament nylon twine, styrofoam boxes in the coating of thermal boxes for fish caught and of diesel motors in boats [1]. Professional fisherman developed from being subsistence fishermen and owing to of such technological improvements became full time fishermen fishing in distant waters to sell to freezing houses and fish markets of urban centers [2, 3].

This commercial fishing initially benefitted from the existing tambaqui (*Colossoma macropomum*) fishing resource, which was of greater market value and availability and which attained the yield landed in Manaus of 14 thousand tons per year until the end of the 1970's [4, 5]. However, this resource reduced progressively during the 1980s. At the same time the jaraquis (*Semaprochilodus* spp.) reached the maximum yield landed [4, 6].

The genera Semaprochilodus belongs to the family Prochilodontidae, with two species in Central Amazon: the jaraqui-escama-fina (*S. taeniurus*) and the jaraqui-escama-grossa (*S. insignis*). Both species were among the most abundant fish landed in Manaus between 2001 and 2003 [7]. In spite of the importance of this and also of the suggestion of overfishing [8, 9, 10], it has not been included in the closed fishing season determined yearly by the national organ of natural resources management (IBAMA). On the other hand, it is affected directly by two legal norms: one aims to protect reproductive areas and other to protect young fish (minimum size of catch of 20 cm in total length - TL). However, considering that the application of such norms is occasional and inefficient, it is necessary that a viable management procedure be developed allowing the sustainability of the Amazon fishing resources.

These scientific knowledge has been used traditionally to evaluate fish stocks but has not been easily accepted by fishermen, particularly artisanal fishermen or even government managers. In the last decade many researchers indicated fishermen culture as important source of information for simple evaluation of natural stocks (e.g. [11]). Fishermen culture also can take an important place in the development and effectiveness of fisheries management [12], particularly when command and control strategies cannot be used. In Brazil,

Citation: Vandick S Batista and Liane G Lima, fisheries co-management: the case of jaraquis Semaprochilodus spp. (Characiformes, Prochilodontidae), DOI: 10.1186/1746-4269-6-15.

Posey, Diegues, Marques and Begossi pioneer researches (e.g. [13, 14, 15, 16]) gave the main references to the actual principles and methods used.

In this context, the use of traditional ecological knowledge associated to scientific knowledge should occupy a fundamental place. Active participation of users of resources in the management conception and actions has been frequently discussed [17, 18]. However, classical fisheries assessment knowledge is not usually understood by stakeholders and fishermen representatives, whose knowledge is fundamental to the efficiency and credibility of any activity to be performed. Traditional fisheries assessment knowledge has been registered in the region on a few occasions (e.g. [19, 20]). As a consequence, the relationship of this knowledge with the scientific knowledge used by the government agencies has rarely been established.

The present study aims to evaluate the potential of traditional bio-ecological knowledge of subsistence and commercial fishermen about the jaraquis in order to contribute to the fisheries management. The results of the assessment were related to similar ones already available in scientific literature evaluating the association of themes to information. Results of fisherman types were also compared, allowing an initial evaluation about the usefulness of the respective sources as a base of information for co-management.

METHODS

The scientific data concerning the biology and ecology of *Semaprochilodus taeniurus* and *S. insignis* were obtained by means of bibliographical revision of published scientific literature, and also by compiling theses, dissertations, monographs and books specialized in the subject. The collection of primary data was made in the fish landing raft in Manaus and in rural communities in a rural zone of the municipal district of Manacapuru, Amazonas, between the latitudes 3° 30› S at 3° 40› S and longitudes 61° 00› W at 60° 45› W. Structured inters were done with 57 fishermen between July and December 2002. Thirty one interees were professional fishermen from Manaus, called city-based fishermen (following [2]), and 26 with the riverine fishermen of the rural areas of Manacapuru. The inters were done individually with each fisherman, using a questionnaire with questions concerning personal characteristics, experience, and social relationships in fishing and bio-ecological knowledge of reproduction, feeding, predatory behavior, growth, migration, mortality and recruitment related to the species in question. The research was done in accordance with the ethical and legal obligations of the institution and the country in which the research took place at the time of the research.

The fishermen were randomly selected for inters of between 45 and 60 minutes. Whole inters were registered in writing, and soon after inserted in a database for analysis.

Dynamic tables, histograms and associated averages and confidence intervals were used to test treatments. The statistical test Z was used in the comparison of proportions with the null hypothesis where the information was the same [21]. The assumption of normality was evaluated by the Kolmogorov-Smirnov test and of the homocedasticity by the Bartlett test [22].

RESULTS

Knowledge related to reproduction

The recorded size and ages that the fish begin to reproduce at were not different for either city-based fishermen or riverine fishermen (P > 0,05). The first maturation length and age was 20 ± 1.5 cm CT and 1,9 ± 0.6 year^{-1}, respectively. The length of 100% sexual maturity was 27 ± 2,4 cm CT, without significant difference among fishermen (P > 0,05). The age of 100% sexual maturity was significantly different between fishermen (P < 0,05), where the city-based fishermen indicated full maturation between 5 ± 2,5 years old and the riverine ones registered 2,3 ± 0,7 years old.

The number of spawning seasons were estimated differently among city-based and riverine fishermen (P < 0,05). It was registered that 93,5% of the city-based fishermen affirmed knowing that the species only spawns once a year, while of the riverine fishermen, just 65% affirmed that the fish spawn only once a year, and 31% said that they would not know how to answer (Figure 1).

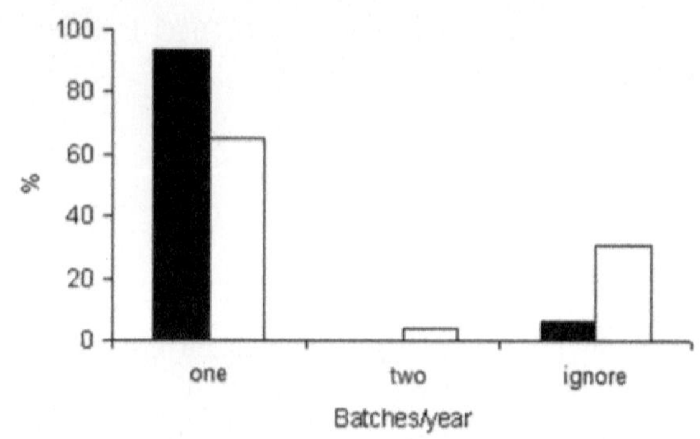

Figure 1: Spawning frequency of jaraquis per year according to city-based (black bars) and riverine (white bars) fishermen.

More common were declarations that jaraquis spawn in White Water Rivers during the beginning of the flood season. No parental care was recorded and the declarations about fecundity usually ranged from a thousand to 50 thousand eggs per batch according to fishermen. However, the fecundity information was very diversified (350 to 1 million eggs according to city-based fishermen and 6,5 thousand to 70 thousand eggs according to riverine fishermen). The difficulties encountered by fishermen for this type of information were so great that just 17% of the riverine and 51% of the city-based fishermen gave an answer on the subject. The declared information is presented concisely along with data of the scientific literature in Table 1 to proceed for further discussion.

Table 1: Cognitive map about jaraquis reproduction according to central Amazon fishermen and the scientific literature

ITEM	CITAÇÕES DOS PESCADORES	CITAÇÕES CIENTÍFICAS
Length of first maturation	"The fish begins to reproduce with 20 cm and 2 years old"	Jaraqui-escama-fina: L_{50}: 24-25 cm FL [47] Age L_{50}: 2,0-2,2 years [10] L_{50}:24,8 cm TL [42]
	"The jaraqui (reproducer) length is 27 cm (CT) with about 5 years old"	Jaraqui-escama-grossa: L_{50}:22,3 cm SL/2,3 years [9] Age L_{50}: 1,9-2,3 years [10] L_{50}:26,4 cm TL [42] Jaraquis: Age L_{50}: 2,0 years [41]
		Jaraquis: Age L_{50}: 2,0 years [41]
Spawning season	"Jaraqui produces youths during the floods "	Jaraqui-escama-fina: Dec-Jan [38] Jaraqui-escama-grossa: Jan-Mar [42] Jaraquis: Dec-Mar/beginning of the flood[1] [41,47]; Nov-Mar [9]
Spawning grounds	" Jaraqui spawns in white waters"	Spawn in the confluence of black and white waters [41, 42]
Parental care	"Jaraqui do not take care"	Parental care not recorded [41, 42, 43, 47, 50]
Spawning type	"Jaraqui spawn once per year"	Total spawners [41, 42, 50]
Fecundity	"The fish has many eggs, thousands"	Jaraqui-escama-fina: 45-105 thousand [57]

Knowledge related to migratory activities

Both city-based fishermen and the riverine fishermen ($P > 0,05$) stated that during the beginning of the flood, the jaraquis were leaving the lakes and small creeks to spawn in the main river (Figure 2). The occurrence of another large scale migration, mainly the dispersion migration at the end of the flood and during the full season was mentioned by about 40% of the professional and riverine fishermen. Lateral migrations are more often mentioned for riverine than for city-based fishermen, and these were the only declarations indicating that jaraquis do not migrate laterally.

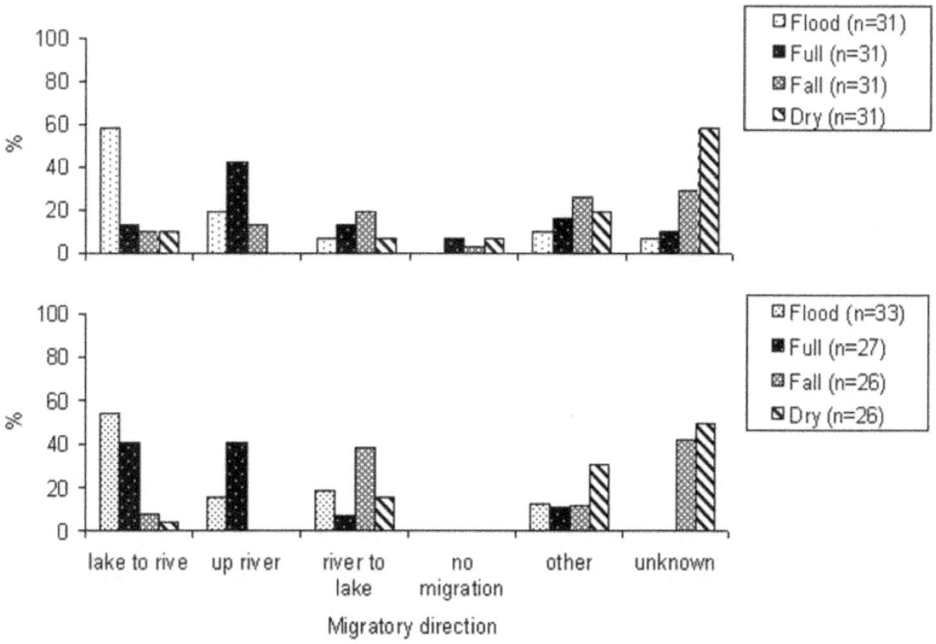

Figure 2: Main migratory activity of jaraqui along an year according to city-based (above) and riverine (below) fishermen, discriminating hydrological cycles: flood, full, fall and dry season.

Knowledge related to trophic ecology

According to the fishermen, the jaraquis feeds preferentially in mud or slime, indicating the lake as a feeding environment during the flood and drain water seasons. They also stated that there is no indiscriminate alteration in the items of the alimentary diet if the waters are rising or falling. Some fishermen emphasized the decrease in food availability during the period of drought, but without significant difference (P > 0,05) among the answers of the types of fishermen (Figure 3).

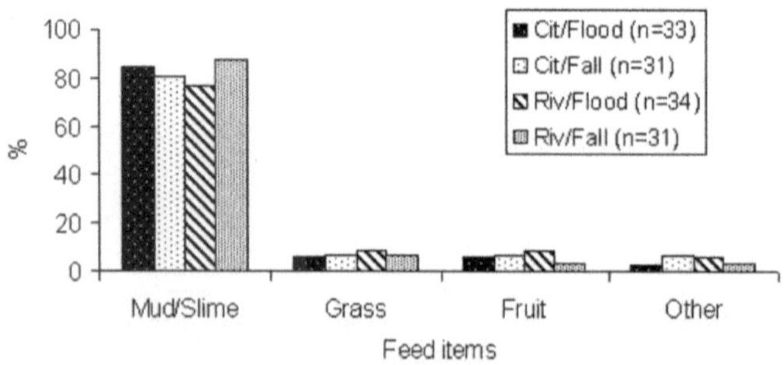

Figure 3: Feeding preferences of jaraqui according to city-based and riverine fishermen, discriminating the flood and fall seasons.

According to fishermen (without significant differences P > 0,05), the main jaraquis predators are other fish, for example, catfish: pirarara (*Phractocephalus hemiliopterus*), piraíba (*Brachyplatystoma filamentosum*), dourada (*Brachyplatystoma rousseauxii*), pacamum (*Zungaro zungaro*) and surubim (*Pseudoplatystoma* spp.). Piranha (Serrasalminae), tucunaré (*Cichla* spp.) and pirarucu (*Arapaima gigas*) are also examples of other fish that feed on jaraquis during all seasons (Figure 4).

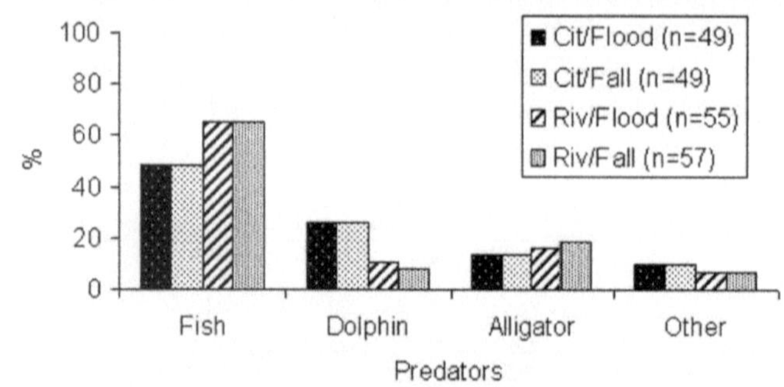

Figure 4: Main predators of jaraqui according to city-based and riverine fishermen, discriminating the flood and fall seasons.

Knowledge related to age and growth

The fishermen stated that the average of size of a one-year-old jaraqui is around 15 cm in total length (TL). For two-years-old fish, the average length measured was 20 cm TL and for 3 years-old-fish the estimation reached 23 cm TL (Table 2), without significant differences among the types of fishermen (P > 0,05).

Table 2: Jaraqui growth according central Amazon fishermen and comparing to that described by the scientific literature

Age	Fishermen	Literature*
1	15,1 ± 2,8 cm CT	14,6 cm SL
2	20,3 ± 2,7 cm CT	22,8 cm SL
3	22,7 ± 5,7 cm CT	25,2 cm SL
Larger/older fish	-	39-41 cm SL

* Jaraqui-escama-grossa [9]

Knowledge related to mortality and fish recruitment

According to the fishermen, mortality by natural causes is the main *causa mortis* of the jaraquis during the larval and juvenile phases, indicating a feeding preference for other fish as the main factor (Figure 5). For the adult phase, the fishermen indicated fishing as the main cause of mortality (without differences between types; P > 0.05).

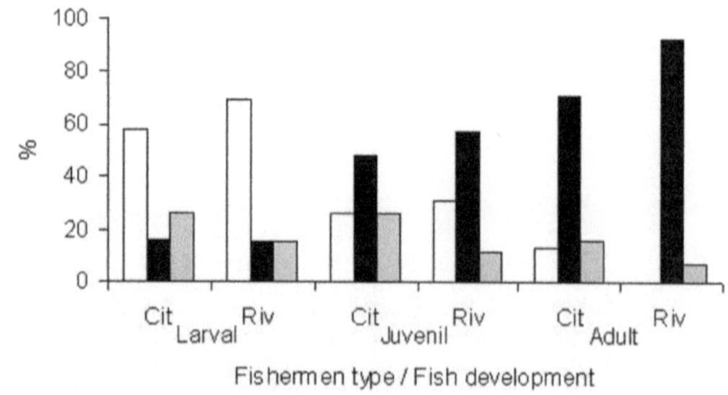

Figure 5: Natural mortality (white bars), fishery mortality (white bars) and ignored causes of mortality (gray bars) of jaraqui for larval, juvenile and adult development phases, according to city-based and riverine fishermen.

The fishery recruitment size of the jaraquis according to both types of fishermen (P > 0,05) was 20,1 ± 2,3 cm TL.

DISCUSSION

The economic and social importance of fishing resources is commonly highlighted in scientific or socioeconomic projects, even though there are few circumstances where there is effective management that allows sustainable and productive use of these resources [23], generating the example of the Tragedy of the Commons [24, 25].

Considering the importance of fishing resources, why is management failure so common? Several elements have been discussed concerning different aspects of these failures (e.g. [26, 27, 28, 29]), and delays associated to centralized management arise as an important contributing factor. Co-management is an alternative proposed by several authors (e.g. [30, 31, 32, 33]) and is being developed in several fisheries (e.g. [31, 34, 35]). However, how is it that managers and resource users obey pre-established norms that they do not understand and that sometimes go against common sense? A typical example of this happens in developing fisheries, where an accumulated stock generates growing yields in response to increases in fishing effort (e.g. [36]). In these circumstances, the political environment does not allow managers to take hard decisions. Thus, it is necessary to understand the users' knowledge and associate it to the technical and scientific knowledge in a comprehensible way in order to justify stakeholders' decisions.

Studies of Brazilian ethnoichthyologist pioneers have already shown that fishermen possess knowledge and carryout practices related to the structure and function of the ecosystems they inhabit, which enables them to be powerful partners in natural resource management [15, 16, 37, 38].

The jaraquis are the principal fishing resource in the Central Amazon, requiring specific management strategies, although they are exploited in a multi-specific system that fishes migrating Characiforms using purse seine nets [23]. There are numerous difficulties in the execution of command-control strategies in the region, which makes co-management an alternative to be developed, requiring that users and governmental managers share knowledge to generate a common knowledge to be used in the management.

In order to find common bioecological knowledge between fishermen and scientific literature, it is necessary to compare them without the objective of evaluating the quality of one against the other, but merely to find common characteristics close enough to facilitate dialogue and the comprehension to share decisions. Hence, this research confirms the existence of kindred knowledge in some areas but strong differences in others.

As regards knowledge related to reproductive behavior, we observed that the information is usually related to the scientific knowledge. However, for fish of larger size and therefore of greater age similarities were fewer.

In the of the fishermen, the size of sexual maturity was similar among city-based and riverine people, however when comparing its results with the scientific knowledge published, the concept of the size of sexual maturity needed to be standardized. In fishing research, the size of sexual maturation is the length in which 50% of a population are capable of reproduction [39, 40], but when asked about the size of sexual maturation or "At which size does it begin to reproduce?", fishermen indicated the smallest fishes that had developed gonads. This may be an answer to the question "At which size do all fishes mature?", which indicates only the sizes of mature individuals. Considering the symmetry of the logistic curve [39], an average of 23,5 cm TL was estimated, which is compatible with sizes registered in the scientific literature. The same occurred with the age of first sexual maturity, estimated to be around 2 years.

The reproductive behavior has common features indicating that the jaraqui begins its reproductive period in the flood season, in downstream shoals, spawning in the confluence of the white and black water [41, 42]. Vieira *et al.* [43] identified the following spawning environments for the jaraqui: black water creek mouth, black river mouth and small lake creeks (a thin water connection between black water lakes and white water rivers).

Vazzoler *et al.* [42] and Ribeiro and Petrere [41] do not indicate that there is any parental care, which was compatible with the answers of the fishermen and with the typical behavior of Characiforms in the Amazon region [44].

Only few fishermen (just 51% of the professional and 17% of the riverine fishermen) were confident to confirm anything about the fecundity. Their answers were very diverse, indicating the difficulty in observing values in an appropriate magnitude. The fecundity is a variable usually correlated with the length of the female based on a potential model [39]. This increases the variability of the fecundity for any species and makes it difficult for fishermen to record. Even in scientific research there are few recordings for the species that reach a magnitude of thousands of eggs. This magnitude increases the difficulties in cultural registration, which may be different for k-strategic species, with few and easily countable eggs.

In relation to patterns of migration, riverine and city-based fishermen pointed out that the jaraquis were leaving the lakes during the flood season. In the scientific literature we found evidence that jaraquis form shoals that perform longitudinal migrations of the order of 1000 to 1300 km every year [41, 45], with a maximum displacement of around 300 km in white water rivers for reproduction, feeding and dispersion. In a general way, the migration of this species is considered very complex, being observed and described in a general model by Goulding [46], and detailed later on by Ribeiro [47]. This description of the migratory circuit was also confirmed by Vazzoler and Amadio [45] and Vieira *et al.* [48]. The fat-fish migration phase described by Ribeiro [47] is particularly important for the dispersion of the jaraquis to other tributaries of the Solimões-Amazonas system and happens in the middle of the full-flood seasons (from March to June). This was also registered as important by fishermen. Lateral migrations are very important type of behavior in the ecology of the Characiformes [49], but it was confirmed more frequently by riverine fishermen, which is natural considering the limited mobility of these people who therefore had more scope to examine aspects of the local dynamics.

The information provided by the fishermen about trophic ecology was similar to that found in the scientific literature, where Ribeiro [47], Vazzoler *et al.* [42] and Barthem and Fabré [50] observed that the jaraqui presents detritivorous habits. More specifically, Ribeiro [47] registered some of the items included in its diet, such as periphyton on the trees, submerged leaves, diatoms, sponge spicules, mushrooms and bacteria, mainly in the full season when the fish are dispersed in the flooded forests. During this period the mud or slime indicated by the fishermen is compatible with scientific information but with fewer details. Fishermen also indicated lakes and flooded areas as being important for its feeding in the flood and fall seasons.

On the other hand, the fishermen of the area pointed out that the jaraquis is predated by alligators, dolphins and other piscivorous fish, particularly catfish of several species. This corresponds to Ribeiro›s research [47], which mentions the great catfish, as the piraíba (*Brachyplatystoma filamentosum*) in the white water rivers, as the main predators of the species along with the dolphins, not to mention the alligators.

In growth analysis, significant divergences were expected, due to the fact that information on age is not available for researchers or fishermen, both facing difficulties in the identification of modal length groups. The count and measurement of age rings in rigid structures also present specific difficulties for scientists [51, 52], but have not been used by fishermen. Such differences contributed to the high dispersion between the values mentioned by fishermen and these found in the scientific literature.

Regarding mortality, fishermen have made clear that for larvae and juveniles, predators are the key factor, but also that fishing is the main factor for the adults. The latter corresponds to the high rate of fishing exploitation observed for the jaraquis in research in the area [8, 10] and the former to the findings of Bayley [53]. However, the fishermen's findings were very superficial and without details particularly for young fish, since these kinds of observation are not useful to them. Hence, fishermen's mortality estimates do not correspond to the instantaneous coefficient of mortality in fishing biology assessments, but rather to the causes of mortality and the variation in intensity.

The mean size of the first catch recorded in the literature indicated that commercial fishing has been concentrated on one year old jaraquis [8]. This fishing recruitment corresponds to jaraquis-escama-grossa usually from 24 to 27 cm CF and jaraquis-escama-fina still smaller (21-22 cm CF), values which are superior to the total length of 20.1 cm indicated by the fishermen. These differences of perception could indicate that the catch is being made of individuals smaller than those observed in the landing harbors, due to the fishermen selecting larger fishes.

Evaluating all the information comparatively, compatibilities in factors were found related to the use of the resource that make both fishermen types into good observers. However, if certain phenomena do not directly affect the interests of the fishermen, these tend to be ignored. The professionals seem to be more efficient in gathering general information, while the riverine fishermen were more efficient in gathering information that required local observations.

Compatibility of knowledge is, however, not enough; it is also necessary to develop a culture of dialogue between stakeholders and users in the establishment of more efficient and sustainable strategies. User knowledge has been gathered by researchers and may contribute as relevant information to government management plans [[33]], particularly for resources with scarce or non-existent data (e.g. [54, 55, 56]). On the other hand, ethnoecological knowledge can also be the basis for the design of management plans, as has been done by some Amazon communities [17, 18], and are the informational basis for fishing agreements recognized by government environmental authorities in Brazil (IBAMA normative instruction 29/2002). Ethnoecological research may help the dialogue between fishermen, researchers and stakeholders, increasing willingness and cooperation to find effective solutions to optimal management of fish resource but also of the environment.

The present paper suggests some themes that can be used to introduce affinities in order to facilitate this dialogue but this only represents one initiative among many others necessary to effectively put in place a co-management regime. Success in this task can change a centralized decision making culture and encourage collaboration in the productive and sustainable use of fishing resources in the region.

CONCLUSIONS

Dialogue between managers and fishermen would develop better if discussions concentrated on information about the reproductive cycle, spawning and nursery grounds, maturity size and feeding preferences of these species. Age and growth as much as mortality were useful information on scientific grounds, but their use in co-management is minimized due to difficulties encountered by fishermen in understanding and appreciating the estimates done. Strategies aimed at developing the dialogue could include short practical courses to encourage fishermen (and also their families) to observe subjects in order to improve their understanding of technical procedures and results and to enable the use of age-related and mortality knowledge. Current thinking suggests concentrating on co-management discussions on the improvement of area and seasonal restrictions.

DECLARATIONS

Acknowledgements

The authors would like to thank the Integrated Program of Aquatic Resources and Várzea of the Federal University of Amazonas (PYRÁ/UFAM) for the infrastructure given and fishermen interees for their collaboration and readiness. CAPES sponsored a ship for the second author in the Post-Graduate Program in Environment Sciences and Sustainability in the Amazon and CNPq by financial support.

Authors' contributions

The authors of this paper were equally responsible for every aspect of the research, the conclusions, and the writing of the paper. All authors read and approved the final manuscript.

REFERENCES

1. McGrath DG, Castro F, Futemma C, Amaral BD, Calabria J: Fisheries and the evolution of resource management on the lower Amazon floodplain. Hum Ecol. 1993, 2: 167-195. 10.1007/BF00889358.
2. Furtado LG: “Reservas pesqueiras”, uma alternativa de subsistência e de preservação ambiental: Reflexões a partir de uma proposta de pescadores do médio Amazonas. Povos das Águas realidade e perspectiva na Amazônia. Edited by: Furtado L, Mello AF, Leitão W. 1993, Belém: MPEG/UFPa, 243-276.
3. Petrere M: As comunidades humanas ribeirinhas da Amazônia e suas transformações sociais. Populações humanas, rios e mares da Amazônia. Edited by: Diégues AC. 1992, São Paulo: Anais do IV Encontro de Ciências Sociais e o Mar no Brasil, 31-68.
4. Merona B, Bittencourt MM: A pesca na Amazônia através dos desembarques no mercado de Manaus: resultados preliminares. Mem Soc Cienc Nat La Salle. 1988, 48: 433-453.
5. Petrere M: A pesca comercial no rio Solimões-Amazonas e seus afluentes: análise dos informes do pescado desembarcado no Mercado Municipal de Manaus (1976-1978). Cien Cult. 1985, 37: 1987-1999.
6. Batista VS, Petrere M: Characterization of the commercial fish production landed at Manaus, Amazonas State, Brazil. Acta Amazonica. 2003, 33: 53-66.
7. IBAMA: Estatística Pesqueira do Amazonas e Pará - 2003. 2006, Manaus: IBAMA/ProVárzea
8. Batista VS: Biologia e administração pesqueira de alguns Caraciformes explotados na Amazônia Central.

Professor titular thesis. 2000, Manaus: UFAM
9. Vieira EF: Determinação da idade e crescimento do jaraqui de escama grossa (*Semaprochilodus insignis*) na Amazônia Central. Master dissertation. 1999, Manaus: INPA/UFAM
10. Vieira EF: Dinâmica sazonal e interanual da estrutura populacional e do impacto da exploração pesqueira dos Jaraquis (Semaprochilodus spp. Schomburgki, 1841) nos subsistemas hidrográficos da Amazônia Central. Doctoral thesis. 2003, Manaus: INPA/UFAM
11. Pomeroy RS, Rivera-Guieb R: Fishery co-management: a practical handbook. 2005, Oxfordshire, CABI Publishing
12. McGoodwin JR: Comprender las culturas de las comunidades pesqueras. 2002, Roma: FAO Doc Técn Pesca, 401:
13. Posey DA: Indigenous knowledge and development: an ideological bridge to the future. Cien Cult. 1983, 35 (7): 18-24.
14. Diegues AC: Pesca Artesanal: Tradição e Modernidade. 1989, São Paulo: IOUSP/FORD/UICN
15. Begossi A, Garavello JC: Notes on the ethnoichthyology from Tocantins River. Acta Amazonica. 1990, 20: 341-351.
16. Marques JGW: Aspectos ecológicos na etnoictioloogia dos pescadores do Complexo Estuário-Lagunas Mundaú-Maguaba - Alagoas. Doctoral Thesis. 1991, Campinas: UNICAMP
17. Ribeiro MOA, Fabré NN, (Eds): Sistemas Abertos Sustentáveis - SAS: uma alternativa de gestão ambiental na Amazônia. 2003, Manaus: EDUA
18. Ruffino ML, Isaac VJ: The fisheries of the lower Amazon: questions of management and development. Acta Biol Venez. 1994, 15: 37-46.
19. Petrere M: Notas sobre a pesca dos índios Kayapó da aldeia de Gorotire, Rio Fresco, Pará. Bol Mus Par Emílio Goeldi, Sér Antrop. 1990, 1: 5-17.
20. Velthem LH: Os Wayana, as águas, os peixes e a pesca. Bol Mus Par Emílio Goeldi, Sér Antrop. 1990, 6: 107-116.
21. Levine DM, Berenson ML, Stephan D: Estatística teoria e aplicações: usando o Microsoft® Excel em português. 2000, Rio de Janeiro: Editora JC
22. Zar JH: Biostatistical Analysis. 1999, New Jersey: Prentice-Hall
23. Batista VS, Isaac VJ, Viana JP: Capítulo 2 - Exploração e manejo dos recursos pesqueiros da Amazônia. A pesca e os recursos pesqueiros na Amazônia. Edited by: Ruffino ML. 2004, Manaus: IBAMA, 63-151.
24. Hardin G: The tragedy of the commons. Science. 1968, 162: 1243-1248. 10.1126/science.162.3859.1243.
25. Hardin G: The tragedy of the unmanaged commons. Tree. 1994, 9: 199-
26. Arlinghaus R, Mehner T, Cowx IG: Reconciling traditional inland fisheries management and sustainability in industrialized countries, with emphasis on Europe. Fish Fish. 2002, 3: 261-316.
27. Caddy J: Fisheries management in the twenty-first century: will new paradigms apply?. Rev Fish Biol Fisher. 1999, 9: 1-43. 10.1023/A:1008829909601.
28. Flaaten O, Salvanes AGV, Schweder T, Ulltang Ø: Fisheries management under uncertainty - an over. Fish Res. 1998, 37: 1-6. 10.1016/S0165-7836(98)00122-2.
29. Stergiou KI: Overfishing, tropicalization of fish stocks, uncertainty and ecosystem management: resharpening Ockham's razor. Fish Res. 2002, 55: 1-9. 10.1016/S0165-7836(01)00279-X.
30. Brown DN, Robert SP: Co-management of Caribbean Community (CARICOM) fisheries. Mar Policy. 1999, 23: 549-570. 10.1016/S0308-597X(98)00040-2.
31. D'Incao F, Reis EG: Community-based management and technical advice in Patos Lagoon estuary (Brazil). Ocean Coast Manage. 2002, 45: 531-539. 10.1016/S0964-5691(02)00084-4.
32. IBAMA: Projeto IARA - Administração dos Recursos Pesqueiros do Médio Amazonas: Estados do Pará e Amazonas. 1995, Brasília: IBAMA, Coleção Meio Ambiente, Série Estudos de Pesca, 15:
33. Pomeroy RS, Rivera-Guieb R: Fishery co-management: a practical handbook. 2005, Oxfordshire: CABI Publishing
34. Johannes RE: The renaissance of community-based marine resource management in Oceania. Annu Rev Ecol Syst. 2002, 33: 317-340. 10.1146/annurev.ecolsys.33.010802.150524.
35. Lane DE, Stephenson RL: Fisheries co-management: organization, process, and decision support. J Northwest Atl Fish Sci. 1998, 23: 251-265. 10.2960/J.v23.a16.
36. Sharp GD: Future climate change and regional fisheries: a collaborative analysis. 2003, Rome: FAO Fisheries Technical Paper, 452:
37. Begossi A, Figueiredo JL: Ethnoichthyology of southern coastal fishermen: cases from Búzios Island and Sepetiba Bay (Brazil). Bull Mar Sci. 1995, 56: 682-689.
38. Marques JGW: Pescando pescadores: etnoecologia abrangente no baixo São Francisco Alagoano. 1995, São Paulo: NUPAUB-USP
39. King M: Fisheries Biology, Assessment and Management. 1995, Oxford: Fishing News Books
40. Vazzoler AEAM: Manual de métodos para estudos biológicos de populações de peixes. 1982, Brasília: CNPq
41. Ribeiro MCL, Petrere M: Fisheries ecology and management of the Jaraqui (*Semaprochildus taenirus, S. insignis*) in the Central Amazonia. Regul River. 1990, 5: 195-215. 10.1002/rrr.3450050302.
42. Vazzoler AEAM, Amadio AS, Malta MCC: Aspectos biológicos dos peixes Amazônicos: XI - Reprodução das espécies do gênero *Semaprochilodus* (Characiformes, Prochilodontidae) no baixo rio Negro, Amazonas, Brasil. Rev Bras Biol. 1989, 49: 165-173.
43. Vieira EF, Fabré NN, Sousa KNS, Araújo LMS: Identificação de novas áreas de desova do jaraqui (*Semaprochilodus spp.*) em microregiões da Amazônia Central. Olam. 2002, 2: 284-308.

44. Vazzoler AEAM, Menezes NA: Síntese de conhecimentos sobre o comportamento reprodutivo dos Characiformes da América do Sul (Teleostei, Ostariophysi). Rev Bras Biol. 1992, 52: 627-640.

45. Vazzoler AEAM, Amadio AS: Aspectos biológicos de peixes Amazônicos. XIII. Estrutura e comportamento de cardumes multiespecíficos de Semaprochilodus (Characiformes, Prochilodontidae) no baixo Rio Negro, Amazonas, Brazil. Rev Bras Biol. 1990, 50: 537-546.

46. Goulding M: Ecologia da pesca do rio Madeira. 1979, Manaus, INPA

47. Ribeiro MCL: As migrações dos jaraquis (Pisces Prochilodontidae) no rio Negro Amazonas Amazonas Brasil. 1983, Master Dissertation, Manaus: INPA/FUA

48. Vieira EF, Isaac VJ, Fabré NN: Biologia reprodutiva do tambaqui *Colossoma macropomum* Cuvier, 1818 (Teleostei, Serrasalmidae), no baixo Amazonas, Brasil. Acta Amazônica. 1999, 29: 625-638.

49. Cox-Fernandes C: Lateral migrations of fishes in Amazon floodplains. Ecol Freshw Fish. 1997, 6: 6-44.

50. Barthem RB, Fabré NN: Capítulo 1 - Biologia e diversidade dos recursos pesqueiros na Amazônia. A pesca e os recursos pesqueiros na Amazônia. Edited by: Ruffino ML. 2004, Manaus: IBAMA, 17-62.

51. Campana SE: Accuracy, precision and quality control in age determination, including a re of the use and abuse of age validation methods. J Fish Biol. 2001, 59: 197-242. 10.1111/j.1095-8649.2001.tb00127.x.

52. Jearld A: Age determination. Fisheries Techniques. Edited by: Nielsen LA, Johnson DL. 1983, Bethesda: AFS, 301-324.

53. Bayley PB: Fish yield from the Amazon in Brazil: comparisons with African river yields and management possibilities. Trans Amer Fish Soc. 1981, 110: 351-359. 10.1577/1548-8659(1981)110<351:FYFTAI>2.0.CO;2.

54. Rosa IL, Alves RRN, Bonifacio KM, Mourao JS, Osorio FM, Oliveira TPR, Nottingham MC: Fishers' knowledge and seahorse conservation in Brazil. J Ethnob Ethnom. 2005, 1: 12-10.1186/1746-4269-1-12.

55. Alves RRN, Nishida AK, Hernandez MIM: Environmental perception of gatherers of the crab caranguejo-uçá (*Ucides cordatus*, Decapoda, Brachyura) affecting their collection attitudes. J Ethnob Ethnom. 2005, 1: 10-10.1186/1746-4269-1-10.

56. Begossi A, Silvano RAM: Ecology and ethnoecology of dusky grouper [garoupa, *Epinephelus marginatus* (Lowe, 1834)] along the coast of Brazil. J Ethnob Ethnom. 2008, 4: 20-10.1186/1746-4269-4-20.

57. Alves MI, Andrade A: Aspectos da reprodução do jaraqui escama fina, *Semaprochilodus taeniurus* (Vallenciennes, 1881). Ciên Agronom. 1992, 23: 123-137.

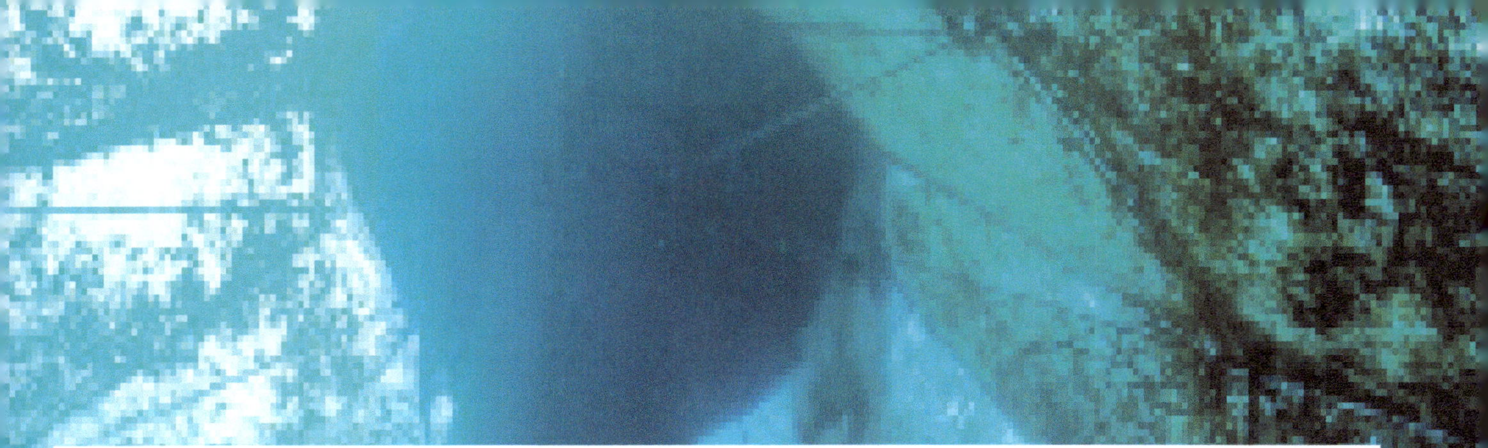

CHAPTER 16

CLIMATE CHANGE IMPACTS AND ADAPTATION OF COMMERCIAL MARINE FISHERIES

Neil J. Holbrook, Johanna E. Johnson

INTRODUCTION

Commercial marine fishing is an important sector of the Australian economy. The Australian Bureau of Agricultural and Resource Economics and Sciences estimate the gross value of production of commercial fisheries and aquaculture in 2009/10 was ~ AUD 2.18 billion. Of this, AUD 1.2 billion was exported in 2009/10 (ABARES 2011), and AUD 1 billion in 2008 (estimate by the Food and Agriculture Organization (FAO); Ridge-Partners 2010). Marine fishing activities (including trapping and Indigenous hunting) provided employment for 7,646 people in 2009/10, a 24 % increase from the previous year, and indirectly employed 6,203 people in seafood wholesaling and processing in 2006 (ABARES 2011).

It is anticipated that marine fisheries will be directly affected by climate-related stresses through changes and access to fisheries resources, and indirectly through habitat changes. Such climate change impacts are already being observed in Australia with changes to fish habitats and distribution (e.g. Hallegraeff 2010; Last et al. 2011; Johnson et al. 2011; Pearce et al. 2011; Poloczanska et al. 2009, 2012), and ability to operate in suitable fishing grounds (Tobin et al.2010). The magnitude of future impacts, and whether marine ecosystems and fisheries can adapt, depends on the rate of change, the response of the marine system to climate change, and other drivers; social (e.g. markets, policy, governance and management) and ecological (e.g. habitat declines and pollution). The adaptation pathways chosen will be critical to the future sustainability of fisheries and will need to consider strategies that are both autonomous (e.g. fish moving to cooler waters) and planned (e.g. factoring climate impacts into stock assessments). Deliberate adaptation that seeks to avoid possible maladaptation requires consideration of: (i) which species, communities or ecosystems are most vulnerable to climate change; (ii) the likelihood of climate-related ecological changes affecting dependent industries; and (iii) what planned adaptations can be implemented to effectively respond to future change.

This review provides a synthesis of the recent published literature relevant to: climate change impacts on, and adaptation of, commercial marine fishing in Australia; how fisheries industries can adapt; and barriers to adaptation. Specifically, it represents key findings of a larger initiative to assess whether research priorities identified in Australia's National Climate Change Adaptation Research Plan for Marine Biodiversity and

Citation: Neil J. Holbrook & Johanna E. Johnson, Climate Change Impacts and Adaptation of Commercial Marine Fisheries, DOI 10.1007/s10584-014-1110-7.

Resources (Mapstone et al. 2010) are being addressed. Further, it complements a review of climate change science for marine biodiversity conservation (Johnson and Holbrook 2014). Ultimately, research will need to remain committed to addressing climate change impacts and adaptation knowledge gaps so appropriate and effective 'no regrets' adaptation strategies can be applied in the future.

OBSERVED AND ANTICIPATED IMPACTS OF CLIMATE CHANGE

Fisheries climate change 'hotspots' have been identified in Australia's southeast and tropical north—with surface and upper ocean temperatures in the southeast increasing at ~3-4 times the global average rate (Holbrook and Bindoff 1997; Ridgway 2007). These represent locations where significant changes for marine and estuarine species are occurring (Stuart-Smith et al.2009; Booth et al. 2011). Temperate fish stocks off southeast Australia are likely to change as a result of climate change, particularly through the direct effects of exposure to increasing ocean temperature affecting species with limited thermal ranges, and indirectly through poleward movement of species (Hobday and Poloczanska 2010; Koehn et al. 2011), habitat changes, and increasing disease outbreaks (Danovaro et al. 2010).

Evidence of climate-related impacts on species distributions is emerging in Australia through a number of studies. Ling et al. (2009) documented the expansion of long-spined sea urchins (*Centrostephanus rodgersii*) in Tasmania from New South Wales (NSW) that are altering benthic habitats critical for valuable rock lobster (*Jasus edwardsii*) and abalone (*Haliotis rubra, H. laevigata*) fisheries. Distributional shifts attributed to warming temperate oceans are documented by: Last et al. (2011), with 45 fish species shifted south since the late 1800s; Stuart-Smith et al. (2009), noting that although Tasmanian rocky reef community structure remains unchanged, there have been southern range shifts (e.g. whiting *Siphonognathus attenuates* and luderick *Girella tricuspidata*) and new records (e.g. rock cale *Aplodactylus lophodon*); Pitt et al. (2010), who document range shifts in 16 species of Tasmanian invertebrates; and Johnson et al. (2011), who show cascading effects of ecological change in benthic (rocky reef) and pelagic systems. Community monitoring in Tasmania has recorded marine species that have extended their ranges, including eastern blue groper (*Achoerodus viridis*), eastern rock lobster (*J. verreauxi*), mahi mahi (*Coryphaena hippurus*) and grey morwong (*Nemadactylus douglasii*).[1]

Dang et al. (2012) note that water temperature and immune response in commercial abalone (*Haliotis rubra, H. laevigata*) in South Australia are correlated, providing some indirect evidence for the possible physiological effects of climate change on fish species. Pecl et al. (2009) conclude that climate change (particularly ocean warming) is expected to have a significant impact on the Tasmanian rock lobster (*J. edwardsii*) fishery, causing declines in biomass and recruitment in northern Tasmania by 2030 under the IPCC SRES A1B ('business-as-usual') emissions scenario, and in southern regions by 2070 under the A1B and A1FI (fossil intensive) scenarios. This is supported by current catch rate monitoring showing a long-term trend of decline, which is expected to continue. A study of banded morwong (*Cheilodactylus spectabilis*) metabolic rate in the Tasman Sea shows increased growth in the middle of their range but reduced growth at the northern edge coinciding with warmer ocean temperatures, potentially leading to declining productivity and range contraction (Neuheimer et al. 2011). Similarly, higher ocean temperatures are shown to result in earlier age at maturity and higher natural mortality of species (Froese and Binohlan 2000; Gardner et al. 2011).

Recent studies document the relationships between catch rates and climate variables. A study of school prawns (*Metapenaeus macleayi*) in northern NSW found that growth and movement were affected by river discharge, with higher discharge rates usually resulting in increased commercial catches (Ives et al. 2009). Meynecke and Lee (2011) show positive correlations between commercial catches of barramundi (*Lates calcarifer*), mud crabs (*Scylla serrata*), sea mullet (*Mugil cephalus*), dusky flathead (*Platycephalus fuscus*), whiting (*Sillago spp.*), tiger prawns (*Penaeus monodon, P. semisulcatus*) and endeavour prawns (*Metapenaeus endeavouri, M. ensis*) with sea surface temperature and rainfall on the Queensland coast. Similarly, NSW commercial fisheries data show that catch-per-unit-effort (CPUE) increased in proportion to freshwater flow for four commercial estuary species (dusky flathead, luderick *G. tricuspidata*, sand whiting *S. ciliata* and sea mullet) and decreased during drought (Gillson et al. 2009). Booth et al. (2011) find similar relationships, with increases in overall CPUE of the northern mud crab fishery interpreted as a response to sea temperature increases.

Alternatively, negative impacts through indirect effects from climate-related habitat loss or degradation are also likely to be very important. For example, studies of seaweed—an important component of temperate rocky reefs—show that bleaching as a result of temperature-mediated disease (Campbell et al. 2011) and southern distributional shifts over the last 40 years (Wernberg et al. 2011) are likely to have implications for habitat-dependent fish.

Tropical fisheries that target species dependent on habitats such as coral reefs, mangroves and seagrass (e.g. prawns, mud crabs, coral trout and aquarium species), are likely to change as a result of climate-related impacts on these habitats (Badjeck et al. 2010; MacNeil et al. 2010; Donnelly 2011; Pratchett et al. 2011a; Bell et al. 2013). For example, barramundi (*L. calcarifer*) landings correlated with an index of climate variability (Balston 2009a) flag the possible effects on early life stages as rainfall, river flow and temperature variability influence nutrient availability and nursery habitat productivity (Balston 2009b). Coral reef fisheries are likely to be affected by future reductions in population connectivity due to ocean warming accelerating larval development, with possible earlier settlement to reefs that will reduce dispersal distances. This is expected to reduce the spatial scale of connectivity and, in conjunction with increased habitat fragmentation due to coral bleaching and ocean acidification, potentially affect catch rates and species availability as reef fish community composition changes (Munday et al. 2009). Further, habitat associations and community dynamics are high priority research gaps that need to be addressed to better understand the impacts of climate change on coral reef fish (Wilson et al. 2010).

Climate change is expected to affect species populations, communities, ecosystem function, and ultimately fishery stocks (Pörtner and Peck 2010; Bell et al. 2013), and recent modelling supports this. Models of future climate change effects on 12 coastal marine food webs in Australia under the A2 emissions scenario over the next 50 years project some increases in coastal primary production in tropical Australia and minor increases in the southeast and southwest, which could benefit fisheries catches and value (Brown et al. 2010). More recent ecosystem model projections suggest an ecosystem regime shift in southeast Australia by 2060 (under the A2 scenario), with primary producers and pelagic systems likely to benefit from climate change, while demersal systems will most likely decline (Fulton 2011). Projections for the wider Pacific region are that oceanic fisheries (primarily tuna) are expected to shift their distribution—following moving productivity hotspots—and become more abundant in the east, while coastal fisheries are likely to decline (Bell et al. 2013). Pelagic fisheries in Australia target 14 large pelagic species using longline that are projected to experience changes in distribution and abundance as their core habitats move south and contract (Hobday 2010).

Projections (A1B scenario) of global catch potential from 2005 to 2055 show an average of 30–70 % increase in high latitude regions and a ~40 % decline in the tropics (Cheung et al. 2010). Further modelling by Cheung et al. (2011) projects that these fishery catch potentials may be reduced by an additional ~10 % with the inclusion of biogeochemical factors (under the A1B scenario by 2050). Although caution should be exercised in translating these results to the Australian context (Fulton 2011), there are implications for fisheries targeting species at the edge of their range, particularly in tropical regions where Australia may become a last refuge for Indo-West Pacific species as the oceans warm (Hobday and Poloczanska 2010). There is now evidence that fisheries distributions are changing due to ocean warming, with catch data from 1970 to 2006 showing that mean temperature of catch has increased globally in 52 large marine ecosystems, and is significantly positively related to increasing ocean temperature (Cheung et al.2013).

Observational and modelling studies suggest that the greatest impacts of climate change on fishery stocks are likely to manifest in southeast Australian marine waters (Hobday and Poloczanska 2010), and for some fisheries in tropical regions (Pratchett et al. 2009, 2011a). The nexus between tropical and temperate systems—the subtropics—is an important zone also likely to experience changes to species abundances and community composition. The fisheries stocks most at risk are those dependent on vulnerable habitats (Koehn et al. 2011), cool temperate endemic species (Hobday and Poloczanska 2010), and demersal species (Barange and Perry2009; Pratchett et al. 2011b), with some shifts in distribution, physiology and catch rates already observed.

WHAT ARE THE INDIRECT EFFECTS OF CLIMATE CHANGE ON FISHERIES?

Coastal areas in tropical Australia are projected to experience more intense weather events that can reduce fishing time, as well as destroy or damage fishery assets and infrastructure such as landing sites, boats and gear (Daw et al. 2009; Badjeck et al. 2010). Increased storm and wave activity may reduce the: (i) number of fishing days, (ii) ability to reach some locations, and (iii) seasonal availability of fish (Hobday and Poloczanska 2010). As an example, the effects of two severe tropical cyclones (Tropical Cyclone (TC) Hamish and TC Justin) on fish abundance, catch composition and catch rates of the coral reef finfish fishery in the Great Barrier Reef (GBR) were that 66 % of the reef structure had been damaged but there was no measurable change in the associated fish community and abundance (Tobin et al. 2010). However, depressed fisheries catches of target species (coral trout *Plectropomus leopardis* and red throat emperor *Lethrinus miniatus*) by >30 % lagged up to 9 months post-cyclone. This reduced 'catchability' of fish is uncorrelated with abiotic data such as reef structural damage or sea temperature. TC Justin, on the other hand, resulted in depressed catch rates by up to 50 % for coral trout, accompanied by ~200 % increase in red throat emperor catch related to a cool-water event following the cyclone. These results demonstrate how extreme weather events can significantly reduce fishing opportunities during the event and catch rates up to 12 months later.

The ability to maintain viable catches is likely to be influenced not only by extreme weather but also distributional shifts of key commercial species away from the major ports/landing sites and economic zones (Booth et al. 2011). Diminished access and property rights as distributions shift may become a significant issue in some regions, while other fishers may gain access to fish as they move (Hobday and Poloczanska 2010). Research on projecting these changes needs to be targeted in locations most likely to experience these shifts (e.g. southeast Australia), and species most likely to expand or contract their ranges. Some range-shifters may become 'locally invasive' as they move south (e.g. the long-spined sea urchin, Ling et al. 2009), impacting on habitats and fishery species. Research is needed on the impacts of these distributional shifts on important fisheries, as they are likely to occur in the short-term.

WHAT ARE THE SOCIO-ECONOMIC IMPLICATIONS?

Grafton (2010) identifies communities with a high proportion of members employed in a particular capture fishery, low overall employment rates, low geographic mobility, and specialised skills as being most vulnerable to changes in their fisheries due to climate change. Such communities are likely to be small remote towns with fisheries as a main income source, e.g. Bowen in tropical Queensland and Kalbarri in Western Australia. As fisheries resources change, small-scale fisheries are less able to adapt due to their limited resources (Daw et al.2009). Based on examples worldwide, fishing communities dependent on local resources of a limited number of species are more vulnerable to fluctuations in stocks, whether due to overfishing, climate change, or other causes (Brander 2010).

Madin et al. (2012) show that the socio-economic implications of range-shifting marine species in Australia include positive and negative effects on economic activities, human health, ecosystem services and management. They explored the potential for utilising different management tools in the Australian context, including spatial protection using no-take and multiple-use protected areas, and standard fisheries management tools. However, they concede that confidence is low in how fisheries management should most effectively deal with range-shifting species. Although approaches to assess the economic and social resilience of fisheries to climate change are limited, one study identifies 12 economic resilience attributes for the rock lobster fishery in Tasmania across the business (e.g. profitability), sectoral (e.g. diversification) and governance (e.g. property rights) levels (van Putten et al. 2013). A more general set of indicators has been developed for adaptation of social systems to climate change that include perception of risk, ability to cope with change (financial and emotional flexibility), and livelihood diversity (Marshall et al. 2010).

The socio-economic impacts of reduced fishing opportunities and depressed catches after tropical cyclones were found to be exacerbated in the GBR by a heavy reliance on high-value live coral trout limiting the industry's ability to adapt to change (Tobin et al. 2010). This reliance on a single species destined for a single market makes the fishery highly vulnerable to declines in catch rates. Sustained reductions in catch rates increased operating costs and reduced profit. In contrast, the recreational and charter fishing sectors employed temporary adaptation strategies (e.g. species diversification and fishing location shifts) and were not significantly impacted by the cyclones.

HOW CAN FISHERIES ADAPT AND WHAT OPPORTUNITIES MIGHT EXIST?

Fishers already live with and adapt to climate variation (Badjeck et al. 2010; Tobin et al. 2010), by moving the location and time that they fish, and the species they target. For example, fishers in the east coast longline fishery use a range of ports to land their catch, and change where they operate as fish distribution and availability changes (Hobday et al. 2009). Analogously, the Indian Ocean tsunami of 2004 impacted India, among other countries, and resulted in a revision of fisheries management in the south with an increased focus on livelihood diversification, coastal rather than offshore fisheries, and post-harvest employment opportunities (FAO 2010).

In Australia, the response of the GBR finfish fishery after tropical cyclones was a short-term shift in effort, with large operators moving more than smaller ones (Tobin et al. 2010). No operators employed long-term adaptations, or diversified their target species due to lack of appropriate gear and the price differential between export live fish and domestic markets. Fisher surveys identified government support to: (i) provide access to locations closed to fishing, (ii) remove other management controls, (iii) provide low interest loans, and (iv) provide relief funds, as the best ways to minimise the impacts of cyclones on the fishery.

Climate change is expected to favour some fisheries species in Australia, such as warm temperate species, changing their distribution and relative abundance. For example, southern fisheries may have increased opportunities where tropical species move south, but lose opportunities where southern species contract (Hobday and Poloczanska 2010). If commercial fishers can change their harvest strategies and processing without incurring significant additional costs, travel time or associated fuel consumption, they may be able to take advantage of these opportunities. Diverting effort to target new or different species will be an important adaptation strategy as distributions shift (MacNeil et al. 2010). The capacity to quickly adapt to changing fisheries resources using new harvest techniques and gear will be a significant factor determining the future success of commercial fisheries (Badjeck et al. 2010). For example, if the distribution of southern bluefin tuna (*Thunnus maccoyii*) contracts south, as predicted, longline fishers would be expected to experience fewer seasonal area restrictions and be able to target other species (Hobday and Poloczanska 2010).

A blueprint for fisher communities recently developed emphasises the importance of understanding the combined role of climate and non-climate pressures in shaping marine sectors in small coastal communities to avoid maladaptation. Case studies (van Putten et al.2014) clearly illustrate that local conditions are important and that adaptation options are mostly location and industry specific, while others are generally applicable across different coastal communities. Coastal management agencies can encourage marine industries to benefit from opportunities that arise from climate change, e.g. from new range shifting species, and create income and labour market opportunities for the charter and recreational sectors. Encouraging community level adaptation to the negative impacts of climate pressures can be found through identifying alternative species, diversification into other sectors, and professional and workforce mobility and flexibility.

HOW CAN MANAGEMENT ADAPT TO MAINTAIN SUSTAINABLE FISHERIES?

The ability of fisheries to adapt and remain viable and sustainable under climate change will depend on the nature of current and future management strategies, with adaptive management frameworks identified as the most likely to succeed (Plaganyi et al. 2011a). In response to projected greater spatial and temporal variability in landings, fishers will have to become more mobile and responsive to fishing opportunities (Badjeck et al. 2010). This will require more flexible management and policy with an adaptive management paradigm, allowing for uncertainty (Brander 2010; Grafton 2010; Johnson and Welch 2010; OECD 2010). It has been suggested that climate change and overfishing can have significant synergistic impacts on fisheries (e.g. North Sea cod fishery, Kirby et al. 2009), while interactions between fishing, ocean warming and ocean acidification over time have shown that all three stressors synergistically impacted top predators and fish biomass in southeast Australia (Griffith et al. 2012). Improved fisheries and ecosystem management will be important for minimising other stressors and promoting more resilient fish stocks (Allison et al. 2009; Perry et al. 2010; Koehn et al. 2011). Many of the management improvements needed do not require new science or understanding; they require development of effective, responsive institutions and tools for achieving adaptive management (Brander 2010), and an ecosystem approach to fisheries management (OECD2010; Hobday et al. 2011).

The range of adaptation options available for fisheries management to minimise the impacts of climate change include, but are not limited to: preserving age and geographic structure of fished populations; protecting key functional groups; multi-jurisdictional management of stocks; integrated management systems that include social, economic and ecological values; reducing overcapacity in fisheries; incorporating a climate change catch quota into stock assessments; and reducing barriers to adaptation, such as resource depletion and resource reliance (Badjeck et al. 2010; Johnson and Welch 2010;

MacNeil et al. 2010). Specific planned adaptations can also reduce or delay impacts, or support autonomous adaptation in the short- to medium-term, such as focused management of target species, habitat protection and restoration, enhancement of native populations, assisted migration, and ex situ conservation (Koehn et al. 2011). Brander (2010) identifies focused fisheries management as being particularly important for populations at the edge of a species range. These populations can be valuable sources of genetic material to climate extremes but also have reduced surplus production, thus increasing their vulnerability to (previously tolerable) levels of fishing. Special protection should therefore be afforded to populations at the trailing edge of their range that may also experience the first impacts of climate change (e.g. increasing ocean temperature).

While the level of resource-dependency and socioeconomic circumstances will drive to some degree the most appropriate adaptation options, adapting to a changing climate will inevitably require trade-offs between biological assets and socioeconomic benefits (Koehn et al. 2011). Ultimately, difficult management decisions may need to be made that consider short-term costs and long-term gains, with preferable adaptations being those that are win:win in the short- and long-term ('no regrets' adaptations), or lose:win that secure future sustainability with some short-term costs (Grafton 2010).

Adaptive management frameworks—such as co-management and ecosystem approaches—focus on building adaptive capacity, reducing vulnerability and increasing socio-ecological resilience while promoting sustainable fisheries. These are all necessary attributes for coping with climate change (Daw et al. 2009). Madin et al. (2012) suggest that the application of frameworks that build resilience and adaptive capacity across the social-ecological system will be the optimal fisheries management response.

In cases where species managed under a quota system move to locations where fishers do not hold quota, designing a flexible framework and markets for trading (OECD 2010), or transferring quotas (McIlgorm et al. 2010), may be options for overcoming the issue of access to a moving resource. Management may also need to consider strategies that allow for future projected changes, e.g. expansions of species ranges or changes in abundance (Koehn et al.2011). Governments have an important role to play in identifying and removing institutional barriers to change, periodically reviewing protection measures to ensure they are still applicable, and ensuring that they do not dilute incentives for fishers to adapt to future climate change (OECD 2010).

WHAT ARE THE BARRIERS TO ADAPTATION?

Adaptation of fisheries to external impacts is possible for even small sectors, and examination of examples of successful adaptations can provide valuable lessons regarding appropriate ways to manage fisheries in an uncertain future, and how to overcome barriers to adaptation (OECD2010). Two reviews (Brander 2010; Johnson and Welch 2010) identify factors expected to limit the ability of fisheries to adapt to climate change: the projected rapid rate of change; the compromised health of overfished stocks; weak social and economic structures; a high dependence on fisheries; inflexible management regimes; and factors that undermine resilience, such as loss of biodiversity, habitat destruction, pollution, invasive species, and pathogens. Uncertainty about future climate change encourages the use of short planning horizons that focus on immediate problems while delaying mitigation actions until more information becomes available (McIlgorm et al. 2010). To avoid this, Miller et al. (2010) propose focusing on integrated science that supports timely institutional responses, a broader planning perspective, and development of resilience-building strategies. Johnson and Welch (2010) propose an integrated assessment approach that can identify highly vulnerable species and targets for action, facilitating prioritisation of management effort and informed application of adaptation options (Koehn et al. 2011; Welch and Johnson 2013). However, despite recent progress towards studying marine social and ecological environments as coupled systems, many issues still challenge full integration (Perry and Ommer 2010).

Socioeconomic constraints are identified as significant barriers to fisheries adaptation (McIlgorm 2010), particularly in the context of oceanic fishers (e.g. tuna) that have made long-term investments in fishing vessels, storage and processing. Future changes in the distribution and abundance of stocks due to climate change and the expected increases in fuel prices are likely to influence fishers' ability to travel greater distances to catch moving stocks, or change gear or practices to target different species. Modelling shows that, from an economic perspective, larger operators have an adaptive advantage that enables them to change their operation in response to fish redistributions (Fulton 2011). In contrast, family-based fishing operations have diminished adaptive capacity due to their long-term association with fishing that connects their identity with the activity and potentially limits their willingness to adopt adaptive strategies that involve occupational diversification (i.e. new income generating activities or leaving fisheries; Coulthard 2009).

Changes in fish stock distribution and the abundance of target and non-target (but potentially "new") species are likely to disrupt existing access and allocation arrangements (Daw et al.2009; OECD 2010). In Australia, where fisheries are managed by a range of jurisdictions, climatic variations that lead to shifts in resource distribution may raise jurisdictional issues or limit flexibility to access cooperatively managed or shared resources. This was experienced in the North Pacific with the wild salmon fishery (Badjeck et al. 2010), and is projected to occur as Australia's east coast tuna stock move south, necessitating changes to the jurisdiction of the Australian fishery (McIlgorm et al. 2010).

Significantly, overcoming barriers to change within the fishing industry will require ongoing involvement of the fishing industry in co-management and self-governance initiatives to assist governments meet the new flexible management paradigm (McIlgorm et al. 2010). Incorporating multi-stakeholder participation, an ecosystem-based approach, a long-term perspective, and flexible livelihood and governance strategies into future fisheries management, will be key to effective adaptation to climate change (Plaganyi et al. 2011b) (Table 1).

Table 1: Summary of observed and anticipated climate change impacts on marine fisheries in Australia, socio-economic implications and possible adaptations in the short-term (5–7 years), including barriers to adaptation. Likelihood of impacts and indirect effects, and confidence in the projections, are colour-coded based on the legend below. [Note: Likelihood and confidence scales follow those of Bell et al. (2011)]

Observed and anticipated impacts	Indirect effects	Socio-economic implications	Adaptation options (next 5-7 years)	Adaptation barriers (next 5-7 years)
Distributional shifts of stocks (most likely poleward)	Reduced species biomass and/or abundance in lower latitudes	Livelihoods at risk, particularly for single species fisheries	Livelihood or species diversification	High resource reliance or single-species fisheries
Expansion of 'locally invasive' species	Reduced fishing days and species catchability	Economic declines or losses (either due to increasing costs or declining catches)	Spatially or temporally flexible management	Inflexible or jurisdiction-driven management
Contraction of thermally sensitive species to smaller ranges	Altered seasonality of catches	Loss of social identity for communities with long association with fishing	Financial support for vulnerable fisheries (e.g. low-interest loans or relief funds)	Weak socio-economic structures (i.e. poor financial or social support systems)
Increased disease outbreaks in marine species	Reduced catch rates of some target species (e.g. demersal fish)	Reduced viability of dependent communities	Multi-jurisdiction stock management (e.g. trading quotas)	Resource depletion or compromised condition of resource
Earlier age at maturity and mortality	Changes to catch compositions, particularly in fisheries that target sensitive or specialist species	Mental health issues in highly dependent communities	Promote opportunities in locations where stocks may expand or increase in abundance	Lack of interest or political will
Habitat loss and/or degradation, particularly of coral reefs and kelp forests	Reduced access to fisheries stocks as distributions shift	Trade-offs between fisheries and other sectors (e.g. water supply and irrigation)	Maintain sustainable fisheries and healthy supporting habitats	Rapidity of the rate of climate change

Likelihood: the probability that a future projection or prediction will occur based on expert judgment

Confidence: a quantitative or qualitative evaluation of the level of scientific understanding in support of a conclusion

CONCLUSIONS

Marine capture fisheries in Australia face significant future challenges associated with the impacts of climate change. However, adaptation options are available, and science can provide valuable information to inform decision-making. Improved projections of: (i) distributional shifts of key commercial fisheries species in locations vulnerable to climate change (e.g. southeast Australia); (ii) species most likely to expand or contract their ranges (e.g. warm temperate species); and, (iii) species that may become 'locally invasive'; are needed to inform management and support immediate adaptation of vulnerable sectors and communities. As species distributions shift, information is required to enable useful dialogue between management agencies that will need to work together in the future, particularly on how community dynamics might change and the likely social and economic impacts. Importantly, information is needed on economic or other factors that limit adaptation, and how these can be addressed for marine capture industries.

This review has considered a range of questions surrounding the impacts of climate change on commercial marine fisheries and adapting to change, with a focus on the Australian context and research over the past few years. Climate change is affecting, and will continue to affect, Australia's commercial marine fishing industries. With greater use of targeted information, industry can be better placed to adapt to future changes and take advantage of opportunities associated with the effects of climate change. It is important however to acknowledge the barriers to adaptation, as reducing these over the next 5–7 years will assist more effective adaptation planning. Efforts to reduce these barriers and synthesise learning from continued research will be critical to improve, or at least preserve, adaptive capacity for the future sustainability of marine capture fisheries.

ACKNOWLEDGMENTS

Funding from the Australian Government Department of Climate Change and Energy Efficiency, through the National Climate Change Adaptation Research Facility, supported this review. The authors thank the three anonymous reviewers for their insightful comments. We acknowledge helpful advice from Stewart Frusher and Ingrid van Putten, and assistance from Clare Brooker in collating the Endnote library.

REFERENCES

1. ABARES (2011) Australian fisheries statistics 2010. Australian Bureau of Agricultural and Resource Economics and Sciences, Canberra
2. Allison EH, Perry AL, Badjeck M-C et al (2009) Vulnerability of national economies to the impacts of climate change on fisheries. Fish Fish 10(2):173–196. doi:10.1111/j.1467-2979.2008.00310.x
3. Badjeck M-C, Allison EH, Halls AS, Dulvy NK (2010) Impacts of climate variability and change on fishery-based livelihoods. Mar Policy 34(3):375–383. doi:10.1016/j.marpol.2009.08.007
4. Balston J (2009a) An analysis of the impacts of long-term climate variability on the commercial barramundi (*Lates calcarifer*) fishery of north-east Queensland, Australia. Fish Res 99(2):83–89. doi:10.1016/j.fishres.2009.05.001
5. Balston J (2009b) Short-term climate variability and the commercial barramundi (*Lates calcarifer*) fishery of north-east Queensland, Australia. Mar Freshw Res 60(9):912–923. doi:10.1071/mf08283
6. Barange M, Perry RI (2009) Physical and ecological impacts of climate change relevant to marine and inland capture fisheries and aquaculture. In: Cochrane K, Young CD, Soto D, Bahri T (eds) Climate change implications for fisheries and aquaculture: overview of current scientific knowledge. FAO Fisheries and Aquaculture Technical Paper No. 530. FAO, Rome, pp 7–106
7. Bell JD, Johnson JE, Hobday AJ (2011) Vulnerability of tropical Pacific fisheries and aquaculture to climate change. Secretariat of the Pacific community. Noumea, New Caledonia
8. Bell JD, Ganachaud AS, Gehrke PC, Griffiths SP, Hobday AJ, Hoegh-Guldberg O, Johnson JE, Le Borgne R, Lehodey P, Lough JM, Matear RJ, Pickering TD, Pratchett MS, Sen Gupta A, Waycott M (2013) Effects of climate change on tropical Pacific fisheries and aquaculture. Nat Clim Chang 3:591–599. doi:10.1038/NCLIMATE1838
9. Booth DJ, Bond N, Macreadie P (2011) Detecting range shifts among Australian fishes in response to climate change. Mar Freshw Res 62(9):1027–1042. doi:10.1071/MF102701323-1650/11/091027
10. Brander K (2010) Impacts of climate change on fisheries. J Mar Syst 79(3–4):389–402. doi:10.1016/j.

jmarsys.2008.12.015

11. Brown CJ, Fulton EA, Hobday AJ et al (2010) Effects of climate-driven primary production change on marine food webs: implications for fisheries and conservation. Glob Chang Biol 16(4):1194–1212. doi:10.1111/j.1365-2486.2009.02046.x
12. Campbell AH, Harder T, Nielsen S, Kjelleberg S, Steinberg PD (2011) Climate change and disease: bleaching of a chemically defended seaweed. Glob Chang Biol 17(9):2958–2970. doi:10.1111/j.1365-2486.2011.02456.x
13. Cheung WWL, Lam VWY, Sarmiento JL, Kearney K, Watson R, Zeller D, Pauly D (2010) Large-scale redistribution of maximum fisheries catch potential in the global ocean under climate change. Glob Chang Biol 16(1):24–35. doi:10.1111/j.1365-2486.2009.01995.x
14. Cheung WWL, Dunne J, Sarmiento JL, Pauly D (2011) Integrating ecophysiology and plankton dynamics into projected maximum fisheries catch potential under climate change in the Northeast Atlantic. ICES J Mar Sci: J Cons 68(6):1008–1018. doi:10.1093/icesjms/fsr012
15. Cheung WWL, Watson R, Pauly D (2013) Signature of ocean warming in global fisheries catch. Nature 497:365–369. doi:10.1038/nature12156
16. Coulthard S (2009) Adaptation and conflict within fisheries: insights for living with climate change. In: Adger WN, Lorenzoni I, O'Brien K (eds) Adapting to climate change: thresholds, values, governance. Cambridge University Press, Cambridge, pp 255–268
17. Dang V, Speck P, Benkendorff K (2012) Influence of elevated temperatures on the antiviral and antibacterial immune response of abalone Haliotis rubra. Fish Shellfish Immunol 32:732–740. doi:10.1016/j.fsi.2012.01.022
18. Danovaro R, Corinaldesi C, Dell'Anno A, Fuhrman JA, Middelburg JJ, Noble RT, Suttle CA (2010) Marine viruses and global climate change. FEMS Microbiol Rev 35:993–1034. doi:10.1111/j.1574-6976.2010.00258.x
19. Daw T, Adger W, Brown K, Badjeck M (2009) Climate change and capture fisheries. Worldfish Centre Report
20. Donnelly R (2011) Climate change vulnerability assessment: Queensland Marine Aquarium Supply Industry, 2010. Research Publication no.108
21. FAO (2010) State of world fisheries and aquaculture 2010. FAO Fisheries and Aquaculture Department, Rome
22. Froese R, Binohlan C (2000) Empirical relationships to estimate asymptotic length, length at first maturity and length at maximum yield per recruit in fishes, with a simple method to evaluate length frequency data. J Fish Biol 56:758–773
23. Fulton EA (2011) Interesting times: winners, losers, and system shifts under climate change around Australia. ICES J Mar Sci 68(6):1329–1342. doi:10.1093/icesjms/fsr032
24. Gardner JL, Peters A, Kearney MR, Joseph L, Heinsohn R (2011) Declining body size: a third universal response to warming? Trends Ecol Evol 26(6):285–291
25. Gillson J, Scandol J, Suthers I (2009) Estuarine gillnet fishery catch rates decline during drought in eastern Australia. Fish Res 99(1):26–37. doi:10.1016/j.fishres.2009.04.007
26. Grafton RQ (2010) Adaptation to climate change in marine capture fisheries. Mar Policy 34(3):606–615. doi:10.1016/j.marpol.2009.11.011
27. Griffith GD, Fulton EA, Gorton R, Richardson AJ (2012) Predicting interactions among fishing, ocean warming, and ocean acidification in a marine system with whole-ecosystem models. Conserv Biol. doi:10.1111/j.1523-1739.2012.01937.x
28. Hallegraeff GM (2010) Ocean climate change, phytoplankton community responses, and harmful algal blooms: a formidable predictive challenge. J Phycol 46(2):220–235. doi:10.1111/j.1529-8817.2010.00815.x
29. Hobday AJ (2010) Ensemble analysis of the future distribution of large pelagic fishes off Australia. Prog Oceanogr 86(1–2):291–301
30. Hobday AJ, Poloczanska ES (2010) Marine fisheries and aquaculture. In: Stokes C, Howden M (eds) Adapting agriculture to climate change: preparing Australian agriculture, forestry and fisheries for the future. CSIRO Publishing, Collingwood, pp 205–228
31. Hobday AJ, Mapstone B, Connolly R, Hughes T, Marshall P, McDonald J, Waschka M (2009) Enhancing species adaptation to climate change. Marine climate change in Australia; Impacts and adaptation responses
32. Hobday AJ, Smith ADM, Stobutzki IC et al (2011) Ecological risk assessment for the effects of fishing. Fish Res 108(2–3):372–384
33. Holbrook NJ, Bindoff NL (1997) Interannual and decadal temperature variability in the southwest Pacific Ocean between 1955 and 1988. J Clim 10:1035–1049
34. Ian Perry R, Ommer RE (2010) Introduction: coping with global change in marine social-ecological systems. Mar Policy 34(4):739–741. doi:10.1016/j.marpol.2010.01.025
35. Ives MC, Scandol JP, Montgomery SS, Suthers IM (2009) Modelling the possible effects of climate change on an Australian multi-fleet prawn fishery. Mar Freshw Res 60(12):1211–1222. doi:10.1071/mf07110
36. Johnson JE, Welch DJ (2010) Marine fisheries management in a changing climate: a review of vulnerability and future options. Rev Fish Sci 18(1):106–124. doi:10.1080/10641260903434557
37. Johnson JE, Holbrook NJ (2014) Adaptation of Australia's marine ecosystems to climate change: Using science to inform conservation management. Int J Ecol. doi:10.1155/2014/140354
38. Johnson CR, Banks SC, Barrett NS et al (2011) Climate change cascades: Shifts in oceanography, species' ranges and subtidal marine community dynamics in eastern Tasmania. J Exp Mar Biol Ecol 400(1–2):17–32. doi:10.1016/j.jembe.2011.02.032
39. Kirby RR, Beaugrand G, Lindley JA (2009) Synergistic effects of climate and fishing in a marine ecosystem. Ecosystems 12(4):548–561. doi:10.1007/s10021-009-

9241-9

40. Koehn JD, Hobday AJ, Pratchett MS, Gillanders BM (2011) Climate change and Australian marine and freshwater environments, fishes and fisheries: synthesis and options for adaptation. Mar Freshw Res 62(9):1148–1164

41. Last PR, White WT, Gledhill DC, Hobday AJ, Brown R, Edgar GJ, Pecl G (2011) Long-term shifts in abundance and distribution of a temperate fish fauna: a response to climate change and fishing practices. Glob Ecol Biogeogr 20:58–72. doi:10.1111/j.1466-8238.2010.00575.x

42. Ling SD, Johnson CR, Ridgway K, Hobday AJ, Haddon M (2009) Climate-driven range extension of a sea urchin: inferring future trends by analysis of recent population dynamics. Glob Chang Biol 15(3):719–731. doi:10.1111/j.1365-2486.2008.01734.x

43. MacNeil MA, Graham NAJ, Cinner JE, Dulvy NK, Loring PA, Jennings S, Polunin NVC, Fisk AT, McClanahan TR (2010) Transitional states in marine fisheries: adapting to predicted global change. Philos Trans R Soc B: Biol Sci 365(1558):3753–3763. doi:10.1098/rstb.2010.0289

44. Madin EMP, Ban NC, Doubleday ZA, Holmes TH, Pecl GT, Smith F (2012) Socio-economic and management implications of range-shifting species in marine systems. Glob Environ Chang-Hum Policy Dimens 22(1):137–146. doi:10.1016/j.gloenvcha.2011.10.008

45. Mapstone B, Appleford P, Broderick K, Connolly R, Higgins J, Hobday A, Hughes T, Marshall P, McDonald J, Waschka M (2010) National climate change adaptation research plan for marine biodiversity and resources. National climate change adaptation research. Facility, Gold Coast

46. Marshall NA, Marshall PA, Tamelander J, Obura D, Malleret-King D, Cinner JE (2010) A framework for social adaptation to climate change; Sustaining tropical coastal communities and industries. IUCN, Gland, v + 36 pp

47. McIlgorm A (2010) Economic impacts of climate change on sustainable tuna and billfish management: insights from the Western Pacific. Prog Oceanogr 86(1–2):187–191. doi:10.1016/j.pocean.2010.04.024

48. McIlgorm A, Hanna S, Knapp G, Le Floc'H P, Millerd F, Pan M (2010) How will climate change alter fishery governance? Insights from seven international case studies. Mar Policy 34(1):170–177. doi:10.1016/j.marpol.2009.06.004

49. Meynecke JO, Lee SY (2011) Climate-coastal fisheries relationships and their spatial variation in Queensland, Australia. Fish Res 110(2):365–376. doi:10.1016/j.fishres.2011.05.004

50. Miller K, Charles A, Barange M et al (2010) Climate change, uncertainty, and resilient fisheries: institutional responses through integrative science. Prog Oceanogr 87(1–4):338–346

51. Munday PL, Leis JM, Lough JM, Paris CB, Kingsford MJ, Berumen ML, Lambrechts J (2009) Climate change and coral reef connectivity. Coral Reefs 28(2):379–395. doi:10.1007/s00338-008-0461-9

52. Neuheimer AB, Thresher RE, Lyle JM, Semmens JM (2011) Tolerance limit for fish growth exceeded by warming waters. Nat Clim Chang. doi:10.1038/NCLIMATE1084

53. OECD (2010) The economics of adapting fisheries to climate change. OECD Publishing, Paris. doi:10.1787/9789264090415-en

54. Pearce A, Lenanton R, Jackson G, Moore J, Feng M, Gaughan D (2011) The "marine heat wave" off Western Australia during the summer of 2010/11. Fisheries Research Report No. 222. Western Australia

55. Pecl G, Frusher S, Gardner C, Haward M, Hobday A, Jennings S, Nursey-Bray M, Punt A, Revill H, van Putten I (2009) The east coast Tasmanian rock lobster fishery – vulnerability to climate change impacts and adaptation response options. Report to the Department of Climate Change, Australia

56. Perry RI, Ommer RE, Barange M, Werner F (2010) The challenge of adapting marine social–ecological systems to the additional stress of climate change. Curr Opin Environ Sustain 2(5–6):356–363. doi:10.1016/j.cosust.2010.10.004

57. Pitt NR, Poloczanska ES, Hobday AJ (2010) Climate-driven range changes in Tasmanian intertidal fauna. Mar Freshw Res 61(9):963–970. doi:10.1071/mf09225

58. Plaganyi EE, Weeks SJ, Skewes TD, Gibbs MT, Poloczanska ES, Norman-Lopez A, Blamey LK, Soares M, Robinson WML (2011a) Assessing the adequacy of current fisheries management under changing climate: a southern synopsis. ICES J Mar Sci 68(6):1305–1317. doi:10.1093/icesims/fsr049

59. Plaganyi EE, Bell JD, Bustamante RH et al (2011b) Modelling climate-change effects on Australian and Pacific aquatic ecosystems: a review of analytical tools and management implications. Mar Freshw Res 62(9):1132–1147

60. Poloczanska ES, Hobday AJ, Richardson AJ (eds) (2009) Report Card of Marine Climate Change for Australia. NCCARF Publication 05/09, ISBN 978-1-921609-03-9

61. Poloczanska ES, Hobday AJ, Richardson AJ (eds) (2012) Marine Climate Change in Australia, Impacts and Adaptation Responses. 2012 Report Card, ISBN 978-0-643-10927-8

62. Pörtner HO, Peck MA (2010) Climate change effects on fishes and fisheries: towards a cause-and-effect understanding. J Fish Biol 77(8):1745–1779. doi:10.1111/j.1095-8649.2010.02783.x

63. Pratchett M, Wilson S, Graham N, Munday M, Jones G, Polunin N (2009) Multi-scale temporal effects of climate-induced coral bleaching on motile reef organisms. In: Oppen MV, Lough J (eds) Coral bleaching: patterns and processes, causes and consequences. Springer, Heidelberg, pp 139–158

64. Pratchett MS, Bay LK, Gehrke PC, Koehn JD, Osborne K, Pressey RL, Sweatman HPA, Wachenfeld D (2011a) Contribution of climate change to degradation and loss of critical fish habitats in Australian marine and freshwater environments. Mar Freshw Res 62(9):1062–1081. doi:10.1071/mf10303

65. Pratchett M, Munday P, Graham N et al (2011b) Vulnerability of coastal fisheries in the tropical Pacific to climate change. In: Bell JD, Johnson JE, Hobday AJ (eds) Vulnerability of tropical pacific fisheries and aquaculture to climate change. Secretariat for the Pacific Community, Noumea

66. Ridge-Partners (2010) Overview of the Australian fishing and aquaculture industry: present and future. A report supporting the development of Working Together: the National Fishing and Aquaculture RD&E Strategy. Copyright Fisheries Research and Development Corporation and Ridge Partners

67. Ridgway KR (2007) Long-term trend and decadal variabiity of the southward penetration of the East Australian Current. Geophys Res Lett 34(13):L13613. doi:10.1029/2007GL030393

68. Stuart-Smith RD, Barrett NS, Stevenson DG, Edgar GJ (2009) Stability in temperate reef communities over a decadal time scale despite concurrent ocean warming. Glob Chang Biol 16(1):122–134. doi:10.1111/j.1365-2486.2009.01955.x

69. Tobin A, Schlaff A, Tobin R, Penny A, Ayling T, Ayling A, Krause B, Welch D, Sutton S, Sawynok B, Marshall N, Marshall P (2010) Adapting to change: minimising uncertainty about the effects of rapidly-changing environmental conditions on the Queensland Coral Reef Finfish Fishery. Final Report to the Fisheries Research & Development Corporation, Project 2008/103. Fishing & Fisheries Research Centre Technical Report No. 11, James Cook University, Townsville, Australia (172pp.)

70. van Putten IE, Jennings S, Frusher S, Gardner C, Haward M, Hobday A, Nursey-Bray M, Pecl G, Punt A, Revill H (2013) Building blocks of economic resilience to climate change: a fisheries example. Reg Environ Chang. doi:10.1007/s10113-013-0456-0

71. van Putten I, Metcalf S, Frusher S, Marshall N, Tull M (2014) Transformation of coastal communities: Where is the marine sector heading? Australasian Journal of Regional Studies (in press)

72. Welch DJ, Johnson JE (2013) Assessing the vulnerability of Torres Strait fisheries and supporting habitats to climate change. Report to the Australian Fisheries Management Authority. C2O Fisheries, Australia, 114pp

73. Wernberg T, Russell BD, Thomsen MS, Gurgel CFD, Bradshaw CJA, Poloczanska ES, Connell SD (2011) Seaweed communities in retreat from ocean warming. Curr Biol 21(21):1828–1832. doi:10.1016/j.cub.2011.09.028

74. Wilson SK, Adjeroud M, Bellwood DR et al (2010) Crucial knowledge gaps in current understanding of climate change impacts on coral reef fishes. J Exp Biol 213(6):894–900. doi:10.1242/jeb.037895

INDEX